Human Biological Variation: A Genetic Perspective

Human Biological Variation:
A Genetic Perspective

Editor: Violet Lawrence

AMERICAN
MEDICAL PUBLISHERS
www.americanmedicalpublishers.com

AMERICAN
MEDICAL PUBLISHERS
www.americanmedicalpublishers.com

Cataloging-in-Publication Data

Human biological variation : a genetic perspective / edited by Violet Lawrence.
    p. cm.
Includes bibliographical references and index.
ISBN 978-1-63927-252-5
1. Human population genetics. 2. Human genetics--Variation. 3. Human evolution.
4. Variation (Biology). I. Lawrence, Violet.
QH431 .H86 2022
599.935--dc23

American Medical Publishers,
41 Flatbush Avenue,
1st Floor, New York,
NY 11217, USA

ISBN 978-1-63927-252-5 (Hardback)

Contents

Preface IX

Chapter 1 **CXCL5 polymorphisms are associated with variable blood pressure in cardiovascular disease-free adults**.. 1
Amber L Beitelshees, Christina L Aquilante, Hooman Allayee,
Taimour Y Langaee, Gregory J Welder, Richard S Schofield and Issam Zineh

Chapter 2 **Complement regulator CD46: genetic variants and disease associations**.......................... 9
M. Kathryn Liszewski and John P. Atkinson

Chapter 3 **Association of genome variations in the renin-angiotensin system with physical performance**.. 22
Argyro Sgourou, Vassilis Fotopoulos, Vassilis Kontos, George P Patrinos and
Adamantia Papachatzopoulou

Chapter 4 **Trans-species polymorphism in humans and the great apes is generally maintained by balancing selection that modulates the host immune response** 28
Luisa Azevedo, Catarina Serrano, Antonio Amorim and David N. Cooper

Chapter 5 **Copy number variation in CEP57L1 predisposes to congenital absence of bilateral ACL and PCL ligaments**... 34
Yichuan Liu, Yun Li, Michael E. March, Kenny Nguyen, Kexiang Xu,
Fengxiang Wang, Yiran Guo, Brendan Keating, Joseph Glessner, Jiankang Li,
Theodore J. Ganley, Jianguo Zhang, Matthew A. Deardorff, Xun Xu and
Hakon Hakonarson

Chapter 6 **The impact of common polymorphisms in CETP and ABCA1 genes with the risk of coronary artery disease**...43
Cyril Cyrus, Chittibabu Vatte, Awatif Al-Nafie, Shahanas Chathoth,
Rudaynah Al-Ali, Abdullah Al-Shehri, Mohammed Shakil Akhtar,
Mohammed Almansori, Fahad Al-Muhanna, Brendan Keating and Amein Al-Ali

Chapter 7 **Copy number variation of human AMY1 is a minor contributor to variation in salivary amylase expression and activity** 49
Danielle Carpenter, Laura M. Mitchell and John A. L. Armour

Chapter 8 **Variants in congenital hypogonadotrophic hypogonadism genes identified in an Indonesian cohort of 46,XY under-virilised boys** 55
Katie L. Ayers, Aurore Bouty, Gorjana Robevska, Jocelyn A. van den Bergen,
Achmad Zulfa Juniarto, Nurin Aisyiyah Listyasari, Andrew H. Sinclair and
Sultana M. H. Faradz

Chapter 9 **In silico prioritization and further functional characterization of SPINK1 intronic variants** 65
Wen-Bin Zou, Hao Wu, Arnaud Boulling, David N. Cooper, Zhao-Shen Li,
Zhuan Liao, Jian-Min Chen and Claude Férec

Chapter 10   **A pipeline combining multiple strategies for prioritizing heterozygous variants for the identification of candidate genes in exome datasets**................................72
Teresa Requena, Alvaro Gallego-Martinez and Jose A. Lopez-Escamez

Chapter 11   **Single nucleotide polymorphisms in the angiogenic and lymphangiogenic pathways are associated with lymphedema caused by Wuchereria bancrofti**..............................83
Linda Batsa Debrah, Anna Albers, Alexander Yaw Debrah, Felix F. Brockschmidt, Tim Becker, Christine Herold, Andrea Hofmann, Jubin Osei-Mensah, Yusif Mubarik, Holger Fröhlich, Achim Hoerauf and Kenneth Pfarr

Chapter 12   **Identification of functional single nucleotide polymorphisms in the branchpoint site** 96
Hung-Lun Chiang, Jer-Yuarn Wu and Yuan-Tsong Chen

Chapter 13   **The peptidylglycine-α-amidating monooxygenase (PAM) gene rs13175330 A>G polymorphism is associated with hypertension**................................102
Hye Jin Yoo, Minjoo Kim, Minkyung Kim, Jey Sook Chae, Sang-Hyun Lee and Jong Ho Lee

Chapter 14   **Identification of compound heterozygous variants in the noncoding RNU4ATAC gene in a Chinese family with two successive foetuses with severe microcephaly**................................110
Ye Wang, Xueli Wu, Liu Du, Ju Zheng, Songqing Deng, Xin Bi, Qiuyan Chen, Hongning Xie, Claude Férec, David N. Cooper, Yanmin Luo, Qun Fang and Jian-Min Chen

Chapter 15   **Nonparametric approaches for population structure analysis** 118
Luluah Alhusain and Alaaeldin M. Hafez

Chapter 16   **Associations between hypertension and the peroxisome proliferator-activated receptor- δ (PPARD) gene rs7770619 C>T polymorphism**................................130
Minjoo Kim, Minkyung Kim, Hye Jin Yoo, Jayoung Shon and Jong Ho Lee

Chapter 17   **Associations of high-altitude polycythemia with polymorphisms in PIK3CD and COL4A3**................................138
Xiaowei Fan, Lifeng Ma, Zhiying Zhang, Yi Li, Meng Hao, Zhipeng Zhao, Yiduo Zhao, Fang Liu, Lijun Liu, Xingguang Luo, Peng Cai, Yansong Li and Longli Kang

Chapter 18   **Architecture of polymorphisms in the human genome reveals functionally important and positively selected variants in immune response and drug transporter genes**................................147
Yu Jin, Jingbo Wang, Maulana Bachtiar, Samuel S. Chong and Caroline G. L. Lee

Chapter 19   **Predicting the combined effect of multiple genetic variants**................................160
Mingming Liu, Layne T. Watson and Liqing Zhang

Chapter 20   **Ranking non-synonymous single nucleotide polymorphisms based on disease concepts** 167
Hashem A Shihab, Julian Gough, Matthew Mort, David N Cooper, Ian NM Day and Tom R Gaunt

Chapter 21   **A study of the role of GATA4 polymorphism in cardiovascular
metabolic disorders** 173
Nzioka P Muiya, Salma M Wakil, Asma I Tahir, Samya Hagos, Mohammed Najai,
Daisy Gueco, Nada Al-Tassan, Editha Andres, Nejat Mazher, Brian F Meyer and
Nduna Dzimiri

Chapter 22   **CER1 gene variations associated with bone mineral density, bone markers, and
early menopause in postmenopausal women** 184
Theodora Koromila, Panagiotis Georgoulias, Zoe Dailiana, Evangelia E Ntzani,
Stavroula Samara, Chris Chassanidis, Vassiliki Aleporou-Marinou and
Panagoula Kollia

Permissions

List of Contributors

Index

Preface

The difference in DNA found among individuals is known as genetic variation. These genetic differences found in individuals and populations forms the basis of human biological variations. The field focuses on identifying the various biological causes that lead to these variations in humans. Some of these include the order of bases in nucleotides present in the genes, variation in enzymes and variation in discrete and quantitative traits. The technique of protein electrophoresis is used to examine the variations in enzymes. Mutation, genetic recombination and segregation are some of the causes of genetic variation studied within this discipline. This book is a valuable compilation of topics, ranging from the basic to the most complex advancements in the field of human biological variations. The various studies that are constantly contributing towards advancing technologies and evolution of this field are examined in detail herein. This book is a vital tool for all researching or studying this discipline as it gives incredible insights into emerging trends and concepts.

Significant researches are present in this book. Intensive efforts have been employed by authors to make this book an outstanding discourse. This book contains the enlightening chapters which have been written on the basis of significant researches done by the experts.

Finally, I would also like to thank all the members involved in this book for being a team and meeting all the deadlines for the submission of their respective works. I would also like to thank my friends and family for being supportive in my efforts.

Editor

CXCL5 polymorphisms are associated with variable blood pressure in cardiovascular disease-free adults

Amber L Beitelshees[1*†], Christina L Aquilante[2†], Hooman Allayee[3], Taimour Y Langaee[4], Gregory J Welder[4], Richard S Schofield[5] and Issam Zineh[4]

Abstract

Objective: Leukocyte count has been associated with blood pressure, hypertension, and hypertensive complications. We hypothesized that polymorphisms in the CXCL5 gene, which encodes the neutrophilic chemokine ENA-78, are associated with blood pressure in cardiovascular disease (CVD)-free adults and that these polymorphisms are functional.

Methods and results: A total of 192 community-dwelling participants without CVD or risk equivalents were enrolled. Two CXCL5 polymorphisms (−156 G > C (rs352046) and 398 G > A (rs425535)) were tested for associations with blood pressure. Allele-specific mRNA expression in leukocytes was also measured to determine whether heterozygosity was associated with allelic expression imbalance. In −156 C variant carriers, systolic blood pressure (SBP) was 7 mmHg higher than in −156 G/G wild-type homozygotes (131 ± 17 vs. 124 ± 14 mmHg; $P = 0.008$). Similarly, diastolic blood pressure (DBP) was 4 mmHg higher in −156 C variant carriers (78 ± 11 vs. 74 ± 11 mmHg; $P = 0.013$). In multivariate analysis of SBP, age, sex, body mass index, and the −156 G > C polymorphism were identified as significant variables. Age, sex, and the −156 G > C SNP were further associated with DBP, along with white blood cells. Allelic expression imbalance and significantly higher circulating ENA-78 concentrations were noted for variant carriers.

Conclusion: CXCL5 gene polymorphisms are functional and associated with variable blood pressure in CVD-free individuals. The role of CXCL5 as a hypertension- and CVD-susceptibility gene should be further explored.

Keywords: CXCL5, ENA-78, Blood pressure, Hypertension, Leukocytes

Introduction

The relationship between inflammation and elevated blood pressure is increasingly being evaluated [1,2]. It has been shown that elevated concentrations of prototypical pro-inflammatory markers such as interleukin-6, C-reactive protein (CRP), and tumor necrosis factor-alpha are associated with increased blood pressure, incidence of hypertension, and the likelihood for hypertensive complications [3-14]. It has been further suggested that this inflammatory-hypertensive relationship results from increased number or activity of common cellular mediators such as white blood cells (WBC) [15,16]. For example, studies have demonstrated elevated WBC count to be associated with increased incident hypertension as well as increased blood pressure within the normal to pre-hypertensive range [17-22].

Although the exact mechanistic relationship between leukocytosis and elevated blood pressure is unknown, it is plausible that low-grade inflammation may be a contributing factor. In this regard, WBC count may be a surrogate marker for increased activation of inflammatory pathways that cause leukocyte recruitment and activation. As such, increased activity of leukocytic chemokines could be related to increased blood pressure.

Epithelial neutrophil activator-78 (ENA-78), a key leukocytic chemokine that is both a neutrophil attractor

* Correspondence: abeitels@medicine.umaryland.edu
†Equal contributors
[1]Division of Endocrinology, Diabetes and Nutrition, University of Maryland School of Medicine, 660 W. Redwood St, HH469, Baltimore, MD 21201, USA
Full list of author information is available at the end of the article

and activator, has been implicated in many diseases with an inflammatory component (e.g., obesity, diabetes, subclinical atherosclerosis, acute coronary syndromes) [23-32]. We have previously reported that two single nucleotide polymorphisms (SNPs), -156 G > C (rs352046) and 398 G > A (rs425535), in the gene encoding ENA-78 (*CXCL5*) occur in sites important for transcription and exon splicing [33]. In our previous work, a relationship existed between these SNPs and both plasma concentrations and leukocyte production of the ENA-78 chemokine protein [33]. We then went on to show an association between the *CXCL5* -156 G > C polymorphism and worse outcomes in patients with acute coronary syndromes [27]. In the present work, to the extent that ENA-78 is important in neutrophil recruitment and degranulation, we hypothesized that one or both of these polymorphisms (–156 G > C and 398 G > A) could be associated with differences in blood pressure in individuals without established cardiovascular disease (CVD). Specifically, we hypothesized that relatively young individuals without known CVD who were carriers of *CXCL5* variant alleles would exhibit higher systolic blood pressure (SBP), diastolic blood pressure (DBP), or pulse pressure (PP) than wild-type homozygotes. Furthermore, to assess whether there was a functional role for these polymorphisms, we measured allele-specific mRNA expression of *CXCL5* in leukocytes obtained from CVD-free individuals who were heterozygous for the SNPs at both loci.

Materials and methods
Study population
The study population has been previously described [33]. Briefly, participants were recruited from two sites in the USA and had to be at least 18 years of age without known CVD or CVD-risk equivalents (e.g., diabetes, peripheral vascular disease, 10-year Framingham Risk ≥20%) as defined by National Cholesterol Education Program criteria [34]. Other exclusions were pregnancy, malignancy, substance abuse, and routine use of medications known to affect WBC counts such as systemic steroids and other anti-inflammatory agents. Individuals were excluded from analysis if they were taking antihypertensive medications for either cardiovascular or non-cardiovascular indications (e.g., migraine). For blood pressure measurement, subjects were seated for at least 5 min in a quiet, temperature-controlled General Clinical Research Center (GCRC) outpatient clinic room, and two blood pressure measurements were taken at least 5 min apart. The average of the duplicate blood pressure measurements was used for this investigation. Blood samples were obtained from participants enrolled in University of Florida- and Colorado Multiple Institutional Review Board (IRB)-approved studies. All subjects

provided written informed consent to specimen and data use in genetic association and related studies.

Genotype and inflammatory biomarker determination
Genomic DNA was isolated from whole blood or buccal cells using previously described methods [35]. *CXCL5* genotypes were determined by polymerase chain reaction (PCR) and pyrosequencing (Qiagen, Valencia, CA, USA) as we have previously described [36]. Circulating high-sensitivity CRP (as a non-specific marker of inflammation) was measured by the Shands Hospital Laboratory at the University of Florida and University of Colorado GCRC. ENA-78 concentrations were measured by cytometric fluorescence detection as previously described (Luminex™100 IS system; Luminex Corp., Austin, TX, USA; Fluorokine® MAP Multiplex Human Cytokine Panel A; R&D Systems, Minneapolis, MN, USA) [37]. Samples were stored at –80°C until CRP and ENA-78 detection was performed.

Allele-specific mRNA quantification
To determine whether variant carrier status results in functional changes at the transcriptional level, we quantified allele-specific mRNA transcripts from leukocytes using pyrosequencing-based methodology [38,39]. Specifically, the presence or absence of allelic expression imbalance was determined using leukocytes obtained from 18 individuals who were heterozygotic for both the –156 G > C and 398 G > A polymorphisms. The 398 G > A SNP was chosen as the genetic biomarker in these experiments because it is located in the coding region of *CXCL5*, while –156 G > C is a promoter polymorphism and as such cannot be quantified at the mRNA level. Because of the near complete linkage of the studied SNPs, we chose individuals who were heterozygotes at both loci so that 398 G > A genotype might serve as a functional surrogate for the upstream promoter locus.

Leukocyte mRNA was prepared from approximately 6×10^6 cells from each individual using the RNeasy mini kit (Qiagen, Valencia, CA, USA). Cells were rinsed, lysed, and homogenized in buffered solutions and subsequently passed through the RNeasy mini column (Qiagen, Valencia, CA, USA). Following a series of washes at room temperature and 15-min incubation with DNase, concentrations were determined by spectrophotometry (NanoDrop Technologies, Wilmington, DE, USA). cDNA was synthesized using approximately 450 ng of cellular RNA from each individual using a High-Capacity cDNA Archive Kit (Applied Biosystems, Foster City, CA, USA) per protocol. Conditions for reverse transcription were 25°C for 10 min followed by 37°C for 2 h. cDNA quality was assessed by comparing cDNA and DNA PCR products generated using intron-spanning primers by gel electrophoresis. For allele-specific transcript quantification, subject DNA and cDNA underwent PCR simultaneously using previously described conditions [36]. PCR products obtained for genotype determination (DNA) and transcript quantification (mRNA) were assayed in parallel pyrosequencing reactions to minimize cycle variability. Pyrosequencing analyses were performed in duplicate on three separate PCR amplification products, and the results were pooled for analysis. Peak heights were determined by the pyrosequencing allele quantification algorithm. In genomic DNA, the ratio of 398A:G alleles for DNA in heterozygotes is expected to be approximately 1, whereas significant deviations from this ratio in mRNA would suggest allele expression imbalance associated with the variant allele.

Statistical analyses

Genotype frequencies were determined by allele counting, and departures from Hardy-Weinberg equilibrium were assessed by chi-square analyses. Differences in blood pressure by genotype groups (0, homozygous for common allele; 1, heterozygous or homozygous for variant allele) were compared using one-way ANOVA. Based on the preexisting sample size and prevalence of variant alleles, we had 80% power with a two-sided α of 0.05 to detect a 6-mmHg difference in SBP, 4-mmHg difference in DBP, and 4-mmHg difference in PP between genotype groups. Multiple regression analysis was performed if blood pressure differences were seen across genotype groups. Covariates for multiple regression were chosen through univariate analyses of age, sex, smoking status (0, non-smoker; 1, current smoker), body mass index (BMI), CRP concentration, ENA-78 concentration, and WBC count. Any variable with a $P \leq 0.1$ on univariate analysis was entered into the multivariable model. Because of small numbers of individuals within racial groups, analyses could not be performed within racial strata. However, race (0, white; 1, non-white) was included in all multivariable analyses, and a race-by-genotype interaction term was considered in the regression models to avoid spurious associations secondary to racial differences in allele frequency. Multiple regression using step-type selection methods was performed to determine the joint effects of CXCL5 genotypes and clinical variables on SBP, DBP, or PP. All statistical analyses were performed using SPSS (version 11.5, SPSS Inc., Chicago, IL, USA) or SAS (version 9.1, SAS Institute Inc., Cary, NC, USA). A P value < 0.05 was considered statistically significant.

Results

Baseline demographic characteristics are shown in Table 1. Participants were on average 39 ± 12 years old

Table 1 Baseline characteristics

Characteristic	N = 192
Age (mean ± SD, years)	39 ± 12
Women (number (%))	124 (65)
Race/ethnicity (number (%))	
White	148 (77)
Black	12 (6)
Hispanic	19 (10)
Other	13 (7)
Family heart disease history (number (%))	29 (15.1)
Smoking (number (%))	35 (18)
Body mass index (mean ± SD, kg/m^2)	29.6 ± 7
Blood pressure (mean ± SD, mmHg)	
Systolic	126 ± 15
Diastolic	75 ± 11
Pulse pressure (mean ± SD, mmHg)	51 ± 10
Cholesterol (mean ± SD, mg/dL^a)	
Total	201 ± 43
LDL	118 ± 36
HDL	55 ± 17
Triglycerides	139 ± 107
White blood cell count, (mean ± SD, ×10^9 cells/L)	6.3 ± 2.0
C-reactive protein (median (range), mg/L^b)	1.78 (0.1–16.9)
ENA-78 (median (range), pg/mL^b)	362 (32.2–3970)

^{a}Total, HDL, and triglycerides available in 94% of subjects; LDL available in 92% of subjects. ^{b}CRP and ENA-78 available for 88% and 91% of subjects, respectively.

with blood pressures of $126/75 \pm 15/11$ mm Hg. -156 G > C and 398 G > A genotypes were determined for 189 and 188 of the 192 individuals, respectively. The overall −156 C and 398A minor allele frequencies were both 15%. Variant allele frequencies differed by race whereby the −156 C allele frequency was 14%, 45%, and 11%, and 398A allele frequency was 13%, 46%, and 9% in Caucasians, blacks, and non-black Hispanics, respectively. Genotype distributions satisfied criteria for Hardy-Weinberg equilibrium (data not shown). The two SNPs were in a high degree of linkage disequilibrium with r^2 for Caucasian, black, and Hispanic individuals of 0.82, 1.0, and 0.51, respectively, in our study population.

Genotype association with blood pressure

In −156 C variant carriers, SBP was 7-mmHg higher than in −156 G/G wild-type homozygotes (131 ± 17 vs. 124 ± 14 mmHg; $P = 0.008$). Similarly, DBP was 4-mmHg higher in −156 C variant carriers (78 ± 11 vs. 74 ± 11 mmHg; $P = 0.013$). PP did not differ between −156 C variant carriers and wild-type homozygotes (53 ± 11 vs. 51 ± 10; $P = 0.22$). Because of the high degree of linkage disequilibrium between the 398 G > A and −156 G > C

SNPs, blood pressure differences were similar when compared by 398 G > A genotypes. For example, SBP was 130 ± 16 and 125 ± 14 mmHg in 398A variant carriers and 398 G/G homozygotes, respectively ($P = 0.033$); DBP was 78 ± 11 and 74 ± 11 mmHg, respectively ($P = 0.038$); and PP was not different between groups (53 ± 11 vs. 51 ± 10 mmHg in 398A carriers and 398 G/G homozygotes, respectively; $P = 0.362$).

Age ($P \le 0.001$), sex ($P \le 0.008$), and BMI ($P \le 0.002$) were common univariate predictors of SBP, DBP, and PP. Furthermore, WBC count ($P = 0.10$ for SBP; $P = 0.076$ for DBP) and both *CXCL5* polymorphisms (range $P = 0.008$ to 0.038) were additional predictors of SBP and DBP, while smoking status was associated with SBP alone ($P = 0.038$). In terms of circulating CRP and ENA-78 levels, both biomarkers were significant for SBP ($P = 0.005$ for CRP and $P = 0.033$ for ENA-78) and PP ($P = 0.001$ for CRP and $P = 0.007$ for ENA-78) in univariate analyses. Consistent with our previous report, *CXCL5* genotype was associated with ENA-78 protein concentrations in the plasma whereby variant carriers at either SNP locus had higher protein concentrations than wild-type homozygotes ($P = 0.003$; Figure 1).

In multivariate analysis of SBP, age, sex, BMI, and the *CXCL5* -156 G > C promoter polymorphism were identified as significant variables (Table 2). The overall model that included these variables explained 32.5% of the variability in SBP ($P < 0.001$). Consideration of the 398 G > A polymorphism rather than the −156 G > C promoter SNP resulted in a model in which only age, sex, and BMI were significantly associated with SBP ($R^2 = 0.301$; $P < 0.001$).

Table 2 Multivariate predictors of systolic blood pressure in cardiovascular disease-free individuals

Variable	β	Standard error	P value
Constant	100	4.86	<0.0001
Age	0.313	0.094	0.001
Sex	−9.84	2.12	<0.0001
BMI	0.637	0.160	<0.0001
−156 C carrier	4.93	2.30	0.034

$R^2 = 0.325$; $P < 0.0001$.

Age, sex, and the −156 G > C SNP were further associated with DBP, along with WBC (Table 3). Consideration of this promoter SNP (model $R^2 = 0.168$; $P < 0.0001$) was slightly more informative than consideration of the 398 G > A SNP ($P = 0.067$) in which case age ($P < 0.0001$), sex ($P = 0.001$), and WBC ($P = 0.02$) still remained significant (model $R^2 = 0.145$; $P < 0.0001$). In multivariable models of PP, only sex ($P < 0.004$) and BMI ($P < 0.0001$) were significant (model $R^2 = 0.247$; $P < 0.0001$).

Allelic expression imbalance

Allele-specific mRNA quantification was performed to determine whether there is a functional basis for the differences seen in blood pressure based on *CXCL5* genotypes (see 'Materials and methods' section for rationale of 398 G > A as marker SNP). Importantly, there was consistently higher expression of *CXCL5* mRNA from the 398A allele compared to the 398 G allele in heterozygous individuals (Figure 2A). For example, individual heterozygotes displayed anywhere from 2.2-fold to 3.4-fold higher expression of 398A variant transcripts compared to the 398 G allele, with a mean ratio of 2.9 (Figure 2B; $P = 7.4\text{E-}15$).

Discussion

Accumulating evidence points to a relationship between inflammation and blood pressure. Data suggest that WBC counts are associated with incident hypertension and correlated with blood pressure concentrations. We hypothesized that WBC count is a surrogate for

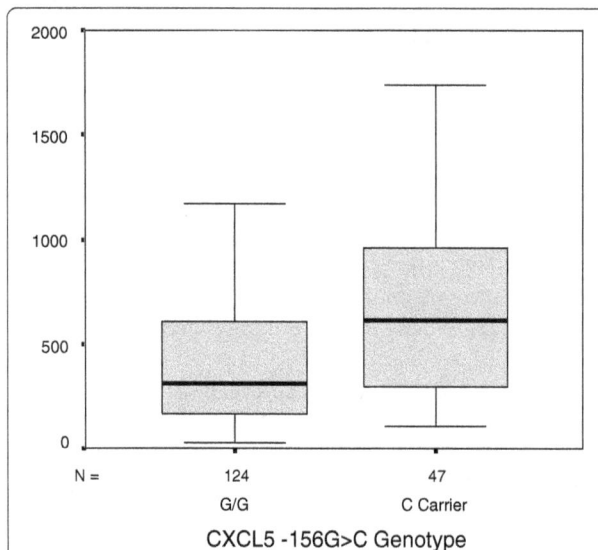

Figure 1 Plasma ENA-78 by *CXCL5* -156 G > C genotype.
$P = 0.003$; data were similar for the exon 2 SNP, data not shown ($P = 0.001$).

Table 3 Multivariate predictors of diastolic blood pressure in cardiovascular disease-free individuals

Variable	β	Standard error	P value
Constant	63.13	3.42	<0.0001
Age	0.247	0.063	<0.0001
Sex	−5.801	1.549	<0.0001
−156 C carrier	3.735	1.630	0.023
WBC	0.768	0.374	0.041

$R^2 = 0.168$; $P < 0.0001$.

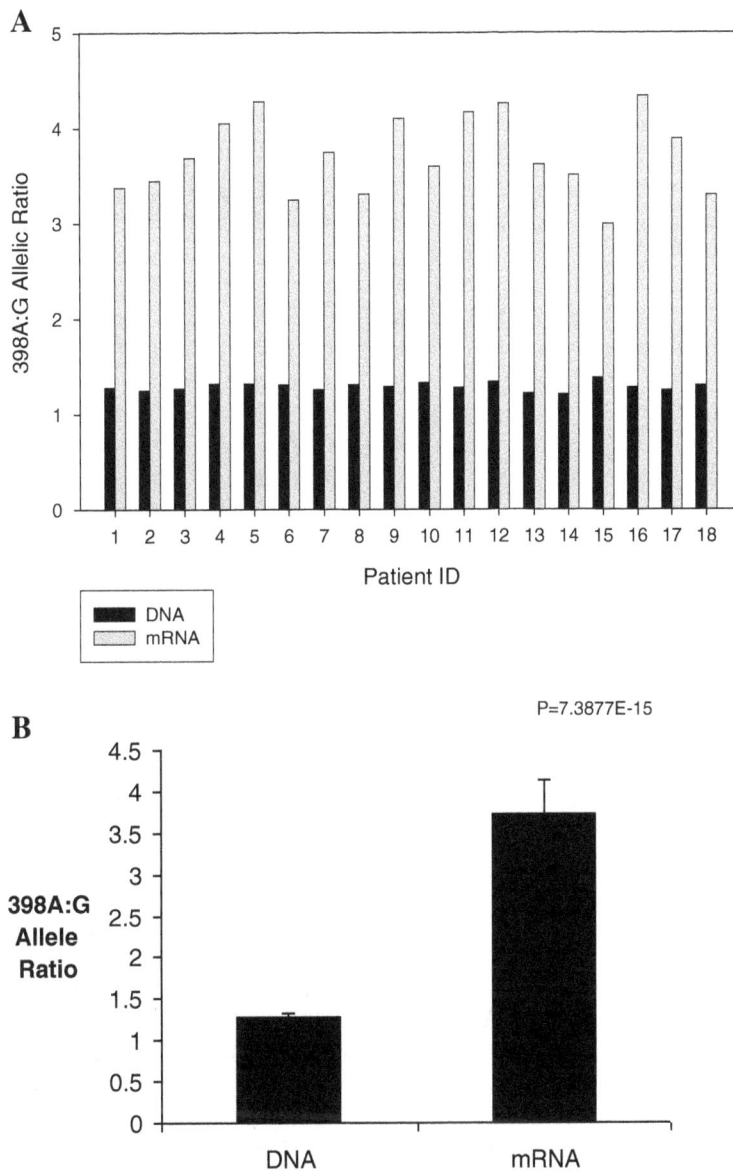

Figure 2 Allele-specific *CXCL5* mRNA expression in leukocytes. (A) Allelic mRNA and DNA ratios were measured in 18 cardiovascular disease-free individuals heterozygous for the 398 G > A SNP. The A/G ratios in DNA were close to 1 suggesting equal abundance of both alleles, whereas there was consistently higher expression of mRNA from the 398A allele compared to the 398 G allele. **(B)** Pooled 398A/G ratios from 18 heterozygous individuals. The sample displayed 2.9-fold higher expression of 398A variant transcripts compared to the 398 G allele (*P* = 7.4E-15). Data are presented as mean ± SD.

leukocytic chemokine activity and that the *CXCL5* gene, which encodes the neutrophil attractor ENA-78, may be an important determinant of blood pressure. We demonstrated a significant, independent relationship between *CXCL5* polymorphisms and SBP and DBP in the overall population of CVD-free individuals. Variant carriers of the −156 G > C promoter SNP had 7-mmHg and 4-mmHg higher SBP and DBP, respectively, than those with the wild-type −156 G/G genotype. Because of the epidemiologically significant difference in CVD risk conferred by

blood pressure differences of this magnitude, and since variant carriers represent approximately 30% of the population studied, *CXCL5* polymorphisms should be considered as a potential novel biomarker of pre-hypertension, hypertension, and CVD risk requiring future study. However, it is important to emphasize that genetic associations are preliminary and will require confirmation in additional populations.

Of particular interest, WBC count (along with traditional variables such as age, sex, smoking status, and

BMI) was significantly associated with SBP and DBP in univariate analysis among CVD-free individuals. This finding supports the report by Orakzai et al. that demonstrated a relationship between WBC counts and SBP among nearly 3,500 white individuals without CVD and with SBP < 140 mmHg on entry [20]. It also supports data from other clinical cohorts showing an association between WBC count, major WBC components (e.g., neutrophils), and blood pressure [21,22,40,41]. However, in our analysis WBC count was no longer a significant predictor of SBP when $CXCL5$ genotype was included in multivariable analysis, suggesting genotype may capture the contribution of inflammation to SBP more effectively than WBC count. WBC did, however, remain a significant predictor of DBP in multivariate analysis, along with age, sex, and $CXCL5$ -156 G > C genotype.

To determine whether there is any functional basis for an observed association between $CXCL5$ variant alleles and blood pressure, we performed allele expression imbalance experiments in a subset of participants. The exonic 398 G > A allele was chosen as the genetic marker given its location in the coding region of the mRNA. However, the 398 G/A heterozygous individuals ($N = 18$) were also heterozygous for the promoter polymorphism, which minimizes confounding of an association by differing genotypes at the upstream locus. It was noted that variant carriers displayed nearly threefold higher expression of variant $CXCL5$ mRNA transcripts from the 398A allele. This novel finding is consistent with our previous observation that variant carriers exhibited higher plasma and leukocyte-produced ENA-78 than wild-type homozygotes and that the promoter and exonic SNPs occur in transcription factor binding and splicing enhancer sites, respectively [33]. Given that the –156 G > C and 398 G > A SNPs are in near perfect linkage disequilibrium, it is unclear which polymorphism is the causal variant and functionally contributes to the blood pressure phenotype. However, the –156 G > C promoter SNP was more significantly correlated with blood pressure in our study. Further functional studies of these SNPs are warranted.

In addition to genotype and traditional covariates, we included plasma CRP and ENA-78 protein concentrations in our analyses. While CRP and ENA-78 were significantly associated with SBP (and PP) in univariate analyses, they fell out of the models when $CXCL5$ genotype was included. This suggests that in our analyses, genotype is more significantly associated with the blood pressure phenotype than systemically circulating concentrations of the non-specific inflammatory mediator CRP and the $CXCL5$ protein product ENA-78. While this observation may appear somewhat contradictory, it can be postulated that $CXCL5$ gene polymorphisms may be better indicators of chemokine activity at the target organ (e.g., endothelium) level than a measurement in the circulation. Because of trans-acting influences on systemic biomarker expression, polymorphisms in $CXCL5$ may be more robustly associated with blood pressure. In fact, we have shown a similar finding in a different population for the endothelial nitric oxide synthase gene where $NOS3$ gene polymorphisms, but not measures of circulating NO activity, were associated with arterial stiffness in children with type 1 diabetes [42,43]. Further support for this observation can be found in a case–control study of the role of ENA-78 in patients with ischemic stroke. Zaremba et al. demonstrated that serum ENA-78 protein concentrations were not different between stroke patients and controls; contrarily, it was demonstrated that ENA-78 concentrations were significantly higher (twofold) in the cerebrospinal fluid of stroke patients compared with controls [44]. Taken in sum, it is possible that genotype more effectively captures the likelihood for local preponderance of chemokine activity than plasma protein level.

In general, there is biological plausibility for the role of $CXCL5$ in CVD. For example, the protein product of $CXCL5$, ENA-78, belongs to the same class of chemokines as IL-8, IP-10, and I-TAC, which have been previously implicated in atherosclerotic inflammation [23,45]. ENA-78 has been shown to be chemotactic for neutrophils and stimulate neutrophilic degranulation causing release of myeloperoxidase and generating reactive oxygen species [24,25]. In addition, ENA-78 is involved in platelet-dependent activation of monocytes, displays angiogenic properties, and has been implicated in diseases such as obesity, diabetes, subclinical atherosclerosis, acute coronary syndromes, ischemic stroke, abdominal aortic aneurysm, and thrombosis [27,28,32,44,46-51]. Hypertension is a risk factor for adverse events such as atherosclerosis, stroke, and abdominal aortic aneurysm, and ENA-78 is overexpressed in these situations. We have shown $CXCL5$ polymorphisms to be associated with ENA-78 concentrations, blood pressure, and prognosis following acute coronary syndromes [27,33]. Thus, the role of $CXCL5$ in CVD should be further explored. As final hypothesis-generating evidence of a link between the $CXCL5$ pathway and blood pressure, statins have been hypothesized to have mild antihypertensive effects, and we have shown that atorvastatin reduces ENA-78 production from human endothelial cells in a dose-dependent fashion [52,53]. Our findings, along with existing data, support the need for future investigation of $CXCL5$ as a hypertension- and CVD-susceptibility gene.

Competing interests

The authors declare that they have no competing interests.

Authors' contributions

ALB performed statistical analyses and drafted the manuscript. CLA enrolled the study subjects and drafted the manuscript. HA assisted in the molecular genetic studies and provided critical revision of the manuscript. TYL assisted in the molecular genetic studies. GJW assisted in the molecular genetic studies. RSS assisted with the clinical study. IZ conceived the manuscript, enrolled the study subjects, and drafted the manuscript. All authors read and approved the final manuscript.

Acknowledgments
We thank Dr. Julie A. Johnson for her thoughtful comments regarding the manuscript. We thank Lauren Burt and Lynda Stauffer for their laboratory assistance. This work was supported by American Heart Association Florida/ Puerto Rico Affiliate Scientist Development Grant 0435278B, American College of Clinical Pharmacy Kos Dyslipidemia Research and Pharmacotherapy New Investigator Awards, American Association of Colleges of Pharmacy New Investigator Program Award, the University of Colorado Denver General Clinical Research Center (RR00051), and NIH C06 Grant RR17568. ALB is supported by K23 HL091120.

Author details
[1]Division of Endocrinology, Diabetes and Nutrition, University of Maryland School of Medicine, 660 W. Redwood St, HH469, Baltimore, MD 21201, USA. [2]Department of Pharmaceutical Sciences, University of Colorado Skaggs School of Pharmacy and Pharmaceutical Sciences, Aurora, CO 80045, USA. [3]Department of Preventive Medicine and Institute for Genetic Medicine, Keck School of Medicine, University of Southern California, Los Angeles, CA 90089, USA. [4]Department of Pharmacotherapy and Translational Research, Center for Pharmacogenomics, University of Florida College of Pharmacy, Gainesville, FL 32610, USA. [5]Division of Cardiovascular Medicine and Department of Veterans Affairs Medical Center, University of Florida College of Medicine, Gainesville, FL 32603, USA.

References
1. Watson T, Goon PK, Lip GY: **Endothelial progenitor cells, endothelial dysfunction, inflammation, and oxidative stress in hypertension.** *Antioxid Redox Signal* 2008, **10**:1079–1088.
2. Harrison DG, Guzik TJ, Lob HE, Madhur MS, Marvar PJ, Thabet SR, Vinh A, Weyand CM: **Inflammation, immunity, and hypertension.** *Hypertension* 2011, **57**:132–140.
3. Chae CU, Lee RT, Rifai N, Ridker PM: **Blood pressure and inflammation in apparently healthy men.** *Hypertension* 2001, **38**:399–403.
4. Abramson JL, Weintraub WS, Vaccarino V: **Association between pulse pressure and C-reactive protein among apparently healthy US adults.** *Hypertension* 2002, **39**:197–202.
5. Sesso HD, Buring JE, Rifai N, Blake GJ, Gaziano JM, Ridker PM: **C-reactive protein and the risk of developing hypertension.** *JAMA* 2003, **290**:2945–2951.
6. Sung KC, Suh JY, Kim BS, Kang JH, Kim H, Lee MH, Park JR, Kim SW: **High sensitivity C-reactive protein as an independent risk factor for essential hypertension.** *Am J Hypertens* 2003, **16**:429–433.
7. Bautista LE, Vera LM, Arenas IA, Gamarra G: **Independent association between inflammatory markers (C-reactive protein, interleukin-6, and TNF-alpha) and essential hypertension.** *J Hum Hypertens* 2005, **19**:149–154.
8. Margolis KL, Manson JE, Greenland P, Rodabough RJ, Bray PF, Safford M, Grimm RH Jr, Howard BV, Assaf AR, Prentice R, Women's Health Initiative Research Group: **Leukocyte count as a predictor of cardiovascular events and mortality in postmenopausal women: the Women's Health Initiative Observational Study.** *Arch Intern Med* 2005, **165**:500–508.
9. Vazquez-Oliva G, Fernandez-Real JM, Zamora A, Vilaseca M, Badimon L: **Lowering of blood pressure leads to decreased circulating interleukin-6 in hypertensive subjects.** *J Hum Hypertens* 2005, **19**:457–462.
10. Horne BD, Anderson JL, John JM, Weaver A, Bair TL, Jensen KR, Renlund DG, Muhlestein JB, Intermountain Heart Collaborative Study Group: **Which white blood cell subtypes predict increased cardiovascular risk?** *J Am Coll Cardiol* 2005, **45**:1638–1643.
11. Lakoski SG, Herrington DM, Siscovick DM, Hulley SB: **C-reactive protein concentration and incident hypertension in young adults: the CARDIA study.** *Arch Intern Med* 2006, **166**:345–349.
12. Sesso HD, Wang L, Buring JE, Ridker PM, Gaziano JM: **Comparison of interleukin-6 and C-reactive protein for the risk of developing hypertension in women.** *Hypertension* 2007, **49**:304–310.

13. Wang TJ, Gona P, Larson MG, Levy D, Benjamin EJ, Tofler GH, Jacques PF, Meigs JB, Rifai N, Selhub J, Robins SJ, Newton-Cheh C, Vasan RS: **Multiple biomarkers and the risk of incident hypertension.** *Hypertension* 2007, **49**:432–438.
14. Lakoski SG, Cushman M, Siscovick DS, Blumenthal RS, Palmas W, Burke G, Herrington DM: **The relationship between inflammation, obesity and risk for hypertension in the Multi-Ethnic Study of Atherosclerosis (MESA).** *J Hum Hypertens* 2011, **25**:73–79.
15. Hoffman M, Blum A, Baruch R, Kaplan E, Benjamin M: *Leukocytes and coronary heart disease. Atherosclerosis* 2004, **172**:1–6.
16. Karthikeyan VJ, Lip GY: **White blood cell count and hypertension.** *J Hum Hypertens* 2006, **20**:310–312.
17. Friedman GD, Selby JV, Quesenberry CP Jr: **The leukocyte count: a predictor of hypertension.** *J Clin Epidemiol* 1990, **43**:907–911.
18. Shankar A, Klein BE, Klein R: **Relationship between white blood cell count and incident hypertension.** *Am J Hypertens* 2004, **17**:233–239.
19. Gillum RF, Mussolino ME: **White blood cell count and hypertension incidence. The NHANES I Epidemiologic Follow-up Study.** *J Clin Epidemiol* 1994, **47**:911–919.
20. Orakzai RH, Orakzai SH, Nasir K, Santos RD, Rana JS, Pimentel I, Carvalho JA, Meneghello R, Blumenthal RS: **Association of white blood cell count with systolic blood pressure within the normotensive range.** *J Hum Hypertens* 2006, **20**:341–347.
21. Schillaci G, Pirro M, Pucci G, Ronti T, Vaudo G, Mannarino MR, Porcellati C, Mannarino E: **Prognostic value of elevated white blood cell count in hypertension.** *Am J Hypertens* 2007, **20**:364–369.
22. Tatsukawa Y, Hsu WL, Yamada M, Cologne JB, Suzuki G, Yamamoto H, Yamane K, Akahoshi M, Fujiwara S, Kohno N: **White blood cell count, especially neutrophil count, as a predictor of hypertension in a Japanese population.** *Hypertens Res* 2008, **31**:1391–1397.
23. Walz A, Burgener R, Car B, Baggiolini M, Kunkel SL, Strieter RM: **Structure and neutrophil-activating properties of a novel inflammatory peptide (ENA-78) with homology to interleukin 8.** *J Exp Med* 1991, **174**:1355–1362.
24. Walz A, Schmutz P, Mueller C, Schnyder-Candrian S: **Regulation and function of the CXC chemokine ENA-78 in monocytes and its role in disease.** *J Leukoc Biol* 1997, **62**:604–611.
25. Walz A, Strieter RM, Schnyder S: **Neutrophil-activating peptide ENA-78.** *Adv Exp Med Biol* 1993, **351**:129–137.
26. Wislez M, Philippe C, Antoine M, Rabbe N, Moreau J, Bellocq A, Mayaud C, Milleron B, Soler P, Cadranel J: **Upregulation of bronchioloalveolar carcinoma-derived C-X-C chemokines by tumor infiltrating inflammatory cells.** *Inflamm Res* 2004, **53**:4–12.
27. Zineh I, Beitelshees AL, Welder GJ, Hou W, Chegini N, Wu J, Cresci S, Province MA, Spertus JA: **Epithelial neutrophil-activating peptide (ENA-78), acute coronary syndrome prognosis, and modulatory effect of statins.** *PLoS One* 2008, **3**:e3117.
28. Chavey C, Lazennec G, Lagarrigue S, Clape C, Iankova I, Teyssier J, Annicotte JS, Schmidt J, Mataki C, Yamamoto H, Sanches R, Guma A, Stich V, Vitkova M, Jardin-Watelet B, Renard E, Strieter R, Tuthill A, Hotamisligil GS, Vidal-Puig A, Zorzano A, Langin D, Fajas L: **CXC ligand 5 is an adipose-tissue derived factor that links obesity to insulin resistance.** *Cell Metab* 2009, **9**:339–349.
29. Dominguez M, Miquel R, Colmenero J, Moreno M, Garcia-Pagan JC, Bosch J, Arroyo V, Ginès P, Caballería J, Bataller R: **Hepatic expression of CXC chemokines predicts portal hypertension and survival in patients with alcoholic hepatitis.** *Gastroenterology* 2009, **136**:1639–1650.
30. Yang Z, Zhang Z, Wen J, Wang X, Lu B, Zhang W, Wang M, Feng X, Ling C, Wu S, Hu R: **Elevated serum chemokine CXC ligand 5 levels are associated with hypercholesterolemia but not a worsening of insulin resistance in Chinese people.** *J Clin Endocrinol Metab* 2010, **95**:3926–3932.
31. Keeley EC, Moorman JR, Liu L, Gimple LW, Lipson LC, Ragosta M, Taylor AM, Lake DE, Burdick MD, Mehrad B, Strieter RM: **Plasma chemokine levels are associated with the presence and extent of angiographic coronary collaterals in chronic ischemic heart disease.** *PLoS One* 2011, **6**:e21174.
32. Chen L, Yang Z, Lu B, Li Q, Ye Z, He M, Huang Y, Wang X, Zhang Z, Wen J, Liu C, Qu S, Hu R: **Serum CXC ligand 5 is a new marker of subclinical atherosclerosis in type 2 diabetes.** *Clin Endocrinol (Oxf)* 2011, **75**(6):766–770.

33. Zineh I, Aquilante CL, Langaee TY, Beitelshees AL, Arant CB, Wessel TR, Schofield RS: **CXCL5 gene polymorphisms are related to systemic concentrations and leukocyte production of epithelial neutrophil-activating peptide (ENA-78).** *Cytokine* 2006, **33:**258–263.

34. Expert Panel on Detection, Evaluation, and Treatment of High Blood Cholesterol in Adults: **Executive Summary of The Third Report of The National Cholesterol Education Program (NCEP) Expert Panel on Detection, Evaluation, And Treatment of High Blood Cholesterol In Adults (Adult Treatment Panel III).** *JAMA* 2001, **285:**2486–2497.

35. Andrisin TE, Humma LM, Johnson JA: **Collection of genomic DNA by the noninvasive mouthwash method for use in pharmacogenetic studies.** *Pharmacotherapy* 2002, **22:**954–960.

36. Zineh I, Welder GJ, Langaee TY: **Development and cross-validation of sequencing-based assays for genotyping common polymorphisms of the CXCL5 gene.** *Clin Chim Acta* 2006, **370:**72–75.

37. Zineh I, Welder GJ, DeBella AE, Arant CB, Wessel TR, Schofield RS: **Atorvastatin effect on circulating and leukocyte-produced CD40 ligand concentrations in people with normal cholesterol levels: a pilot study.** *Pharmacotherapy* 2006, **26:**1572–1577.

38. Shiao YH, Crawford EB, Anderson LM, Patel P, Ko K: **Allele-specific germ cell epimutation in the spacer promoter of the 45 S ribosomal RNA gene after Cr(III) exposure.** *Toxicol Appl Pharmacol* 2005, **205:**290–296.

39. Sun A, Ge J, Siffert W, Frey UH: **Quantification of allele-specific G-protein beta3 subunit mRNA transcripts in different human cells and tissues by Pyrosequencing.** *Eur J Hum Genet* 2005, **13:**361–369.

40. Tian N, Penman AD, Mawson AR, Manning RD Jr, Flessner MF: **Association between circulating specific leukocyte types and blood pressure: the atherosclerosis risk in communities (ARIC) study.** *J Am Soc Hypertens* 2010, **4:**272–283.

41. Angeli F, Angeli E, Ambrosio G, Mazzotta G, Cavallini C, Reboldi G, Verdecchia P: **Neutrophil count and ambulatory pulse pressure as predictors of cardiovascular adverse events in postmenopausal women with hypertension.** *Am J Hypertens* 2011, **24:**591–598.

42. Haller MJ, Pierce GL, Braith RW, Silverstein JH: **Serum superoxide dismutase activity and nitric oxide do not correlate with arterial stiffness in children with type 1 diabetes mellitus.** *J Pediatr Endocrinol Metab* 2006, **19:**267–269.

43. Zineh I, Beitelshees AL, Haller MJ: **NOS3 polymorphisms are associated with arterial stiffness in children with type 1 diabetes.** *Diabetes Care* 2007, **30:**689–693.

44. Zaremba J, Skrobanski P, Losy J: **The level of chemokine CXCL5 in the cerebrospinal fluid is increased during the first 24 hours of ischaemic stroke and correlates with the size of early brain damage.** *Folia Morphol (Warsz)* 2006, **65:**1–5.

45. Libby P: *Inflammation in atherosclerosis.* Nature 2002, **420:**868–874.

46. Damas JK, Gullestad L, Ueland T, Solum NO, Simonsen S, Froland SS, Aukrust P: **CXC-chemokines, a new group of cytokines in congestive heart failure–possible role of platelets and monocytes.** *Cardiovasc Res* 2000, **45:**428–436.

47. Hamid C, Norgate K, D'Cruz DP, Khamashta MA, Arno M, Pearson JD, Frampton G, Murphy JJ: **Anti-beta2GPI-antibody-induced endothelial cell gene expression profiling reveals induction of novel pro-inflammatory genes potentially involved in primary antiphospholipid syndrome.** *Ann Rheum Dis* 2007, **66:**1000–1007.

48. Holm T, Damas JK, Holven K, Nordoy I, Brosstad FR, Ueland T, Währe T, Kjekshus J, Frøland SS, Eiken HG, Solum NO, Gullestad L, Nenseter M, Aukrust P: **CXC-chemokines in coronary artery disease: possible pathogenic role of interactions between oxidized low-density lipoprotein, platelets and peripheral blood mononuclear cells.** *J Thromb Haemost* 2003, **1:**257–262.

49. Middleton RK, Lloyd GM, Bown MJ, Cooper NJ, London NJ, Sayers RD: **The pro-inflammatory and chemotactic cytokine microenvironment of the abdominal aortic aneurysm wall: a protein array study.** *J Vasc Surg* 2007, **45:**574–580.

50. Hasani Ranjbar S, Amiri P, Zineh I, Langaee TY, Namakchian M, Heshmet R, Sajadi M, Mirzaee M, Rezazadeh E, Balaei P, Tavakkoly Bazzaz J, Gonzalez-Gay MA, Larijani B, Amoli MM: **CXCL5 gene polymorphism association with diabetes mellitus.** *Mol Diagn Ther* 2008, **12:**391–394.

51. Turner NA, Das A, O'Regan DJ, Ball SG, Porter KE: **Human cardiac fibroblasts express ICAM-1, E-selectin and CXC chemokines in response to proinflammatory cytokine stimulation.** *Int J Biochem Cell Biol* 2011, **43:**1450–1458.

52. Milionis HJ, Liberopoulos EN, Achimastos A, Elisaf MS, Mikhailidis DP: **Statins: another class of antihypertensive agents?** *J Hum Hypertens* 2006, **20:**320–335.

53. Zineh I, Luo X, Welder GJ, Debella AE, Wessel TR, Arant CB, Schofield RS, Chegini N: **Modulatory effects of atorvastatin on endothelial cell-derived chemokines cytokines, and angiogenic factors.** *Pharmacotherapy* 2006, **26:**333–340.

2

Complement regulator CD46: genetic variants and disease associations

M. Kathryn Liszewski[*] and John P. Atkinson

Abstract

Membrane cofactor protein (MCP; CD46) is an ubiquitously expressed complement regulatory protein that protects host cells from injury by complement. This type-I membrane glycoprotein serves as a cofactor for the serine protease factor I to mediate inactivation of C3b and C4b deposited on host cells. More than 60 disease-associated mutations in *MCP* have now been identified. The majority of the mutations are linked to a rare thrombotic microangiopathic-based disease, atypical hemolytic uremic syndrome (aHUS), but new putative links to systemic lupus erythematosus, glomerulonephritis, and pregnancy-related disorders among others have also been identified. This review summarizes our current knowledge of disease-associated mutations in this complement inhibitor.

Keywords: CD46, Membrane cofactor protein, Complement, Complement regulation, Atypical hemolytic uremic syndrome

Introduction

The complement system is one of the most ancient components of innate immunity. It likely evolved from a C3-like protein that was cleaved by proteases into biologically active self-defense fragments to counteract invading microbes, particularly bacteria, and to clear biologic debris (self and foreign) [1, 2]. The development of a circulatory system may have been the key evolutionary pressure that drove the need for a rapidly acting (within seconds) innate immune host-defense process with the destructive power to prevent invasion and multiplication of bacteria in the blood stream [3]. The complement system is sometimes referred to as "the guardian of the intravascular space".

The vertebrate-complement system consists of a set of sequentially interacting proteins featuring three major pathways that provide a swift and powerful host-defense system. It promotes the inflammatory response and mediates the identification and destruction of pathogens. This is accomplished in two major ways (Fig. 1). First, complement modifies the membranes of microbes. Activated fragments are covalently deposited in large amounts on microorganisms, immune complexes, and damaged tissue. For example, several million C3-derived activation fragments can attach in clusters to a bacterial surface in less

than five minutes. The most important function of these covalently deposited complement proteins is to opsonize the target, thus serving as ligands for complement receptors on peripheral blood cells; specifically, erythrocytes, neutrophils, B lymphocytes and monocytes, as well as dendritic cells, and tissue monocytes. The receptors serve to bind (immune adherence), internalize (in some cases), transport, and clear the microbe. Additionally, the membranes of some pathogens (especially gram-negative bacteria) can be disrupted by the terminal-complement components (membrane-attack complex, MAC) leading to osmotic lysis.

The second function of complement is to promote the inflammatory response. Thus, peptides, released by proteolysis during complement activation, bind to receptors to elicit an inflammatory reaction. These peptides are termed anaphylatoxins because they can trigger the release of mediators such as histamine to cause shock.

Due to its proinflammatory and destructive capabilities, it is no surprise that nearly half of the complement proteins serve in its regulation. Unimpeded, the complement system fires to exhaustion, a point illustrated by inherited deficiencies of its regulatory proteins [4, 5]. These inhibitors also participate in "self" versus "nonself" discrimination in that foreign surfaces, usually *lacking* such regulators, are recognized and attacked, while healthy self-tissues expressing the regulators are protected.

* Correspondence: kliszews@dom.wustl.edu
Division of Rheumatology, Department of Medicine, Washington University School of Medicine, 660 South Euclid, Saint Louis, MO 63110, USA

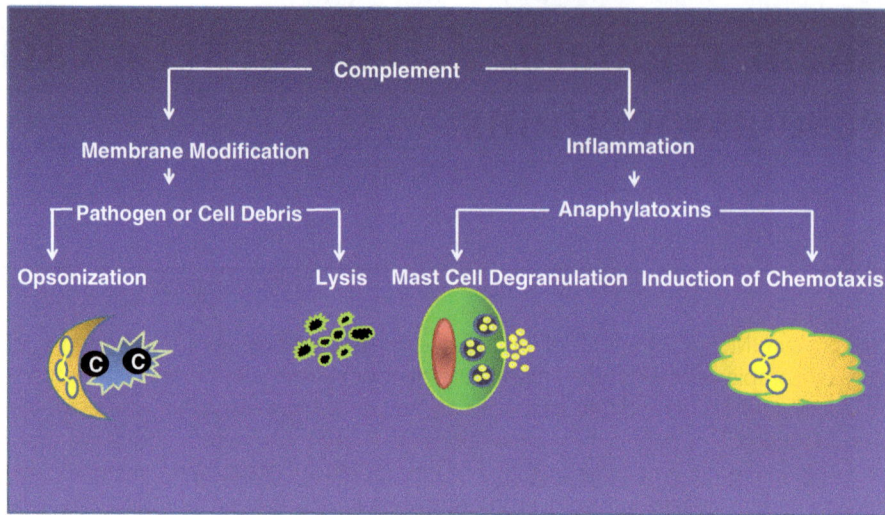

Fig 1 Complement function. The two primary functions of the complement system are to modify pathogens and self-debris with clusters of complement fragments (opsonization). This, in turn, facilitates interaction with complement receptors and, in some bacteria and viruses, induces lysis. The second function is to promote the inflammatory response. Complement fragments C3a and C5a generated during activation of the cascades stimulate many cell types. In the case of mast cells, release of immunomodulatory granules also attracts phagocytic cells into the area of inflammation (chemotaxis)

The alternative pathway (AP) continuously turns over, generating C3 fragments. If a C3b lands on host cells, it must be inactivated by regulatory proteins. One such control protein, CD46 (membrane cofactor protein; MCP), is a member of a group of genetically-, structurally-, and functionally-related proteins called the regulators of complement activation (RCA) [6, 7]. As its name implies, the goal of this gene cluster of receptors and inhibitors is to provide homeostasis by tightly controlling the rapid and powerful amplification process of the AP in order to focus complement attack, in both time and space, on pathogens and, in a more homeostatic manner, injured tissue.

Recently identified associations of human disease featuring excessive AP activation with heterozygous mutations in its components and regulators and the development of a novel therapeutic agent to block C5 cleavage have reignited interest in the field [8, 9].

This review focuses on the ubiquitously expressed inhibitor of C3b and C4b, CD46, and the primary diseases associated with its dysfunction. Citations included are not meant to be exhaustive, but rather to provide key review articles.

Complement pathways

More than a billion years ago, primitive elements of the complement system arose to form a humoral immune system likely derived by proteolysis of a primeval protein whose fragmentation released one piece to mediate opsonization and a second one to elicit an inflammatory response [1, 10–12]. This original pathway (that remains today with enhanced sophistication and inappropriately called the AP) provided a simple protein-based recognition and effector scheme against pathogens. Lectins and antibodies, representing subsequent evolutionary developments, became connected to the complement-dependent effector mechanisms of opsonization and membrane perturbation. As the system grew in capacity and efficiency, control mechanisms were required to maintain homeostasis and to focus attack on pathogens while minimizing damage to self.

The contemporary human-complement system now consists of an efficient, interacting set of nearly 60 blood (serum) and cellular components that include components of the activating cascades, receptors, and positive and negative regulators. Complement systems similar to that in mammals also have been identified in birds, fish, amphibians, and reptiles. An AP is also found in more primitive species, even those lacking a circulatory system [13]. The complement system consists of three major activating pathways that are independently triggered, yet all have the common goal of modifying the target membrane by depositing C3 activation products and then engaging a common terminal membrane-attack complex (Fig. 2).

The AP is the most ancient cascade. It does not require an antibody, a lectin, or prior contact with a pathogen to become engaged. Indeed, it serves as a rapid, self-

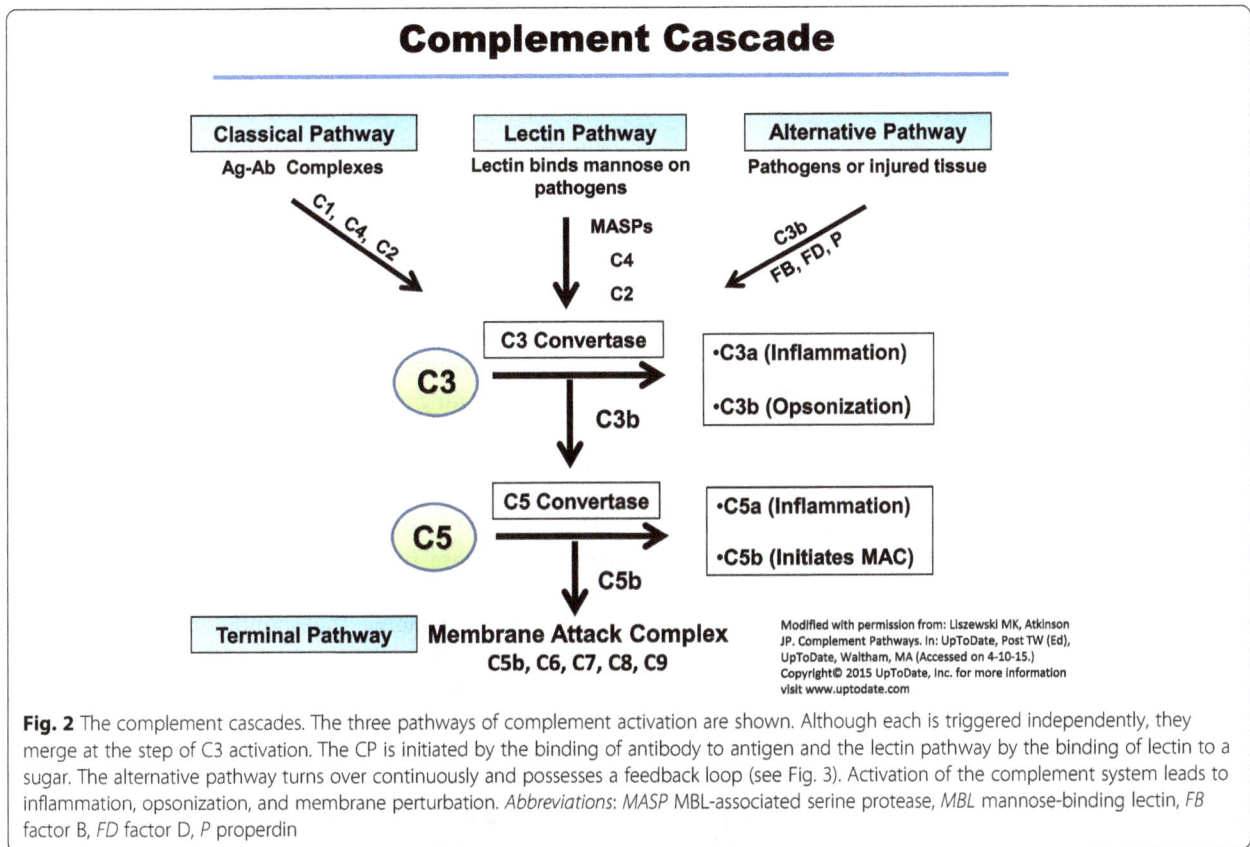

Fig. 2 The complement cascades. The three pathways of complement activation are shown. Although each is triggered independently, they merge at the step of C3 activation. The CP is initiated by the binding of antibody to antigen and the lectin pathway by the binding of lectin to a sugar. The alternative pathway turns over continuously and possesses a feedback loop (see Fig. 3). Activation of the complement system leads to inflammation, opsonization, and membrane perturbation. *Abbreviations: MASP* MBL-associated serine protease, *MBL* mannose-binding lectin, *FB* factor B, *FD* factor D, *P* properdin

amplifying, and exceptionally powerful innate immune system capable of independently recognizing and destroying foreign targets and promoting an inflammatory response. A small amount of auto-activated C3 (so called, C3 tickover) is constantly generated in blood secondary to engagement of its labile thioester bond. This C3 turnover mechanism serves as a surveillance system. If it deposits on healthy self, it is inactivated. If it deposits on a microbe, it can be rapidly amplified. Thus, in the latter case, C3b sequentially engages two proteases, factor B (FB) and factor D (FD), and the stabilizing protein properdin (P). These interact to form an AP C3 convertase (C3bBbP) that cleaves C3 to C3b and C3a. This system, therefore, represents a powerful feedback loop for the generation of C3b (Fig. 3).

The classical pathway (CP) was discovered first in the late 1800s, hence its name. It is primarily triggered by antibodies (IgM and IgG, subclasses 1 and 3) binding to antigens. This initiates a cascade featuring multiple proteolytic cleavage steps beginning with the C1s component of C1 that cleaves C4 and C2. The newly generated C4b and C2a fragments then interact to form the C3 convertase, a proteolytic complex that activates C3. This convertase "converts" C3 into the C3a fragment (an inflammatory modulator) and the larger fragment, C3b, which serves as an opsonin and nidus for a feedback

loop to generate more C3b. In an analogous process, the LP generates the same C3 convertase, but in this scheme, lectins substitute for antibodies in binding antigen-like sugar (mannose) residues, and mannose binding lectin-associated serine proteases (MASPs) replace C1r and C1s [14].

All three pathways merge at the step of the generation of a C3 convertase (Fig. 2). The C3b that deposits on pathogens serves both as a ligand for receptors in addition to being a central component of the feedback loop. Addition of a second C3b to a C3 convertase generates a C5 convertase. The latter cleaves C5 into a potent anaphylatoxin (C5a) and the larger C5b fragment. The C5b binds to C6 and C7 (without proteolysis) and this complex attaches to a membrane. Next, the C5b67 complex binds C8 followed by multiple C9s (~10–15) to form the MAC (C5b-9) (reviewed in [15]).

CD46 and the "regulators of complement activation" gene cluster

Control of complement occurs in the fluid phase (plasma) and on self-tissue at each of the major steps in the pathways: initiation; amplification leading to C3 and then C5 cleavage; and, formation of C5b-9.

Originally identified as a C3b- and C4b-binding protein of human peripheral blood cells, CD46 is expressed

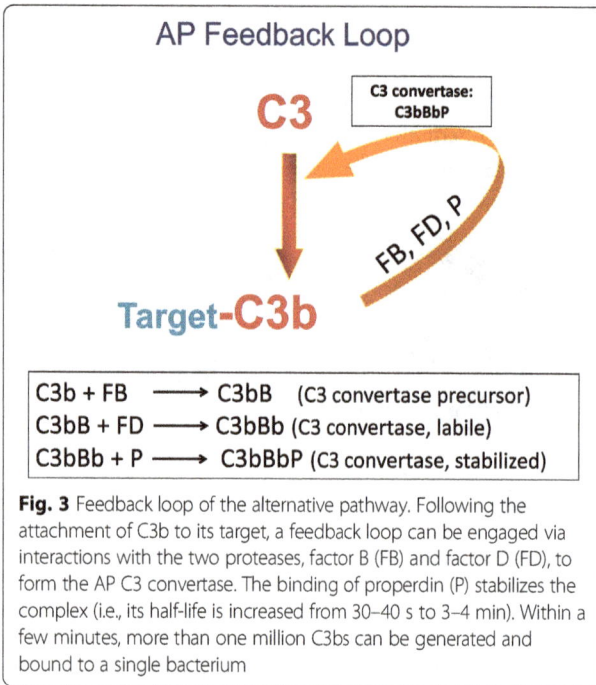

AP Feedback Loop

C3 convertase:
C3bBbP

C3

Target-C3b

FB, FD, P

C3b + FB	⟶ C3bB	(C3 convertase precursor)
C3bB + FD	⟶ C3bBb	(C3 convertase, labile)
C3bBb + P	⟶ C3bBbP	(C3 convertase, stabilized)

Fig. 3 Feedback loop of the alternative pathway. Following the attachment of C3b to its target, a feedback loop can be engaged via interactions with the two proteases, factor B (FB) and factor D (FD), to form the AP C3 convertase. The binding of properdin (P) stabilizes the complex (i.e., its half-life is increased from 30–40 s to 3–4 min). Within a few minutes, more than one million C3bs can be generated and bound to a single bacterium

as a type-I transmembrane protein on nucleated cells [16–18]. The *MCP* gene is located in the regulators of complement activation (RCA) cluster at 1q3.2 [6, 19]. In addition to CD46, other proteins in this C3/C4-interacting family are CD35 (complement receptor one; CR1); CD21 (complement receptor 2, CR2); CD55 (decay-accelerating factor, DAF); C4b binding protein (C4BP); factor H (FH) and its family of proteins, factor H-like protein 1 (FHL-1), and factor H-related proteins 1–5 (FHR-1 to 5) [7, 20, 21].

The human RCA gene cluster spans a total of 21.45 cM on the long arm of chromosome 1 and includes more than 60 genes of which 15 are related to complement [21]. One group of genes is telomeric in a 900-kb DNA segment and the other is a centromeric 650-kb fragment. These are separated by a 14–59 cM segment that includes a number of genes unrelated to complement [21]. Complement regulatory genes share a common

ancestral motif from which they arose by multiple gene duplication events [13, 19, 21].

CD46 and other members of the RCA are composed of a repeating unit that begins at the amino-terminus and comprises most or all of the protein. This structural feature, called a complement control protein (CCP) module (also a short consensus repeat or a sushi domain), consists of ~60 amino acids with four invariant cysteines and 10–18 other highly conserved amino acids. The CCP modules house the sites for regulatory activities, i.e., binding sites for C3b, C4b and/or factor I (FI) that foster cofactor activity (CD46, FH, C4BP, CR1), and decay-accelerating activity (CD55, FH, C4BP, CR1).

Cofactor activity is the process whereby C3b and C4b are proteolytically inactivated by FI, a plasma serine protease. A cofactor protein such as CD46 or FH must bind to the substrate (C4b or C3b) before FI can cleave these two large fragments (Fig. 4). Decay-accelerating activity (DAA) refers to a regulatory process for enhancing the spontaneous decay of the convertases. The active protease fragment is dissociated. CD46 possesses cofactor activity for C4b and C3b but does *not* possess DAA.

The gene for CD46 is ~43 kb and contains 14 exons [6]. A partial duplicate (exons 1–4; *MCP-like*) has been identified, but there is no evidence for its expression. CD46 consists of four major alternatively spliced transcripts [6, 16]. Most cell types express all four of the proteins in approximately the same ratio as is observed on peripheral blood cells (although cell-specific isoform expression has been identified [16, 19]). Each isoform shares an identical amino-terminal portion consisting of four CCPs followed by an alternatively spliced region for O-glycosylation. Although there are three exons coding for this region (A, B, and C), the four most commonly expressed isoforms carry B + C or C alone. Next is the flanking juxtamembraneous 12 amino-acid domain of unknown significance followed by a hydrophobic transmembrane region and a charged cytoplasmic anchor. The carboxyl-terminal tail is also alternatively spliced. It contains one of two nonhomologous cytoplasmic tails of

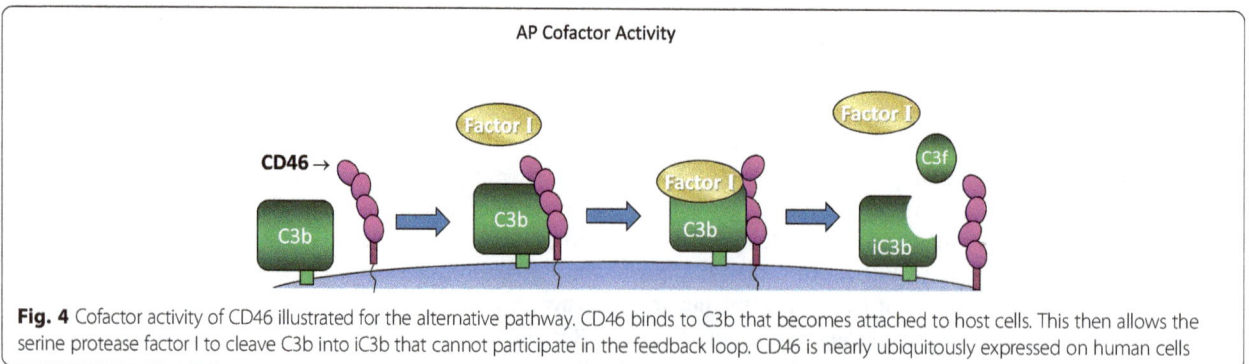

AP Cofactor Activity

CD46 →

Factor I

Factor I

Factor I

C3f

C3b

C3b

C3b

iC3b

Fig. 4 Cofactor activity of CD46 illustrated for the alternative pathway. CD46 binds to C3b that becomes attached to host cells. This then allows the serine protease factor I to cleave C3b into iC3b that cannot participate in the feedback loop. CD46 is nearly ubiquitously expressed on human cells

16 or 23 amino acids each of which houses signaling motifs. The four major isoforms are named CD46-BC1,–BC2,–C1, and – C2.

CD46 function

In addition to its primary role in the regulation of C3b and C4b as a membrane cofactor protein, CD46 has been called the "multitasker" [18] and a "pathogen magnet" [22]. CD46 is increasingly recognized for its roles in linking innate and adaptive immune responses [17, 18, 23].

Pathogen magnet

The nearly universal expression of CD46 may particularly explain its exploitation by a number of pathogens who employ it as a receptor, possibly to co-opt one or more of its complement regulatory activities, signaling capabilities, or internalization mechanisms (reviewed in [22]). Nine human-specific pathogens target CD46 including four viruses (measles virus, adenovirus groups B and D, and herpesvirus 6) and five bacterial species (Neisseria gonorrhoeae, Neisseria meningitidis, Streptococcus pyogenes, Escherichia coli, and Fusobacterium nucleatum). Additionally, bovine CD46 is a receptor for bovine viral diarrhea virus, an enveloped RNA virus.

Pathogens target different domains of CD46 for attachment and subsequent reactions. The adenovirus fiber-knob protein and measles virus hemagglutinin attach to CCPs 1 and 2. Measles-virus binding elicits internalization and alters intracellular processing and antigen presentation (reviewed in [24, 25]). An envelope glycoprotein complex from herpesvirus 6 binds CCPs 2 and 3 (reviewed in [26]). Following clathrin-mediated endocytosis, the viral nucleic acid transits to the nucleus and replicates [25]. The type-IV pilus of the Neisseria species targets CCPs 3 and 4 and the STP segment. Neisseria infection causes the phosphorylation of the cytoplasmic domain (CYT-2) of CD46 by c-Yes, a member of the Src tyrosine kinase family. Studies suggest this phosphorylation is essential for Neisseria attachment and cytoskeletal rearrangement and that Neisseria also stimulates proteolytic cleavage of CD46 tails during infection (reviewed in [18, 27]). The M protein of S. pyogenes, which facilitates invasion of epithelial cells, attaches to CCPs 3 and 4. On epithelial cells, infection leads to the shedding of CD46 at the same time as the bacteria induce apoptosis and cell death [28]. Furthermore, interaction of S. pyogenes and CD46 triggers cell signaling pathways that lead to an immunosuppressive/regulatory phenotype in T cells (reviewed in [29] and [22]). The binding site for the periodontal disease-associated F. nucleatum is unknown [30]. However, the binding of F. nucleatum to CD46 on epithelial cells contributes to increasing levels of proinflammatory mediators and matrix metalloproteinases likely involved in periodontal tissue destruction [30].

Pathogenic microbes also develop their own human-like proteins to subvert host defenses. Poxviruses express CD46-like proteins (30–40 % homologous) to control host complement [31, 32]. Called poxviral inhibitors of complement enzymes (PICES), proteins from variola, and monkeypox are named SPICE (smallpox inhibitor of complement enzymes) and MOPICE (monkeypox inhibitor of complement enzymes), respectively [31–34]. They consist of three or four CCPs that are structural and functional mimics of CD46 and CD55. These virulence factors attach to glycosaminoglycans on the cell surface via their heparin-binding sites and, thus, down-regulate the complement system's ability to attack the virus [31].

Better understanding of the mechanism by which pathogens hijack (e.g., poxviruses) and usurp CD46 function may also provide greater insights relative to its normal functional repertoire. Additionally, replication-defective forms of adenovirus are being utilized in gene transfer and vaccine clinical trials necessitating a better understanding of its attachment mechanisms (reviewed in [25]).

Immunomodulatory functions

Over the last 15 years, it has become increasingly apparent that CD46 plays important and surprising immunomodulatory roles. These studies have been extensively reviewed particularly in relationship to T-cell biology (reviewed in [17, 18, 23]). The co-engagement of the T-cell receptor and CD46 by CD4$^+$ T cells leads to the induction of interferon-γ (IFN-γ)-secreting effector T-helper type-1 (Th1) cells (reviewed in [17, 18, 23]). These subsequently switch predominantly into interleukin-10 (IL-10)-secreting regulatory T cells (Tregs). Thus, a time-ordered functionally-relevant sequence occurs following stimulation through CD46; activation initially induces a rapid burst of interleukin-2 (IL-2) secretion and the generation of a proinflammatory IFN-γ^+/IL-10$^-$ T cell phenotype followed by an intermediate step with IFN-γ^+/IL-10$^+$ cells. The latter then switches to an IFN-γ^-/IL-10$^+$ self-regulatory phenotype (Treg). Recent studies suggest a possible mechanism for this via the binding by CD46 to a newly discovered extracellular ligand, Jagged1, a member of the Notch family of proteins that is regulated by CD46 [18].

Thus, one could envision that at the outset of an immune response, complement-induced Tregs provide a supportive role by facilitating B-cell activation via high IL-10 and sCD40L production (factors required for optimal B-cell activation and Ig class switching). As the immune response progresses, especially with antibody production and complement-mediated clearance of infectious pathogens, complement-induced Tregs might then constrain and/or deactivate the immune response.

Development of a Treg response is important for maintaining peripheral tolerance and control of immune responses. Further, T-cell subsets are increasingly recognized to possess different levels of plasticity in which they acquire new features and functions relative to secondary or chronic immune responses [35]. Thus, disruption or alterations of any of these pathways may contribute to susceptibility to infections or to the development of autoimmunity. Human diseases implicating defects in CD46-mediated signaling (reviewed in [17]) are multiple sclerosis, rheumatoid arthritis, asthma, IPEX-like syndrome, and primary C3 deficiency. Further, a subset of CD46-deficient patients develops common variable immunodeficiency (CVID), a syndrome characterized by hypogammaglobulinemia [36]. One study demonstrated that CD46-activated T cells support B-cell activation and that T cells from a CD46-deficient patient are impaired in promoting IgG production by B cells [37]. However, it is as yet unclear how these CD46-signaling functions play out in human disease.

CD46 variants and disease association

Linkage analyses, genome-wide association studies, and the recent dramatic progress in next-generation sequencing have revealed an expanding number of disease-associated genetic alterations in CD46. Specifically, MCP variants have been increasingly associated with inflammatory disorders particularly characterized by the development of thrombi in small blood vessels (thombotic microangiopathy), especially atypical hemolytic uremic syndrome (aHUS) [38–43].

As expected, a majority of aHUS and other disease-associated mutations in CD46 occur in the four CCPs, the extracellular domains responsible for its complement regulatory activity (Fig. 5 and Tables 1 and 2). An earlier publication dissecting many of the active sites of CD46 by mutation modeling demonstrated how critical amino acids at binding sites lead to a loss-of-function [44].

Note that there exists confusion in the literature detailing CD46-protein numbering of mutations since some references include the 34 amino acid signal peptide and all exons of the STP domain (for example, [38, 39, 45, 46]). Others do not include the signal peptide and/or may leave out the exon of the STP domain that is a rare protein product (for example, [36, 47, 48]). In this review, we have adopted the format of numbering from the translated protein, i.e., the signal peptide and including all exons of the protein as recommended by the Human Genome Variation Society.

Overall, two mutations have been identified in the promoter, four in the signal peptide, thirteen in CCP1, nine in CCP2, fourteen in CCP3, thirteen in CCP4, one in the STP region, four in the transmembrane domain, and one in cytoplasmic tail one (CYT-1) (Tables 1 and 2). While 52 mutations are associated with the development of aHUS, 13 may be associated with other diseases, and four mutations have been described in several diseases (see below).

Atypical hemolytic uremic syndrome

Mutation of CD46 has been linked most often to development of aHUS. The overall incidence of this rare disorder was estimated to be 2 in 1 million (10^6) in a North American population (reviewed in [43]). Although aHUS is characterized by the triad of microangiopathic hemolytic anemia, thrombocytopenia (lowered platelet count) and acute renal failure, other organs such as brain, lung, and gastrointestinal tract can also be affected [38–43, 49] . Most typical HUS cases (~90 %) are epidemic in nature featuring diarrhea in association with an enteric infection from a verocytotoxin-secreting bacteria (e.g., *Escherichia coli* O157:H7). Following gastrointestinal infection, most patients recover, although 5–10 % will progress into enteropathic or Shigatoxin-producing *E. coli* (STEC) HUS, which has a good prognosis for recovery.

In contrast, atypical HUS (aHUS) is a more severe, non-diarrheal type that results from alternative pathway over-reaction on endothelial cells, particularly in the kidney. Penetrance is ~50 % with a relapsing and remitting course resulting in a post-mortality rate in the acute phase of ~25 %. For survivors, ~50 % will remain dialysis-dependent. Approximately 60 % of aHUS cases occur during childhood, and in a majority, the initial episode occurs before the age of 2 years (reviewed in [50]). In contrast to mutations in FH or FI, kidney transplantation for CD46-deficient individuals has a nearly normal success rate since the transplanted organ carries a normal level of CD46 expression [43, 46, 49].

Factors that are reported to precipitate aHUS include infections, pregnancy, trauma, or drugs. Why the kidney endothelium is the major site of organ damage is unknown. What is increasingly clear is that the fundamental defect in this disease is an inability to control the AP on damaged or stressed cell surfaces resulting in excessive and harmful activation on "altered self". Dysfunction of mutated proteins can result from loss-of-function (in regulators responsible for cofactor activity) or gain-of-function (activating components, hyperactive C3 convertases). Both lead to inefficiently degraded C3b and abnormal persistence of C3 and C5 convertases that, in turn, generate excessive amounts of complement-pathway effectors. Further, C5b initiates the assembly of the MAC, leading to membrane injury, while C5a recruits and activates leukocytes and upregulates vascular adhesiveness. With the delicate balance between complement activation versus complement regulation

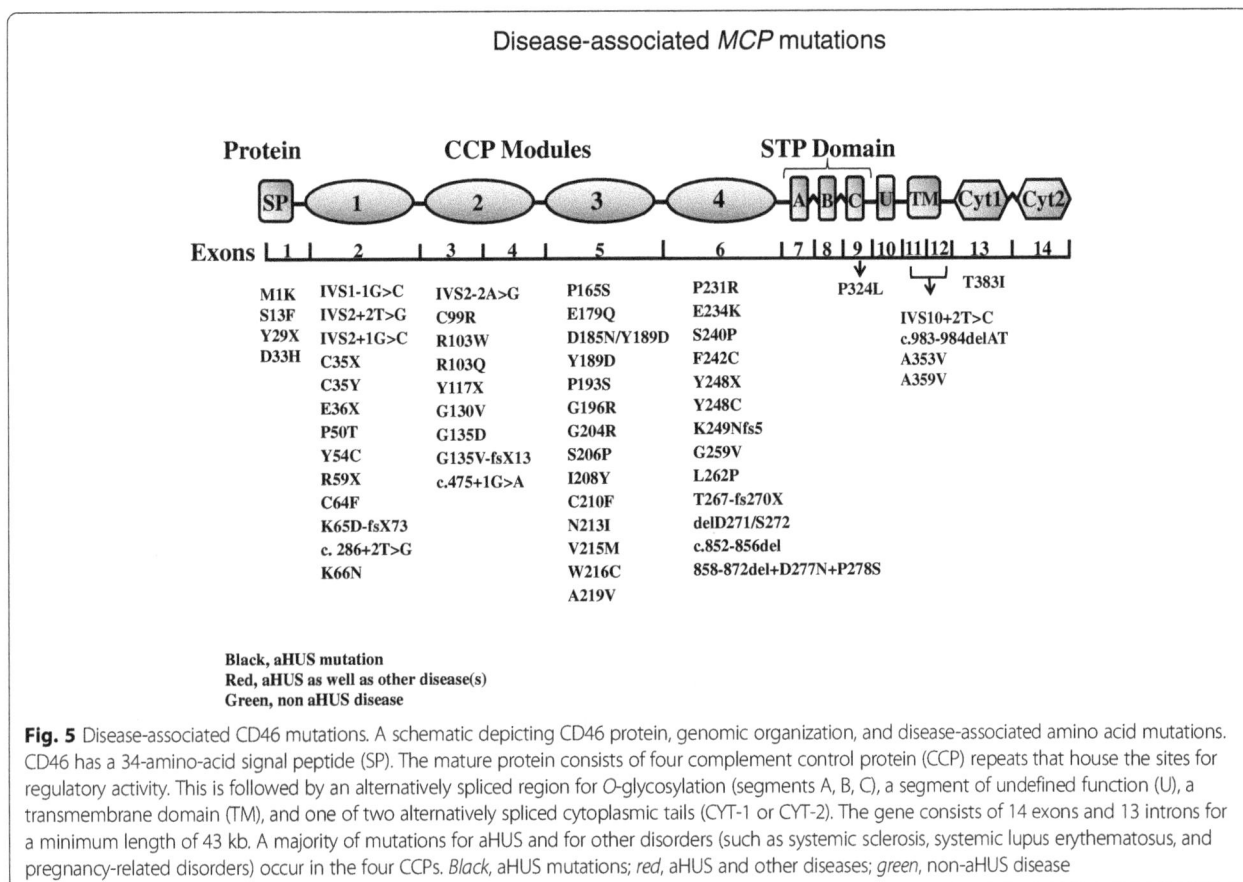

Fig. 5 Disease-associated CD46 mutations. A schematic depicting CD46 protein, genomic organization, and disease-associated amino acid mutations. CD46 has a 34-amino-acid signal peptide (SP). The mature protein consists of four complement control protein (CCP) repeats that house the sites for regulatory activity. This is followed by an alternatively spliced region for *O*-glycosylation (segments A, B, C), a segment of undefined function (U), a transmembrane domain (TM), and one of two alternatively spliced cytoplasmic tails (CYT-1 or CYT-2). The gene consists of 14 exons and 13 introns for a minimum length of 43 kb. A majority of mutations for aHUS and for other disorders (such as systemic sclerosis, systemic lupus erythematosus, and pregnancy-related disorders) occur in the four CCPs. *Black*, aHUS mutations; *red*, aHUS and other diseases; *green*, non-aHUS disease

perturbed on endothelial cells, the thrombotic microangiopathy ensues with vessel-wall thickening, cell engorgement, and destruction.

Mutations in *MCP* that predispose to aHUS were first identified in 2003 [51, 52]. *MCP* mutations were evaluated in three families [51] with a second group reporting a mutation in one family [52]. At present, at least 52 mutations linked to development of aHUS and 13 to other diseases have been reported (Tables 1 and 2). Mutations in *MCP* are found in ~10–20 % of aHUS patients. Most mutations are missense but nonsense, and splice-site variants are also observed (reviewed in [40, 43, 46, 48, 53–55]). The majority are also commonly rare, novel, and deleterious.

In about 75 % of cases, the mutant protein is not expressed. In the remainder, the aberrant protein is expressed but has a reduced or absent function, i.e., C3b- or C4b-binding and/or cofactor activity. Reduced cofactor activity for C3b impairs proper regulation of the AP. In addition to *MCP* mutations, a specific SNP haplotype block termed the *MCPggaac* haplotype in the *MCP* promoter region may be associated in vitro with reduced transcriptional activity (reviewed in [43]). This has been linked with an increased risk of aHUS but only in the setting of a causative variant in another AP component or regulator.

Other diseases

Mutations in *MCP* associated with diseases other than aHUS have been noted (Table 2). These studies involve a small number of patients and all will require confirmation by further investigations.

Systemic sclerosis is an autoimmune disease characterized by immune system activation, microvascular dysfunction, and tissue fibrosis. Scambi et al. reported an association between abnormally low CD46 expression in skin vessels in a subset of patients with two polymorphic variants (−366A > G, rs2796268 and −652A > G, rs2796267) in the *MCP* promoter [56]. These two SNPs have also been linked to enhanced severity of aHUS [38, 39].

Nonsynonymous *MCP* mutations (S13F and A219V) were implicated in earlier development of nephritis, but were not predisposing to systemic lupus erythematosus (SLE) or nephritis [57]. The A353V (rs353665573) is an uncommon polymorphism (1–5 %) that has been reported to be associated with aHUS and several other diseases discussed below. The S13F *MCP* mutation is also associated with development of the HELLP syndrome that features a combination of hemolysis, elevated liver enzymes, and low platelets [58]. This disorder occurs in about 0.5–0.9 % of all pregnancies and in 5–10 % of patients who develop severe preeclampsia [58].

Table 1 CD46 mutations associated with aHUS and functional consequences

Mutation[#]	Domain	Functional studies	Notes	Refs
M1K	SP	Reduced expression		[69]
S13F	SP	ND		[70]
Y29X	SP	ND	Compound heterozygote with factor H mutation	[43, 53]
D33H	SP	ND		[46]
IVS1 − 1G > C	CCP1	Reduced expression	Aberrant splicing of 2 bp after normal splice site; premature stop codon C35X	[36, 43, 47, 48]
IVS2 + 2 T > G	CCP1	Reduced expression	Deleted 144 bp and 48 amino acids in phase with wild-type	[36, 43, 47, 48, 71]
IVS2 + 1G > C	CCP1	Reduced expression		[39, 38, 71, 48, 43]
C35X	CCP1	Reduced expression		[43, 72]
C35Y	CCP1	Reduced expression		[43, 47, 48, 71]
E36X	CCP1	Reduced expression		[43, 71]
P50T	CCP1	ND		[43, 72]
Y54C	CCP1	ND	Successful treatment with eculizumab; transplanted	[73]
R59X	CCP1	Reduced expression		[43, 46, 47, 71, 74]
C64F	CCP1	Reduced expression	Varicella trigger; 16 y/o successfully treated with plasma exchange	[43, 75, 76]
K65D-fsX73	CCP1	ND		[43, 53]
c. 286 + 2 T > G	Between CCP1-2	ND	Frameshift leads to stop in CCP2	[45]
IVS2 − 2A > G	CCP2	Reduced expression	c. 287 − 2A > G; Lacks exon 3 and creates stop codon at L133X;	[36, 43, 45, 47, 48, 71, 77–79]
C99R	CCP2	Reduced expression		[43, 47, 48]
R103W	CCP2	Normal expression & C3/C4 regulatory activity		[38, 39, 43, 48, 61, 71]
R103Q	CCP2	ND	Compound heterozygote with factor H mutation	[43, 53]
Y117X	CCP2	Reduced expression		[46, 69]
G130V	CCP2	ND		[43, 53]
G135D	CCP2	Reduced expression		[69]
G135V-fsX13	CCP2	ND		[43, 74]
P165S	CCP3	Reduced expression		[38, 39, 43, 48, 53]
E179Q	CCP3	Normal to higher expression; 50 % loss regulatory activity		[36, 43, 48]
D185N/Y189D	CCP3	Reduced expression		[36, 43, 48]
Y189D	CCP3	Reduced expression		[43, 45, 48, 70–72, 80]
G196R	CCP3	Reduced expression; decreased C4b CA only (FI interaction site)		[36, 48, 53, 73, 81]
G204R	CCP3	ND		[43, 53]
S206P	CCP3	ND		[43, 53]
I208Y	CCP3	ND	Compound heterozygote with factor H mutation	[43, 53]
C210F	CCP3	ND	Compound heterozygote with factor-I mutation	[43, 53]
W216C	CCP3	ND	Near functional site per [44]	[43, 45]
P231R	CCP4	ND	Functional site per [44]	[43, 45]
E234K	CCP4	ND		[46]
S240P	CCP4	Normal expression; loss of C3b CA & binding	Originally numbered S206P	[43, 48, 51]
F242C	CCP4	Normal expression; reduced C3b/C4b binding and CA		[43, 45, 48, 53, 72]

Table 1 CD46 mutations associated with aHUS and functional consequences *(Continued)*

Y248X	CCP4	Reduced expression		[43, 48, 52, 71, 72]
Y248C	CCP4	ND		[46]
K249N-fsX5	CCP4	ND		[46]
G259V	CCP4	Reduced expression; reduced C3b/C4b binding and CA	Compound heterozygote with FH mutations	[82]
L262P	CCP4	Reduced expression	aHUS pts successfully treated with eculizumab;	[83]
T267-fs270X	CCP4	Reduced expression	delA843-C844	[43, 48, 52, 72]
Del D271/S272	CCP4	Reduced expression	Originally numbered D237/S238	[43, 48, 51, 84]
c.852-856del	CCP4	Reduced expression	Originally numbered as 903-907del	[38, 39, 43, 48]
858-872del + D277N + P278S	CCP4	Reduced expression		[43, 47, 48]
IVS10 + 2 T > C	TM	Reduced expression	Exon 10 skipped changing aa 316–321 & adding a stop at 322	[43, 71, 74]
c.983-984delAT	TM	ND	Frameshift with stop	[43, 45]
A353V	TM	Normal expression and complement regulatory function	Uncommon variant; numerous studies and several disease implications; sometimes termed A304V	[43, 47, 48, 61]
A359V	TM	ND	Japanese pt; compound heterozygote with Y189D in CCP3	[70]
T383I	CYT-1	ND	Fatal infections triggered aHUS in 2 patients. Mother massive viral infection; son by *E. coli* O157:H7	[43, 53, 69]

#Numbered from SP and using ABC1 isoform or as noted for CYT-1. *Abbreviations*: *SP* signal peptide (34 amino acids), *CCP1-4* complement control protein modules 1–4, *UN* Segment of unknown function proximal to membrane (12 amino acids), *TM* transmembrane domain, *CYT-1* cytoplasmic tail 1 (16 amino acids), *CA* cofactor activity, *ND* not done

Preeclampsia complicates 4–5 % of pregnancies worldwide, causing significant maternal and neonatal mortality. Pregnancy in women with systemic lupus erythematosus (SLE) and/or the antiphospholipid syndrome (APS) may be particularly susceptible to complement-mediated injury with increased risk of preeclampsia, placental insufficiency, retardation, fetal growth issues, and miscarriage. A study analyzing a SLE and/or APL Ab cohort (PROMISSE)-sequenced genes for complement *FI*, *FH*, and *MCP* and found heterozygous mutations in seven (18 %) [59]. Five had risk variants that had been previously identified in aHUS, and one had a novel mutation in CD46, K66N (identified as K32N in the paper since the 34 amino acid signal peptide was not counted) that impairs only regulation of C4b. The study suggested a linkage between excessive complement activation and disease pathogenesis in patients with SLE and/or APL Ab who develop preeclampsia.

Idiopathic, recurrent miscarriage has also been associated with mutations both in *MCP* and C4b-binding protein (*C4BP*) [60]. All exons coding for CD46, C4BP, and decay-accelerating factor (DAF; CD55) were sequenced in a cohort of 384 childless women with at least two miscarriages. In addition to the first-time identification of a disease association of *C4BP* mutation, four *MCP* variants were identified. One of the rare variants, P324L, had decreased expression, while N213I had both impaired expression and function. Two mutations that did

not appear to affect complement regulatory function were located in the transmembrane domain, and a third one was in the cytoplasmic tail that could impair signaling function [6, 60, 61].

Of the ~ eight known aHUS cases of homozygous CD46 deficiency, three also developed common variable immunodeficiency [36, 37]. The remaining five patients all presented with subnormal IgG1 levels. T cells from CD46-deficient patients were not capable of promoting B-cell responses suggesting a defect in the ability to optimize B cell responses could account for this disease [37].

Mutation screening was also undertaken in 19 patients with C3 glomerulonephritis (a form of glomerulonephritis characterized by mesangial C3 deposits) [62]. One patient was a compound heterozygote for *MCP* mutations in both exon 5 (V215M in CCP3) and exon 11 (A353V in the transmembrane domain). The V215M (termed V181M in the paper) mutation occurred in a site determined to be functionally important in a previous investigation [44].

Studies in the last decade have demonstrated how dysregulation of the AP contributes to age-related macular degeneration (AMD) (reviewed in [63–67]). It is a leading worldwide cause of central vision loss in individuals over the age of 50. Hypomorphs (i.e., genetically-based changes resulting in functionally deficient complement inhibitors that control the alternative pathway) account for ~50 % of the attributable genetic risk for AMD. Although no mutations have yet been identified linking

Table 2 CD46 mutations associated with aHUS and/or other diseases

Mutation[#]	Domain	Disease association	Functional studies	Notes	References
−366A > G	Promoter	Systemic sclerosis	Reduced expression	Polymorphic variant; rs2796268	[56]
−652A > G	Promoter	Systemic sclerosis	Reduced expression	Polymorphic variant; rs2796267	[56]
S13F	SP	HELLP; SLE; aHUS	ND		[57, 58, 70]
R59X	CCP1	aHUS and common variable immunodeficiency	Reduced expression	Homozygote	[36, 37]
K66N	CCP1	PE & SLE	Normal expression; reduced ability to regulate C4b	Dimerization site on s tructural model	[59]
c.475 + 1G > A	CCP2	TTP	Reduced expression	Splice-site single nucleotide variant; deletes G152-C157	[81]
P193S	CCP3 (indel)	Miscarriage	Normal expression & C3b/C4b regulatory activity		[60]
N213I	CCP3	Miscarriage	Reduced expression & C3b/C4b regulatory activity		[60]
V215M	CCP3	Glomerulonephritis	Expression normal	Patient also has A353V mutation	[62]
A219V	CCP3	SLE	ND		[57]
P324L	STP-C	Miscarriage	Reduced expression; Normal C3b/C4b regulatory activity		[60]
A353V	TM	Miscarriage; C3-glomerulonephritis; HELLP Syndrome; aHUS	Reduced complement control on cell surface		[47, 48, 60, 64]
T383I	CYT-1	Miscarriage; aHUS	Normal expression & C3b/C4b regulatory activity; but could disrupt phosphorylation site on tail		[60, 69]

Abbreviations: See Table 1 abbreviations, *SLE* systemic lupus erythematosus, *PE* preeclampsia, *TTP* thrombotic thrombocytopenic purpura, *HELLP* syndrome, hemolysis, elevated liver enzymes and low platelets

AMD and CD46, several studies suggest it may have a role (reviewed in [67]). CD46 is found in drusen, the hallmark of dry-type AMD. Lower expression is observed in the monocytes of patients with AMD [67]. Further, smoking has been directly linked to development of AMD, and cigarette smoke extract decreased CD46 expression in retinal epithelial cells [67]. A mouse model that knocked out CD46 ($Cd46^{-/-}$) found increased levels of membrane-attack complex, and vascular endothelial growth factors were increased in the retina and choroid of mice deficient in CD46 [67]. Further, these mice also developed more severe retinal damage in a laser induced model of AMD [67].

Conclusion

Rare and uncommon mutations in *MCP* lead to development of aHUS and possibly several other diseases such as pregnancy-related disorders and SLE. The pathophysiological implications of the defective functioning of CD46 in aHUS relate to an inability to sufficiently control the alternative pathway of complement. Incomplete penetrance of mutations is 50 %, indicating that additional genetic or environmental triggers are involved. For aHUS, the outcome for renal transplantation with a normal CD46 kidney is much more favorable than for other mutations in complement proteins (reviewed in

[43, 47]). Additionally, the availability of a new therapeutic option for aHUS, the treatment with a mAb to C5 (eculizumab), induces a remission in most patients (reviewed in [49, 68]). With the advent of whole exome and whole genome sequencing reaching more reasonable costs, and in view of other putative conditions associated with CD46, additional disease associations are likely on the horizon.

Abbreviations

aHUS: atypical hemolytic uremic syndrome; AP: alternative pathway of complement; AMD: age-related macular degeneration; CCP: complement control protein module; CR1: complement receptor one (CR1 CD35); C4BP: C4b binding protein; DAA: decay-accelerating activity; DAF: decay-accelerating factor (CD55); FH: factor H; FI: factor I; MAC: membrane-attack complex; MCP: membrane cofactor protein (CD46); RCA: regulators of complement activation; STP: serine-threonine-proline-enriched domain of CD46.

Competing interests

MKL declares that she has no competing interests. JPA Disclosures: Grant/Research/Clinical Trial Support: Alexion (Targeted Deep Sequencing of Complement Genes in Human Disease (Complement system)); NIH (Interactions, Homeostasis And Translational Implications (Complement system)); NIH (CD46 And Risk Variants: Expanding Roles In Disease And Dissecting Membrane Dynamics (Complement system)); NIH (Genetic Predisposition To The Thrombomicroangiopathies (GP-TMAs) (Complement system)). Midwest Strategic Pharma-Academic Research Consortium; Consultant/Advisory Boards: Celldex Therapeutics (Complement system); Biothera (Complement system); Clinical Pharmacy Services, CDMI (Complement system); Kypha, Inc (Complement system); Stock, equity or options: Compliment Corporation; scientific advisory board.

Author's contributions

Both authors contributed equally to this review. MKL prepared the text and graphics and JPA modified as needed. Both authors read and approved the final manuscript.

Acknowledgements

MKL & JPA were supported by a) the National Institute of General Medical Sciences of the National Institutes of Health under Award Number R01GM099111 and by b) the National Heart, Lung, and Blood Institute of the National Institutes of Health under Award Number U54HL112303. The content is solely the responsibility of the authors and does not necessarily represent the official views of the National Institutes of Health.

References

1. Zhu Y, Thangamani S, Ho B, Ding JL. The ancient origin of the complement system. EMBO J. 2005;24(2):382–94.
2. Ariki S, Takahara S, Shibata T, Fukuoka T, Ozaki A, Endo Y, et al. Factor C acts as a lipopolysaccharide-responsive C3 convertase in horseshoe crab complement activation. J Immunol. 2008;181 (11):7994–8001. doi:181/11/7994.
3. Markiewski MM, Nilsson B, Ekdahl KN, Mollnes TE, Lambris JD. Complement and coagulation: strangers or partners in crime? Trends Immunol. 2007;28(4):184–92.
4. Grumach AS, Kirschfink M. Are complement deficiencies really rare? Overview on prevalence, clinical importance and modern diagnostic approach. Mol Immunol. 2014;61(2):110–7. doi:10.1016/j.molimm.2014.06.030.
5. Pettigrew HD, Teuber SS, Gershwin ME. Clinical significance of complement deficiencies. In: Shoenfeld Y, Gershwin ME, editors. Annals of the New York Academy of Sciences. Boston: Blackwell Publishing; 2009. p. 108–23.
6. Liszewski MK, Post TW, Atkinson JP. Membrane cofactor protein (MCP or CD46): newest member of the regulators of complement activation gene cluster. Annu Rev Immunol. 1991;9:431–55.
7. Zipfel PF, Skerka C. Complement regulators and inhibitory proteins. Nat Rev Immunol. 2009;9(10):729–40. doi:10.1038/nri2620.
8. Holers VM. Complement and its receptors: new insights into human disease. Annu Rev Immunol. 2014;32:433–59. doi:10.1146/annurev-immunol-032713-120154.
9. Mayilyan KR. Complement genetics, deficiencies, and disease associations. Protein & cell. 2012;3(7):487–96. doi:10.1007/s13238-012-2924-6.
10. Nonaka M, Kimura A. Genomic view of the evolution of the complement system. Immunogenetics. 2006;58(9):701–13.
11. Gros P, Milder FJ, Janssen BJ. Complement driven by conformational changes. Nat Rev Immunol. 2008;8(1):48–58.
12. Walport MJ. Complement. First of two parts N Engl J Med. 2001;344(14):1058–66. doi:10.1056/NEJM200104053441406.
13. Tsujikura M, Nagasawa T, Ichiki S, Nakamura R, Somamoto T, Nakao M. A CD46-like molecule functional in teleost fish represents an ancestral form of membrane-bound regulators of complement activation. J Immunol. 2015;194(1):262–72. doi:10.4049/jimmunol.1303179.
14. Wallis R, Mitchell DA, Schmid R, Schwaeble WJ, Keeble AH. Paths reunited: initiation of the classical and lectin pathways of complement activation. Immunobiology.2010;215(1):1–11. doi:S0171-2985(09)00146-6 [pii] 10.1016/j.imbio.2009.08.006
15. Bubeck D. The making of a macromolecular machine: assembly of the membrane attack complex. Biochemistry (Mosc). 2014;53(12):1908–15. doi:10.1021/bi500157z.
16. Liszewski MK, Kemper C, Price JD, Atkinson JP. Emerging roles and new functions of CD46. Springer Semin Immunopath. 2005;27(3):345–58.
17. Cardone J, Le Friec G, Kemper C. CD46 in innate and adaptive immunity: an update. Clin Exp Immunol. 2011;164(3):301–11. doi:10.1111/j.1365-2249.2011.04400.x.
18. Yamamoto H, Fara AF, Dasgupta P, Kemper C. CD46: the 'multitasker' of complement proteins. Int J Biochem Cell Biol. 2013;45(12):2808–20. doi:10.1016/j.biocel.2013.09.016.
19. Hourcade D, Holers VM, Atkinson JP. The regulators of complement activation (RCA) gene cluster. Adv Immunol. 1989;45:381–416.
20. Kim DD, Song WC. Membrane complement regulatory proteins. Clin Immunol. 2006;118(2–3):127–36.
21. Rodriguez De Cordoba S, Goicoechea De Jorge E. Translational mini-review series on complement factor H: genetics and disease associations in human complement factor H. Clin Exp Immuno. 2008;151:1–13.
22. Cattaneo R. Four viruses, two bacteria, and one receptor: membrane cofactor protein (CD46) as pathogens' magnet. J Virol. 2004;78(9):4385–8.
23. Kemper C, Atkinson JP. T-cell regulation: with complements from innate immunity. Nat Rev Immunol. 2007;7(1):9–18.
24. Kemper C, Atkinson JP. Measles virus and CD46. Curr Top Microbiol Immunol. 2009;329:31–57.
25. Nemerow GR, Pache L, Reddy V, Stewart PL. Insights into adenovirus host cell interactions from structural studies. Virology. 2009;384(2):380–8. doi:10.1016/j.virol.2008.10.016.
26. Tang H, Mori Y. Human herpesvirus-6 entry into host cells. Future Microbiol. 2010;5(7):1015–23. doi:10.2217/fmb.10.61.
27. Weyand NJ, Calton CM, Higashi DL, Kanack KJ, So M. Presenilin/gamma-secretase cleaves CD46 in response to Neisseria infection. J Immuno. 2010;184(2):694–-701. doi:10.4049/jimmunol.0900522. jimmunol.0900522.
28. Lovkvist L, Sjolinder H, Wehelie R, Aro H, Norrby-Teglund A, Plant L, et al. CD46 contributes to the severity of Group A streptococcal infection. Infect Immun. 2008;76(9):3951–8.
29. Price JD, Schaumburg J, Sandin C, Atkinson JP, Lindahl G, Kemper C. Induction of a regulatory phenotype in human CD4⁺ T cells by streptococcal M protein. J Immunol. 2005;175:677–84.
30. Mahtout H, Chandad F, Rojo JM, Grenier D. Fusobacterium nucleatum binding to complement regulatory protein CD46 modulates the expression and secretion of cytokines and matrix metalloproteinases by oral epithelial cells. J Periodontol. 2011;82(2):311–9. doi:10.1902/jop.2010.100458.
31. Liszewski MK, Leung MK, Hauhart R, Fang CJ, Bertram P, Atkinson JP. Smallpox inhibitor of complement enzymes (SPICE): dissecting functional sites and abrogating activity. J Immunol. 2009;183:3150–9.
32. Ojha H, Panwar HS, Gorham Jr RD, Morikis D, Sahu A. Viral regulators of complement activation: structure, function and evolution. Mol Immunol. 2014;61(2):89–99. doi:10.1016/j.molimm.2014.06.004.
33. Liszewski MK, Leung MK, Hauhart R, Buller RM, Bertram P, Wang X, et al. Structure and regulatory profile of the monkeypox inhibitor of complement: comparison to homologs in vaccinia and variola and evidence for dimer formation. J Immunol. 2006;176(6):3725–34.
34. Yadav VN, Pyaram K, Mullick J, Sahu A. Identification of hot spots in the variola virus complement inhibitor (SPICE) for human complement regulation. J Virol. 2008;82:3283–9.
35. Murphy KM, Stockinger B. Effector T cell plasticity: flexibility in the face of changing circumstances. Nat Immunol. 2010;11(8):674–80. doi:10.1038/ni.1899.
36. Fremeaux-Bacchi V, Moulton EA, Kavanagh D, Dragon-Durey M-A, Blouin J, Caudy A, et al. Genetic and functional analyses of membrane cofactor protein (CD46) mutations in atypical hemolytic uremic syndrome. J Am Soc Nephrol. 2006;17:2017–25.
37. Fuchs A, Atkinson JP, Fremeaux-Bacchi V, Kemper C. CD46-induced human Treg enhance B-cell responses. Eur J Immunol. 2009;39(11):3097–109. doi:10.1002/eji.200939392.
38. Esparza-Gordillo J. Goicoechea De Jorge E, Buil A, Berges LC, Lopez-Trascasa M, Sanchez-Corral P, et al. Predisposition to atypical hemolytic uremic syndrome involves the concurrence of different susceptibility alleles in the regulators of complement activation gene cluster in 1q32. Hum Mol Genet. 2005;14(5):703–12.
39. Esparza-Gordillo J, Goicoechea De Jorge E, Buil A, Berges LC, Lopez-Trascasa M, Sanchez-Corral P, et al. Predisposition to atypical hemolytic uremic syndrome involves the concurrence of different susceptibility alleles in the regulators of complement activation gene cluster in 1q32. Hum Mol Gene. 2005;14(8):1107. CORRIGENDUM.
40. Saunders RE, Abarrategui-Garrido C, Fremeaux-Bacchi V. Goicoechea De Jorge E, Goodship TH, Lopez Trascasa M, et al. The interactive Factor H-atypical hemolytic uremic syndrome mutation database and website: update and integration of membrane cofactor protein and Factor I mutations with structural models Hum Mutat. 2007;28(3):222–34.
41. Sullivan M, Erlic Z, Hoffmann MM, Arbeiter K, Patzer L, Budde K, et al. Epidemiological approach to identifying genetic predispositions for atypical

hemolytic uremic syndrome. Ann Hum Genet. 2010;74(1):17–26. doi:10.1111/j.1469-1809.2009.00554.x.

42. Meri S. Complement activation in diseases presenting with thrombotic microangiopathy. Eur J Intern Med. 2013;24(6):496–502. doi:10.1016/j.ejim.2013.05.009.

43. Kavanagh D, Goodship TH, Richards A. Atypical hemolytic uremic syndrome. Semin Nephrol. 2013;33(6):508–30. doi:10.1016/j.semnephrol.2013.08.003.

44. Liszewski MK, Leung M, Cui W, Subramanian VB, Parkinson J, Barlow PN, et al. Dissecting sites important for complement regulatory activity in membrane cofactor protein (MCP; CD46). J Biol Chem. 2000;275(48):37692–701.

45. Maga TK, Nishimura CJ, Weaver AE, Frees KL, Smith RJ. Mutations in alternative pathway complement proteins in American patients with atypical hemolytic uremic syndrome. Hum Mutat. 2010;31(6):E1445–60. doi:10.1002/humu.21256.

46. Rodriguez De Cordoba S, Hidalgo MS, Pinto S, Tortajada A. Genetics of atypical hemolytic uremic syndrome (aHUS). Semin Thromb Hemost. 2014;40:422–30. doi:10.1055/s-0034-1375296.

47. Caprioli J, Noris M, Brioschi S, Pianetti G, Castelletti F, Bettinaglio P, et al. Genetics of HUS: the impact of MCP, CFH and IF mutations on clinical presentation, response to treatment, and outcome. Blood. 2006;108:1267–79.

48. Richards A, Liszewski MK, Kavanagh D, Fang CJ, Moulton EA, Fremeaux-Bacchi V, et al. Implications of the initial mutations in membrane cofactor protein (MCP; CD46) leading to atypical hemolytic uremic syndrome. Mol Immunol. 2007;44:111–22.

49. Riedl M, Fakhouri F, Le Quintrec M, Noone DG, Jungraithmayr TC, Fremeaux-Bacchi V, et al. Spectrum of complement-mediated thrombotic microangiopathies: pathogenetic insights identifying novel treatment approaches. Semin Thromb Hemost. 2014;40(4):444–64. doi:10.1055/s-0034-1376153.

50. Loirat C, Fremeaux-Bacchi V. Atypical hemolytic uremic syndrome. Orphanet J Rare Dis. 2011;6:60. doi:10.1186/1750-1172-6-60.

51. Richards A, Kemp EJ, Liszewski MK, Goodship JA, Lampe AK, Decorte R, et al. Mutations in human complement regulator, membrane cofactor protein (CD46), predispose to development of familial hemolytic uremic syndrome. Proc Natl Acad Sci U S A. 2003;100:12966–71.

52. Noris M, Brioschi S, Caprioli J, Todeschini M, Bresin E, Porrati F, et al. Familial haemolytic uraemic syndrome and an MCP mutation. Lancet. 2003;362(9395):1542–7.

53. Bresin E, Rurali E, Caprioli J, Sanchez-Corral P, Fremeaux-Bacchi V. Rodriguez De Cordoba S, et al. Combined complement gene mutations in atypical hemolytic uremic syndrome influence clinical phenotype J Am Soc Nephro. 2013;24(3):475–86. doi:10.1681/ASN.2012090884.

54. Bu F, Maga T, Meyer NC, Wang K, Thomas CP, Nester CM, et al. Comprehensive genetic analysis of complement and coagulation genes in atypical hemolytic uremic syndrome. J Am Soc Nephrol. 2014;25(1):55–64. doi:10.1681/ASN.2013050453.

55. Rodriguez E, Rallapalli PM, Osborne AJ, Perkins SJ. New functional and structural insights from updated mutational databases for complement factor H, Factor I, membrane cofactor protein and C3. Biosci Rep. 2014;34(5):art:e00146. doi:10.1042/BSR20140117.

56. Scambi C, Ugolini S, Jokiranta TS, De Franceschi L, Bortolami O, La Verde V, et al. The local complement activation on vascular bed of patients with systemic sclerosis: a hypothesis-generating study. PLoS One. 2015;10(2), e0114856. doi:10.1371/journal.pone.0114856.

57. Jonsen A, Nilsson SC, Ahlqvist E, Svenungsson E, Gunnarsson I, Eriksson KG, et al. Mutations in genes encoding complement inhibitors CD46 and CFH affect the age at nephritis onset in patients with systemic lupus erythematosus. Arthritis Res Ther. 2011;13(6):R206. doi:10.1186/ar3539.

58. Crovetto F, Borsa N, Acaia B, Nishimura C, Frees K, Smith RJ, et al. The genetics of the alternative pathway of complement in the pathogenesis of HELLP syndrome. J Matern Fetal Neonatal Med. 2012;25(11):2322–5. doi:10.3109/14767058.2012.694923.

59. Salmon JE, Heuser C, Triebwasser M, Liszewski MK, Kavanagh D, Roumenina L, et al. Mutations in complement regulatory proteins predispose to preeclampsia: a genetic analysis of the PROMISSE cohort. PLoS Med. 2011;8(3), e1001013. doi:10.1371/journal.pmed.1001013.

60. Mohlin FC, Mercier E, Fremeaux-Bacchi V, Liszewski MK, Atkinson JP, Gris JC, et al. Analysis of genes coding for CD46, CD55, and C4b-binding protein in patients with idiopathic, recurrent, spontaneous pregnancy loss. Eur J Immunol. 2013;43(6):1617–29. doi:10.1002/eji.201243196.

61. Fang CJ, Fremeaux-Bacchi V, Liszewski MK, Pianetti G, Noris M, Goodship TH, et al. Membrane cofactor protein mutations in atypical hemolytic uremic syndrome (aHUS), fatal Stx-HUS, C3 glomerulonephritis, and the HELLP syndrome. Blood. 2008;111(2):624–32. doi:10.1182/blood-2007-04-084533.

62. Servais A, Fremeaux-Bacchi V, Lequintrec M, Salomon R, Blouin J, Knebelmann B, et al. Primary glomerulonephritis with isolated C3 deposits: a new entity which shares common genetic risk factors with haemolytic uraemic syndrome. J Med Genet. 2007;44(3):193–9.

63. Gehrs KM, Jackson JR, Brown EN, Allikmets R, Hageman GS. Complement, age-related macular degeneration and a vision of the future. Arch Ophthalmol. 2010;128(3):349–58. doi:10.1001/archophthalmol.2010.18. 128/3/349.

64. Khandhadia S, Cipriani V, Yates JR, Lotery AJ. Age-related macular degeneration and the complement system. Immunobiology. 2012;217(2):127–46. doi:10.1016/j.imbio.2011.07.019. S0171-2985(11) 00159-8.

65. Sobrin L, Seddon JM. Nature and nurture- genes and environment- predict onset and progression of macular degeneration. Prog Retin Eye Res. 2014;40:1–15. doi:10.1016/j.preteyeres.2013.12.004.

66. Schramm EC, Clark SJ, Triebwasser MP, Raychaudhuri S, Seddon JM, Atkinson JP. Genetic variants in the complement system predisposing to age-related macular degeneration: a review. Mol Immunol. 2014;61(2):118–25. doi:10.1016/j.molimm.2014.06.032.

67. Bora NS, Matta B, Lyzogubov VV, Bora PS. Relationship between the complement system, risk factors and prediction models in age-related macular degeneration. Mol Immunol. 2015;63(2):176–83. doi:10.1016/j.molimm.2014.07.012.

68. Legendre CM, Licht C, Muus P, Greenbaum LA, Babu S, Bedrosian C, et al. Terminal complement inhibitor eculizumab in atypical hemolytic-uremic syndrome. N Engl J Med. 2013;368(23):2169–81. doi:10.1056/NEJMoa1208981.

69. Provaznikova D, Rittich S, Malina M, Seeman T, Marinov I, Riedl M, et al. Manifestation of atypical hemolytic uremic syndrome caused by novel mutations in MCP. Pediatr Nephrol. 2012;27(1):73–81. doi:10.1007/s00467-011-1943-5.

70. Fan X, Yoshida Y, Honda S, Matsumoto M, Sawada Y, Hattori M, et al. Analysis of genetic and predisposing factors in Japanese patients with atypical hemolytic uremic syndrome. Mol Immunol. 2013;54(2):238–46. doi:10.1016/j.molimm.2012.12.006.

71. Fremeaux-Bacchi V, Fakhouri F, Garnier A, Bienaime F, Dragon-Durey MA, Ngo S, et al. Genetics and outcome of atypical hemolytic uremic syndrome: a nationwide French series comparing children and adults. CJASN. 2013;8(4):554–62. doi:10.2215/CJN.04760512.

72. Noris M, Caprioli J, Bresin E, Mossali C, Pianetti G, Gamba S, et al. Relative role of genetic complement abnormalities in sporadic and familial aHUS and their impact on clinical phenotype. CJASN. 2010;5(10):1844–59. doi:10.2215/CJN.02210310.

73. Reuter S, Heitplatz B, Pavenstadt H, Suwelack B. Successful long-term treatment of TMA with Eculizumab in a transplanted patient with atypical hemolytic uremic syndrome due to MCP mutation. Transplantation. 2013;96(10):e74–6. doi:10.1097/01.TP.0000435705.63428.1f.

74. Fremeaux-Bacchi V, Sanlaville D, Menouer S, Blouin J, Dragon-Durey MA, Fischbach M, et al. Unusual clinical severity of complement membrane cofactor protein-associated hemolytic-uremic syndrome and uniparental isodisomy. Am J Kidney Dis. 2007;49(2):323–9. doi:10.1053/j.ajkd.2006.10.022.

75. Kwon T, Belot A, Ranchin B, Baudouin V, Fremeaux-Bacchi V, Dragon-Durey MA, et al. Varicella as a trigger of atypical haemolytic uraemic syndrome associated with complement dysfunction: two cases. Nephrol Dial Transplant. 2009;24(9):2752–4. doi:10.1093/ndt/gfp166.

76. Reid VL, Mullan A, Erwig LP. Rapid recovery of membrane cofactor protein (MCP; CD46) associated atypical haemolytic uraemic syndrome with plasma exchange. BMJ case reports. 2013;2013(sep04 1):bcr2013200980–bcr2013200980. doi:10.1136/bcr-2013-200980.

77. Bento D, Mapril J, Rocha C, Marchbank KJ, Kavanagh D, Barge D, et al. Triggering of atypical hemolytic uremic syndrome by influenza A (H1N1). Ren Fail. 2010;32(6):753–6. doi:10.3109/0886022×.2010.486491.

78. Fakhouri F, Roumenina L, Provot F, Sallee M, Caillard S, Couzi L, et al. Pregnancy-associated hemolytic uremic syndrome revisited in the era of complement gene mutations. J Am Soc Nephrol. 2010;21(5):859–67. doi:10.1681/ASN.2009070706.

79. Pabst WL, Neuhaus TJ, Nef S, Bresin E, Zingg-Schenk A, Sparta G. Successful long-term outcome after renal transplantation in a patient with atypical

haemolytic uremic syndrome with combined membrane cofactor protein CD46 and complement factor I mutations. Pediatr Nephrol. 2013. doi:10.1007/s00467-013-2450-7.

80. Khan S, Tarzi MD, Dore PC, Sewell WA, Longhurst HJ. Secondary systemic lupus erythematosus: an analysis of 4 cases of uncontrolled hereditary angioedema. Clin Immunol. 2007;123(1):14–7.

81. Rossio R, Lotta LA, Pontiggia S, Borsa Ghiringhelli N, Garagiola I, Ardissino G, et al. A novel CD46 mutation in a patient with microangiopathy clinically resembling thrombotic thrombocytopenic purpura and normal ADAMTS13 activity. Haematologica. 2015;100:e87–9. doi:10.3324/haematol.2014.111062.

82. Mohlin FC, Nilsson SC, Levart TK, Golubovic E, Rusai K, Muller-Sacherer T, et al. Functional characterization of two novel non-synonymous alterations in CD46 and a Q950H change in factor H found in atypical hemolytic uremic syndrome patients. Mol Immunol. 2015;65(2):367–76. doi:10.1016/j.molimm.2015.02.013.

83. Gulleroglu K, Fidan K, Hancer VS, Bayrakci U, Baskin E, Soylemezoglu O. Neurologic involvement in atypical hemolytic uremic syndrome and successful treatment with eculizumab. Pediatr Nephrol. 2013;28(5):827–30. doi:10.1007/s00467-013-2416-9.

84. Westra D, Volokhina E, van der Heijden E, Vos A, Huigen M, Jansen J, et al. Genetic disorders in complement (regulating) genes in patients with atypical haemolytic uraemic syndrome (aHUS). Nephrol Dial Transplant. 2010;25(7):2195–202. doi:10.1093/ndt/gfq010.

3

Association of genome variations in the renin-angiotensin system with physical performance

Argyro Sgourou[1], Vassilis Fotopoulos[1], Vassilis Kontos[2], George P Patrinos[3] and Adamantia Papachatzopoulou[2*]

Abstract

Background: The aim of this study was to determine the genotype distribution and allelic frequencies of *ACE* (I/D), *AGTR1* (A +1166 C), *BDKRB2* (+9/−9) and *LEP* (G−2548A) genomic variations in 175 Greek athletes who excelled at a national and/or international level and 169 healthy Greek adults to identify whether some particular combinations of these loci might serve as predictive markers for superior physical condition.

Results: The D/D genotype of the *ACE* gene ($p = 0.034$) combined with the simultaneous existence of *BDKRB2* (+9/−9) ($p = 0.001$) or *LEP* (G/A) ($p = 0.021$) genotypes was the most prevalent among female athletes compared to female controls. A statistical trend was also observed in *BDKRB2* (+9/−9) and *LEP* (G−2548A) heterozygous genotypes among male and female Greek athletes, and in *ACE* (I/D) only in male athletes. Finally, both male and female athletes showed the highest rates in the *AGTR1* (A/A) genotype.

Conclusions: Our results suggest that the co-existence of *ACE* (D/D), *BDKRB2* (+9/−9) or *LEP* (G/A) genotypes in female athletes might be correlated with a superior level of physical performance.

Keywords: Genetic variations, Renin-angiotensin system, Physical performance

Introduction

Genetic polymorphisms that act as potential mediators of the human health and physical performance are targets for many research groups attempting to unravel their role to the genetic predisposition for a superior performance and endurance. There are up to 170 gene variant sequences, 17 mitochondrial DNA markers and 25 additional nuclear genetic markers in the human genetic map which are related to physical performance phenotypes as well as to good physical fitness [1].

One of the most extensively studied genome variations, widely associated with the human performance over the last decade, was the insertion (I) or deletion (D) of 287-bp *Alu* repeats within intron 16 of the angiotensin-converting enzyme (*ACE*) gene [rs1799752] [2,3]. *ACE* plays a key role along the biochemical pathway of the renin-angiotensin system (RAS), which controls the homoeostasis of the human circulatory system. Renin is a low molecular weight enzyme that is released by juxtaglomerular cells of the kidney in response to

blood pressure failure. Renin converts its substrate angiotensinogen to angiotensin I, which is almost immediately converted by *ACE* to angiotensin II (AT II). AT II is a potent vasoconstrictor substance that acts mainly via AT II type-1 receptors. Also, *ACE* hydrolyses bradykinin which is a vasodilator, thus reduces peripheral resistance and hence blood pressure [4]. Additionally, RAS acts through other tissues as a paracrine/autocrine system [5], and its local activity in the cardiac [6], adipose [7] and skeletal muscular tissues [8] has been reported. It has been currently verified that the local adipose RAS is capable of functioning independently of the plasma RAS and it is up-regulated in obesity [9,10], where the presence of AT II stimulates leptin gene expression and secretion from adipocytes [11], revealing a considerable cross-interaction between leptin expression and RAS components. In particular, the leptin G−2548A promoter polymorphism (*LEP* G−2548A) [rs7799039] has been strongly associated with the serum leptin levels in overweight individuals and obesity and an increased risk for obesity [12-14]. A study with obese Zucker rats treated with *ACE* inhibitors have shown decreased leptin release [11], which supports the cross-interaction between leptin and *ACE* gene products. Recent results have shown

* Correspondence: apapacha@med.upatras.gr
[2]Laboratory of General Biology, Faculty of Medicine, University of Patras, Patras 265 04, Greece
Full list of author information is available at the end of the article

that alterations in adipocyte production of the RAS components may contribute to disorders of the metabolic syndrome, including obesity and obesity-related hypertension and diabetes [15,16].

The presence of other polymorphisms, like the *AGTR1* (A +1166 C) allele in 3' UTR of the AT II type-1 receptor gene [rs12721276], results in increased expression of the receptor gene [17], and the 9-bp deletion in exon 1 of the *BDKRB2* (β2 receptor of bradykinin) gene [rs72348790] results in a higher transcriptional activity and, consequently, to a quicker receptor's response to bradykinin molecules [18,19]. Additionally, the coexistence of the latter polymorphism with the *ACE* D/D genotype responsible for elevated *ACE* enzyme activity might counterbalance this activity, preventing bradykinin's hydrolysation, by withdrawing it in a higher rate.

In this study, we have investigated the presence of known polymorphisms named *ACE* (I/D), *AGTR1* (A +1166 C) and *BDKRB2* (+9/−9) along the RAS biochemical pathway as well as the one in the promoter of the *LEP* gene (G−2548A). Leptin exhibits a cross-interaction with RAS components. The presence of all the above polymorphisms in specific combination showed to play a role not only

in the blood pressure control, but also in other metabolic pathways that might affect fitness and physical performance in humans.

Results

Genotypic distribution

For all four polymorphisms studied, among male athletes versus male controls, the highest percentages of male athletes appeared as heterozygotes for *ACE* (I/D), *BDKRB2* (+9/−9) and *LEP* (G/A) genes and homozygotes (A/A) for *AGTR1* gene polymorphism. In total, there were no statistically significant differences in genotypes between male athletes versus male controls (Table 1). However, *ACE* (I/I) genotype was absent in the international male athlete group of 39 out of 102 male athletes.

In both female athletes and female controls, their genotypic distribution is shown in Table 1. In female groups, significant differences were apparent, such as the higher score in female athletes (47.95%) versus female controls (31.33%) (*p* = 0.034) of the *ACE* (D/D) genotype, while the *ACE* (I/D) genotype exhibited a higher score in female controls (51.81%) versus the

Table 1 Genotype distributions and allele frequencies of the four polymorphisms in athletes and control groups

		Male controls (n = 88)	Male athletes (n = 102)	p value	Female controls (n = 83)	Female athletes (n = 73)	p value
Genotype distributions							
ACE	II	12 (13.64%)	7 (6.86%)	0.121	14 (16.87%)	13 (17.81%)	0.877
	ID	45 (51.14%)	60 (58.82%)	0.288	*43 (51.81%)*	*25 (34.25%)*	*0.027*
	DD	31 (35.23%)	35 (34.31%)	0.859	*26 (31.33%)*	*35 (47.95%)*	*0.034*
AGTR1 (A +1166 C)	AA	53 (60.23%)	57 (55.88%)	0.545	45 (54.22%)	41 (56.16%)	0.807
	AC	32 (36.36%)	39 (38.24%)	0.79	35 (42.17%)	26 (35.62%)	0.403
	CC	3 (3.41%)	6 (5.88%)	0.424	3 (3.61%)	6 (8.22%)	0.218
BDKRB2 (+9/−9)	(+9/+9)	30 (34.09%)	33 (32.35%)	0.757	24 (28.92%)	20 (27.40%)	0.833
	(+9/−9)	46 (52.27%)	58 (56.86%)	0.53	45 (54.22%)	45 (61.64%)	0.349
	(−9/−9)	12 (13.64%)	11 (10.78%)	0.528	14 (16.87%)	8 (10.96%)	0.29
LEP (G−2548A)	AA	16 (18.18%)	19 (18.63%)	0.937	12 (14.46%)	15 (20.55%)	0.297
	GA	46 (57.27%)	58 (56.86%)	0.526	46 (55.42%)	41 (56.16%)	0.926
	GG	26 (29.55%)	25 (24.51%)	0.435	25 (30.12%)	17 (23.29%)	0.363
Allele frequencies							
I allele (*ACE* gene)		57 (64.8%)	67 (65.7%)	0.895	*57 (68.7%)*	*38 (52.1%)*	*0.034*
D allele (*ACE* gene)		76 (86.4%)	95 (93.1%)	0.121	69 (83.1%)	60 (82.2%)	0.877
A allele (*AGRT1* gene)		85 (96.6%)	96 (94.1%)	0.424	80 (96.4%)	67 (91.8%)	0.218
C allele (*AGRT1* gene)		35 (39.8%)	45 (44.1%)	0.545	38 (45.8%)	32 (43.8%)	0.807
+9 allele (*BDKRB2* gene)		76 (86.4%)	91 (89.2%)	0.548	69 (83.1%)	65 (89%)	0.29
−9 allele (*BDKRB2* gene)		58 (65.9%)	69 (67.6%)	0.800	59 (71.1%)	53 (72.6%)	0.833
A allele (*LEP* gene)		62 (70.5%)	77 (75.5%)	0.435	58 (69.9%)	56 (76.7%)	0.337
G allele (*LEP* gene)		72 (81.8%)	83 (81.4%)	0.937	71 (85.5%)	58 (79.5%)	0.316

Values in italics have significant *p* values.

female athlete group (34.25%) (p = 0.027). All other genotypic distributions did not reach statistical significance (Table 1). Furthermore, in the female athlete group that were homozygous for the ACE (D/D) genotype, the BDKRB2 (+9/−9) or LEP (G/A) genotypes were more prevalent (p = 0.001 and p = 0.021, respectively), compared to the female control group (Table 2). Also, a significant difference was revealed in female athletes in the distribution of the BDKRB2 (+9/+9) genotype (27.40%) versus that of the BDKRB2 (−9/−9) genotype (10.96%) (p = 0.042). This trend was also observed in male athletes but did not reach statistical significance. Finally, the distribution of the LEP (A/A) and LEP (G/G) genotypes was similar among female athletes (20.55% and 23.29%, respectively).

Allele frequencies

Allele frequencies concerning both cohorts of male athletes versus male controls and female athletes versus female controls are shown in Table 1. The frequency of each allele resulted from the sum of the homo- and the heterozygotes carrying the counted allele in their genotypes. A statistically significant higher percentage was noticed for the ACE I allele in female controls (68.7%) versus female athletes (52.1%) (p = 0.034).

Allelic combinations

The impact of the allele frequencies was further assessed by analysing the 16 possible allelic combinations coming from the four different studied polymorphisms (ACE (I/D), LEP (G/A), BDKRB2 (+9/−9) and AGTR1 (A/C); Table 3). Polymorphisms are referred as I or D, G or A, +9 or −9 and A or C. The percentages of each allelic combination quartet resulted from the use of the SPSS statistical program and the implementation of appropriate functions. Once more, significant differences were only observed in the female group (Figure 1). Specifically, the allelic combinations compared between female athletes and female controls revealed a significantly decreased frequencies of the IG+9A (32.9% versus 51.8%) and of the IG−9A (20.5% versus 42.2%) (p = 0.017 and 0.004, respectively) among female athletes.

The same tests were applied throughout the groups of male athletes versus controls and showed no statistical significance (data not shown).

Discussion

Physical performance seems to be controlled by many genetic factors that interact with the environment to affect complex interactions in human physical performance characteristics. All these render such investigations quite complex, and extended studies are needed to unravel possible interactions. To date, genetic studies, attempting to ascertain the role of genetic variants involved in the human superior physical performance, have focused on candidate genes mainly associated with cardiovascular functions, including ACE and proteins participating in skeletal muscle activity such as a-actinins [20,21]. Much of the mechanisms underlying the human athletic performance remain unexplored, despite 12 years of research on the most widely studied candidate polymorphic site of ACE I/D [3].

Both I and D alleles have been so far successfully associated in sports and with superior athletic performance in South African triathletes [22], British distance runners [23], swimmers [24] and sprinters [25]. The relationships between genotypes and performance, however, remain ambiguous. A recent study on 230 elite Jamaican and American sprinters found no association of either allele with sprint athlete status [26]. It has been suggested that at least part of the association of ACE with high athletic performance phenotypes is mediated through changes in kinin activity and is related to the existence of the BDKRB2(−9) allele as it provides a higher expression and abundance of the bradykinin receptor [27]. Besides the contribution of bradykinin to an impaired blood pressure, it also enhances insulin-stimulated tyrosine kinase activity of the insulin receptor, with subsequent GLUT-4 translocation in skeletal muscle tissue, thus giving a theoretical boosting effect during exercise [28]. Alternatively, the effects of the ACE (I/D) genotype may be mediated through changes in AT II activity, which acts via the AGTR1 receptor. AGTR1 (A +1166 C) gene polymorphism appears functional, with the C allele acting as an enhancer of the receptor activity; however, it does not seem to be associated with differences in high-level human performance [29].

In our study, we observed a strong statistical trend towards ACE (D/D) polymorphism among female athletes, which is in line with previous publications [24,25]. The highest rates of the corresponding group of male athletes appeared as heterozygotes (I/D) for the same polymorphism. Our results might not be in contrast with another study, which included 101 elite Greek track and field athletes [30]; this study suggested weak evidence that the presence of the ACE (D/D) genotype could influence sprint performance in Greek athletes. This might be due to both, i.e. the limited number of the participating athletes and the different kinds of sports they are considered elite.

Table 2 Cross-tabulation: female athletes/female controls and ACE (D/D)/other genotypes

		Female controls ACE (D/D)	Female athletes ACE (D/D)	p value
BDKRB2 (+9/−9)	(+9/−9)	11 (13.3%)	26 (35.62%)	0.001
LEP (G−2548A)	GA	15 (18.1%)	25 (34.2%)	0.021

Values in italics have significant p values.

Table 3 Allele combination frequencies of the four polymorphisms in athletes and control groups

ACE (I, D), LEP (G, A), BDKRB2 (+9, −9), AGTR1 (A, C)	Male controls (n = 88)	Male athletes (n = 102)	p value	Female controls (n = 83)	Female athletes (n = 73)	p value
IA+9A	31 (35.22%)	42 (41.2%)	0.435	32 (38.6%)	24 (32.9%)	0.461
IG+9A	36 (40.9%)	45 (44.1%)	0.705	*43 (51.8%)*	*24 (32.9%)*	*0.017*
IA−9A	26 (29.54%)	30 (29.4%)	0.943	27 (32.5%)	17 (23.3%)	0.201
IG−9A	28 (31.8%)	38 (37.3%)	0.466	*35 (42.2%)*	*15 (20.5%)*	*0.004*
IA+9C	11 (12.5%)	15 (14.7%)	0.682	14 (16.9%)	7 (9.6%)	0.184
IG+9C	15 (17%)	20 (19.6%)	0.676	20 (24.1%)	11 (15.1%)	0.159
IA−9C	13 (14.77%)	15 (14.7%)	0.964	13 (15.7%)	6 (8.2%)	0.156
IG−9C	15 (17%)	20 (19.6%)	0.676	15 (18.1%)	6 (8.2%)	0.072
DA+9A	46 (52.27%)	61 (59.8%)	0.338	38 (45.8%)	38 (52.1%)	0.434
DG+9A	50 (56.8%)	65 (63.7%)	0.380	49 (59%)	39 (53.4%)	0.481
DA−9A	38 (43.2%)	42 (41.2%)	0.729	31 (37.3%)	32 (43.8%)	0.410
DG−9A	38 (43.2%)	53 (52%)	0.256	44 (53%)	31 (42.5%)	0.188
DA+9C	16 (18.18%)	24 (23.5%)	0.389	16 (19.3%)	17 (23.3%)	0.541
DG+9C	18 (20.45%)	31 (30.4%)	0.129	24 (28.9%)	25 (34.2%)	0.474
DA−9C	15 (17%)	22 (21.6%)	0.455	14 (16.9%)	14 (19.2%)	0.707
DG−9C	16 (18.18%)	28 (27.5%)	0.142	20 (24.1%)	18 (24.7%)	0.935

Values in italics have significant p values.

Similar studies have shown that the *BDKRB2* (−9/−9) genotype was associated with the actual performance of 701 South African males who completed an Ironman Triathlon [19], but there were controversial results among Israeli athletes [31]. Likewise, a current study on the contribution of leptin gene promoter polymorphism *LEP* (G−2548A) in the human capacity for athletic performance indicated that the G allele provides an advantage to the reduction of body mass index as a response to physical training [32]. On the contrary, our data showed that female athletes of the *BDKRB2* (+9/+9) genotype were statistically higher than those of the *BDKRB2* (−9/−9) homozygotes, while a statistically significant increase was obtained for the co-existence of the *ACE* (D/D) genotype together with the heterozygosity of the *LEP* (G/A) or *BDKRB2* (+9/−9) genotypes (Table 2). Among female athletes, significant reduction of the I allele frequencies (Table 1) and of both IG+9A and IG−9A allelic frequency combinations (Table 3, Figure 1) were demonstrated, which are consistent

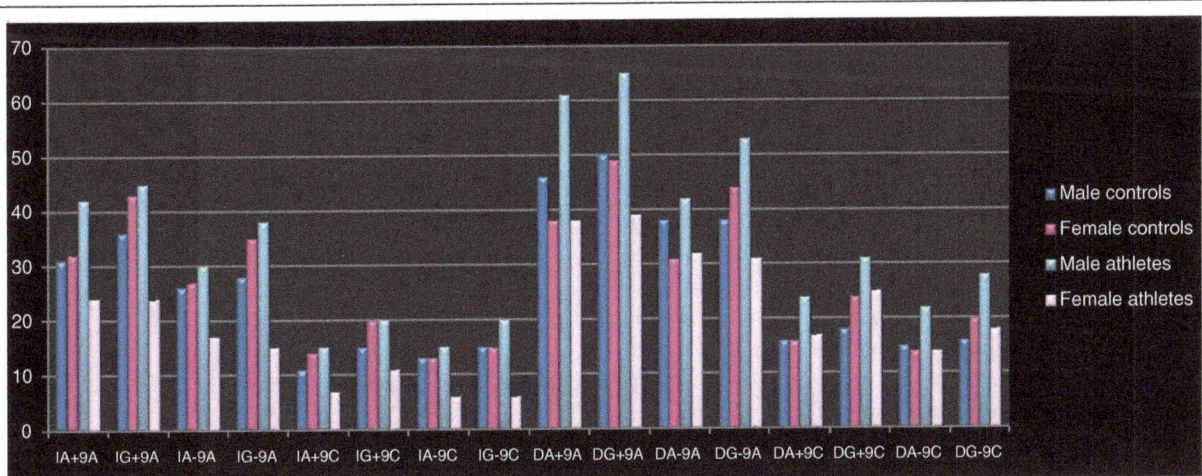

Figure 1 Diagrammatic representation of the 16 allelic combinations among male/female controls and athletes. The 16 allelic combinations resulted from all possible combinations of the eight different polymorphic alleles studied (*ACE* (I, D), *LEP* (G, A), *BDKRB2* (+9, −9) and *AGTR1* (A, C)).

with results coming from genotype distributions within the same group of athletes.

An interesting trend towards the heterozygous state of ACE (I/D), LEP (G/A) and BDKRB2 (+9/−9) genotypes was also highlighted among male Greek athletes, which may require a larger sample size of athletes in order to be demonstrated. Finally, for the AT II type-1 receptor gene, both male and female athletes showed the highest rates in the AGTR1 (A/A) genotype, but with no statistical significant results compared to the corresponding control groups.

New approaches should be identified to evaluate the impact of DNA polymorphisms in human fitness and high-level performance. One possible approach is conducting genome-wide association studies as the one performed by De Moor et al. [33], yet it is unclear whether their findings can be extrapolated to actual elite athletic status. Williams and Folland [34] in 2008 computed the 'total genotype score' (TGS), ranging from 0 to 100, resulting from the accumulated combination of 23 polymorphisms that are candidates to explain individual variations in endurance performance. Using a similar model, limited to seven well-studied polymorphisms associated with endurance capacity in Caucasians, Ruiz et al. [35] determined the actual TGS of the best Spanish male distance runners and road cyclists and suggested an overall more 'favourable' polygenic endurance profile in the athlete group than in Spanish normal individuals.

The limitation of our study is the small number of elite athletes, both as a whole and also in each sport. As such, this study can be considered a pilot investigation that can be expanded with the inclusion of further athletes. For the same reason, the abovementioned genomic markers have to be considered with caution and under no circumstances can be exploited to predict ones athletic performance. A meta-analysis study is currently under way to confirm or to overrule the predictive value of these biomarkers to assess athletic performance.

Conclusions

The tendency for the heterozygous state in three: ACE (I/D), BDKRB2 (+9/−9) and LEP (G/A), out of the four gene polymorphic sites studied was shown, but not proved by the statistical analysis in male athletes. Among female athletes, the co-existence of ACE (D/D) with BDKRB2 (+9/−9) or LEP (G/A) genotypes and the reduction of I allele frequency and of both IG+9A and IG−9A allelic combinations were proved to be significant, compared to the female control group. Probably, a broader and more homogeneous sampling of athletes would demonstrate how strong the results highlighted in this study are and examine the effects of multiple genetic variants and allele combinations in superior physical performance.

Methods
Subjects and genotyping

One hundred and seventy-five Greek athletes and 169 Greek normal individuals who served as the control group were recruited. The athlete group consisted of 102 males and 73 females, whereas the control group consisted of 88 males and 83 females. Inclusion criteria for athletes were their participation and excellence at least once at international and/or national competitions, respectively. In particular, 39 out 102 men and 35 out 73 women had represented Greece at an international level in various sports: swimming (44 males:25 females), volleyball (16 males), handball (29 females) and athletics (long distance runners; 42 males:19 females).

DNA was collected with consent from 10 ml of peripheral blood or 10 ml of saline mouth rinse samples, and the DNA was isolated by QIAamp DNA Blood Kit (QIAGEN GmbH, Hilden, Germany). DNA amplification was performed with polymerase chain reaction, and subsequent restriction fragment length polymorphism analysis was carried out, according to the protocols described by Rigat et al. [36], Di Mauro et al. [37], Fischer et al. [38] and Mammes et al. [12] for ACE (I/D), AGTR1 (A +1166 C), BDKRB2 (+9/−9) and LEP (G−2548A) polymorphisms, respectively.

Statistical analysis

Statistical analysis was performed with SPSS statistical software package (IBM SPSS Statistics version 19.0, Chicago, IL, USA). The statistical differences between groups in genotype distribution, allele frequencies and allelic combination frequencies are presented in Tables 1 and 3. The p values less than 0.05 were considered significant, and they were further assessed by Fisher's exact test.

We have statistically cross-tested genotypic distribution, allele frequencies and furthermore the frequencies of the 16 allelic combinations derived from the four genes examined (two different alleles per gene) between all groups. The tests were initially applied to the whole athlete group in comparison to the non-athlete group, and then each of the above groups were subdivided into cohorts of males and females where the same tests were applied again.

Competing interests
The authors declare that they have no competing interests.

Authors' contributions
AS designed the study, collected the samples from (male/female) athletes and international teams, performed the experiments and drafted the manuscript. VF collected the data, carried out the statistical analysis and gave technical support and conceptual advice for the manuscript. VK participated in the design of the study, collected the control (male/female) samples and performed the initial experiments. GPP and AP developed the concept, designed the experiments and compiled the manuscript. AP also administered the experiments and supervised the statistical analysis. All authors read and approved the final manuscript.

Acknowledgements
This work was supported by grants from the Hellenic Open University (HOU) and the Medical Faculty, University of Patras. The participation of all athletes from the Greek Federations of Volleyball, Handball, Swimming and Athletics (long-distance runners) is gratefully acknowledged.

Author details
[1]School of Science and Technology, Hellenic Open University, Patras 262 22, Greece. [2]Laboratory of General Biology, Faculty of Medicine, University of Patras, Patras 265 04, Greece. [3]Department of Pharmacy, School of Health Sciences, University of Patras, Patras 265 04, Greece.

References
1. Rankinen T, Bray MS, Hagberg JM, Pérusse L, Roth SM, Wolfarth B, Bouchard C: The human gene map for performance and health-related fitness phenotypes: the 2005 update. *Med Sci Sports Exerc* 2006, **38**:1863–1888.
2. Rigat B, Hubert C, Alhenc-Gelas F, Cambien F, Corvol P, Soubrier F: An insertion/deletion polymorphism in the angiotensin I-converting enzyme gene accounting for half the variance of serum enzyme levels. *J Clin Invest* 1990, **86**:1343–1356.
3. Puthucheary Z, Skipworth JR, Rawal J, Loosemore M, Van Someren K, Montgomery HE: The ACE gene and human performance: 12 years on. *Sports Med* 2011, **41**:433–448.
4. Guyton AC, Hall JE: *Dominant role of the kidney in the long-term regulation of arterial pressure and in hypertension: the integrated system for pressure control. In Guyton & Hall: Textbook of Medical Physiology.* 10th edition. Philadelphia: Saunders; 2000:195–209.
5. Lavoie JL, Sigmund CD: Minireview: overview of the renin-angiotensin system–an endocrine and paracrine system. *Endocrinology* 2003, **144**:2179–2183.
6. Dzau VJ: Tissue angiotensin and pathobiology of vascular disease: a unifying hypothesis. *Hypertension* 2001, **37**:1047–1052.
7. Jonsson JR, Game PA, Head RJ, Frewin DB: The expression and localisation of the angiotensin-converting enzyme mRNA in human adipose tissue. *Blood Press* 1994, **3**:72–85.
8. Jones A, Woods DR: Skeletal muscle RAS and exercise performance. *Int J Biochem Cell Biol* 2003, **35**:855–866.
9. Giacchetti G, Faloia E, Mariniello B, Sardu C, Gatti C, Camilloni MA, Guerrieri M, Mantero F: Overexpression of the renin-angiotensin system in human visceral adipose tissue in normal and overweight subjects. *Am J Hypertens* 2002, **15**:381–388.
10. Boustany CM, Bharadwaj K, Daugherty A, Brown DR, Randall DC, Cassis LA: Activation of the systemic and adipose renin-angiotensin system in rats with diet-induced obesity and hypertension. *Am J Physiol Regul Integr Comp Physiol* 2004, **287**:943–949.
11. Cassis LA, English VL, Bharadwaj K, Boustany CM: Differential effects of local versus systemic angiotensin II in the regulation of leptin release from adipocytes. *Endocrinology* 2004, **145**:169–174.
12. Mammès O, Betoulle D, Aubert R, Herbeth B, Siest G, Fumeron F: Association of the G-2548A polymorphism in the 5' region of the LEP gene with overweight. *Ann Hum Genet* 2000, **64**:391–404.
13. Nieters A, Becker N, Linseisen J: Polymorphisms in candidate obesity genes and their interaction with dietary intake of n-6 polyunsaturated fatty acids affect obesity risk in a sub-sample of the EPIC-Heidelberg cohort. *Eur J Nutr* 2002, **41**:210–221.
14. Hoffstedt J, Eriksson P, Mottagui-Tabar S, Arner P: A polymorphism in the leptin promoter region (−2548 G/A) influences gene expression and adipose tissue secretion of leptin. *Horm Metab Res* 2002, **34**:355–369.
15. Hilzendeger AM, Morais RL, Todiras M, Plehm R, da Costa Goncalves A, Qadri F, Araujo RC, Gross V, Nakaie CR, Casarini DE, Carmona AK, Bader M, Pesquero JB: Leptin regulates ACE activity in mice. *J Mol Med* 2010, **88**:899–907.
16. Jayasooriya AP, Mathai ML, Walker LL, Begg DP, Denton DA, Cameron-Smith D, Egan GF, McKinley MJ, Rodger PD, Sinclair AJ, Wark JD, Weisinger HS, Jois M, Weisinger RS: Mice lacking angiotensin-converting enzyme have increased energy expenditure, with reduced fat mass and improved glucose clearance. *Proc Natl Acad Sci USA* 2008, **105**:6531–6546.
17. Bonnardeaux A, Davies E, Jeunemaitre X, Féry I, Charru A, Clauser E, Tiret L, Cambien F, Corvol P, Soubrier F: Angiotensin II type 1 receptor gene polymorphisms in human essential hypertension. *Hypertension* 1994, **24**:63–69.
18. Lung CC, Chan EK, Zuraw BL: Analysis of an exon 1 polymorphism of the B2 bradykinin receptor gene and its transcript in normal subjects and patients with C1 inhibitor deficiency. *J Allergy Clin Immunol* 1997, **99**:134–146.
19. Saunders CJ, de Milander L, Hew-Butler T, Xenophontos SL, Cariolou MA, Anastassiades LC, Noakes TD, Collins M: Dipsogenic genes associated with weight changes during Ironman Triathlons. *Hum Mol Genet* 2006, **15**:2980–2987.
20. Bray MS, Hagberg JM, Pérusse L, Rankinen T, Roth SM, Wolfarth B, Bouchard C: The human gene map for performance and health-related fitness phenotypes: the 2006–2007 update. *Med Sci Sports Exerc* 2009, **41**:35–73.
21. Macarthur DG, North KN: Genes and human elite athletic performance. *Hum Genet* 2005, **116**:331–339.
22. Collins M, Xenophontos SL, Cariolou MA, Mokone GG, Hudson DE, Anastasiades L, Noakes TD: The ACE gene and endurance performance during the South African Ironman Triathlons. *Med Sci Sports Exerc* 2004, **36**:1314–1320.
23. Myerson S, Hemingway H, Budget R, Martin J, Humphries S, Montgomery H: Human angiotensin I-converting enzyme gene and endurance performance. *J Appl Physiol* 1999, **87**:1313–1326.
24. Woods D, Hickman M, Jamshidi Y, Brull D, Vassiliou V, Jones A, Humphries S, Montgomery H: Elite swimmers and the D allele of the ACE I/D polymorphism. *Hum Genet* 2001, **108**:230–242.
25. Amir O, Amir R, Yamin C, Attias E, Eynon N, Sagiv M, Sagiv M, Meckel Y: The ACE deletion allele is associated with Israeli elite endurance athletes. *Exp Physiol* 2007, **92**:881–886.
26. Scott RA, Irving R, Irwin L, Morrison E, Charlton V, Austin K, Tladi D, Deason M, Headley SA, Kolkhorst FW, Yang N, North K, Pitsiladis YP: ACTN3 and ACE genotypes in elite Jamaican and US sprinters. *Med Sci Sports Exerc* 2010, **42**:107–112.
27. Williams AG, Dhamrait SS, Wootton PT, Day SH, Hawe E, Payne JR, Myerson SG, World M, Budgett R, Humphries SE, Montgomery HE: Bradykinin receptor gene variant and human physical performance. *J Appl Physiol* 2004, **96**:938–942.
28. Taguchi T, Kishikawa H, Motoshima H, Sakai K, Nishiyama T, Yoshizato K, Shirakami A, Toyonaga T, Shirontani T, Araki E, Shichiri M: Involvement of bradykinin in acute exercise-induced increase of glucose uptake and GLUT-4 translocation in skeletal muscle: studies in normal and diabetic humans and rats. *Metabolism* 2000, **49**:920–930.
29. Alvarez R, Terrados N, Ortolano R, Iglesias-Cubero G, Reguero JR, Batalla A, Cortina A, Fernández-García B, Rodríguez C, Braga S, Alvarez V, Coto E: Genetic variation in the renin-angiotensin system and athletic performance. *Eur J Appl Physiol* 2000, **82**:117–120.
30. Papadimitriou ID, Papadopoulos C, Kouvatsi A, Triantaphyllidis C: The ACE I/D polymorphism in elite Greek track and field athletes. *J Sports Med Phys Fitness* 2009, **49**:459–463.
31. Eynon N, Meckel Y, Alves AJ, Nemet D, Eliakim A: Is there an interaction between BDKRB2−9/+9 and GNB3 C825T polymorphisms and elite athletic performance? *Scand J Med Sci Sports* 2011, **21**:242–246.
32. Huuskonen A, Lappalainen J, Tanskanen M, Oksala N, Kyröläinen H, Atalay M: Genetic variations of leptin and leptin receptor are associated with body composition changes in response to physical training. *Cell Biochem Funct* 2010, **28**:306–312.
33. De Moor MH, Spector TD, Cherkas LF, Falchi M, Hottenga JJ, Boomsma DI, De Geus: Genome-wide linkage scan for athlete status in 700 British female DZ twin pairs. *Twin Res Hum Genet* 2007, **10**:812–820.
34. Williams AG, Folland JP: Similarity of polygenic profiles limits the potential for elite human physical performance. *J Physiol* 2008, **586**:113–121.
35. Ruiz JR, Gómez-Gallego F, Santiago C, González-Freire M, Verde Z, Foster C, Lucia A: Is there an optimum endurance polygenic profile? *J Physiol* 2009, **587**:1527–1534.
36. Rigat B, Hubert C, Corvol P, Soubrier F: PCR detection of the insertion/deletion polymorphism of the human angiotensin converting enzyme gene (DCP1) (dipeptidyl carboxypeptidase 1). *Nucleic Acids Res* 1992, **20**:1433.

Trans-species polymorphism in humans and the great apes is generally maintained by balancing selection that modulates the host immune response

Luisa Azevedo[1,2,3*], Catarina Serrano[1,2,3], Antonio Amorim[1,2,3] and David N. Cooper[4]

Summary

Known examples of ancient identical-by-descent genetic variants being shared between evolutionarily related species, known as trans-species polymorphisms (TSPs), result from counterbalancing selective forces acting on target genes to confer resistance against infectious agents. To date, putative TSPs between humans and other primate species have been identified for the highly polymorphic major histocompatibility complex (MHC), the histo-blood ABO group, two antiviral genes (ZC3HAV1 and TRIM5), an autoimmunity-related gene LAD1 and several non-coding genomic segments with a putative regulatory role. Although the number of well-characterized TSPs under long-term balancing selection is still very small, these examples are connected by a common thread, namely that they involve genes with key roles in the immune system and, in heterozygosity, appear to confer genetic resistance to pathogens. Here, we review known cases of shared polymorphism that appear to be under long-term balancing selection in humans and the great apes. Although the specific selective agent(s) responsible are still unknown, these TSPs may nevertheless be seen as constituting important adaptive events that have occurred during the evolution of the primate immune system.

Introduction

Trans-species polymorphisms (TSPs) are ancient genetic variants whose origin predates speciation events, resulting in shared alleles between evolutionarily related species [1]. Shared polymorphisms are only considered to be TSPs sensu stricto when there is convincing evidence to show that they are identical-by-descent rather than recurrent mutations occurring independently in different lineages (i.e. identical-by-state); the latter often involve the CpG dinucleotide [2, 3] whose hypermutability is directly attributable to its role as the major site of cytosine methylation, with the attendant risk of spontaneous deamination of 5-methylcytosine to yield thymine. It follows that, even in a genomic region which manifests signals of balancing selection, a specific shared polymorphism that

is located within a CpG dinucleotide is, by its very nature, more likely to be identical-by-state due to recurrent mutation in distinct lineages rather than it is to be identical-by-descent. In such cases, any of the strong signals of balancing selection observed may emanate from functionally relevant balanced polymorphisms that are closely linked to the CpG site in question but which are not themselves shared across lineages.

The long-term (i.e. post-speciation) preservation of bona fide identical-by-descendent TSPs is inherently unlikely under a purely neutralist model of evolution [4, 5], and hence the action of selection must invariably be assumed. In passing, it should be noted that positive selection is also implicit in the case of independently occurring identical-by-state polymorphisms in different lineages. Here, we review the small number of relatively well-characterized examples of TSPs that are shared between the genomes of great apes (human, common chimpanzee, bonobo, gorilla and orangutan), focusing

* Correspondence: lazevedo@ipatimup.pt
[1]Instituto de Investigação e Inovação em Saúde, Universidade do Porto, Porto, Portugal
[2]IPATIMUP-Institute of Molecular Pathology and Immunology, University of Porto, Rua Dr. Roberto Frias s/n, 4200-465, Porto, Portugal

specifically on those for which there is a body of evidence for long-standing balancing selection (Fig. 1).

TSPs maintained by balancing selection: the MHC and ABO loci

Evidence for the common ancestral origin of an extant TSP shared by humans and one or more of the great apes was first documented about three decades ago when orthologous sequences from the highly polymorphic major histocompatibility complex (MHC) loci were compared between species [4, 6]. MHC (known as HLA, human leucocyte antigen, in humans) loci play a key role in the adaptive response to pathogens [7, 8], and some of the allelic lineages of the human HLA-DQ alpha locus (HLA-DQA1, HLA-DQA3 and HLA-DQA4) were deduced to have been present in the most recent common ancestor of the human, chimpanzee and gorilla lineages [4] and must therefore have survived for at least 8 million years [9, 10]. More recently, the human MHC lineage most strongly associated with delayed HIV-1 progression (*HLA-B*57*) was found to exhibit a high degree of similarity to a lineage frequently found among SIV-infected chimpanzees [11]. Other comparisons have reached a similar conclusion, namely that many alleles at the MHC locus have survived for an extended period of evolutionary time and hence are currently shared by multiple primate lineages. One example is provided by the MHC-DQB1*06 allele which predates the separation of the hominid and Old World monkey lineages more than 35 million years ago [12]. Shared alleles at the MHC locus have also been found in other primates such as the rhesus and cynomolgous macaques [13], between Madagascan lemurs in which some alleles appear to have been maintained for more than 40 million years [14] and between several non-primate lineages such as mice and rats [15], the brown bear (*Ursus arctos*) and the giant panda (*Ailuropoda melanoleuca*) [16], South American mouse opossums (*Gracilinanus microtarsus* and *Marmosops incanus*) [17] and between equines [18] among others (reviewed in [19]). The maintenance of these shared polymorphisms over such extended periods of evolutionary time implies strong selective pressure on the host immune response elicited by the pathogenic agent. However, it should be borne in mind that the potential contribution of recurrent mutation to the origin of these shared variants has not always been unequivocally excluded. It may therefore be that some of these 'shared variants' are actually identical-by-state rather than identical-by-descent. At the same time, direct evidence for a functional role for the balanced variant, or group of variants, is lacking in most cases. Although these *caveats* do not necessarily challenge the now well-established role of balancing selection acting on MHC loci, it precludes the acquisition of a clear picture of how often the TSPs are actually identical-by-descent as opposed to simply being identical-by-state.

Another well-established example of long-term balancing selection operating in the primate genome is provided by the ABO blood group locus. In humans, three main alleles account for the diversity at this locus, corresponding to the A, B and O blood groups. The A and B

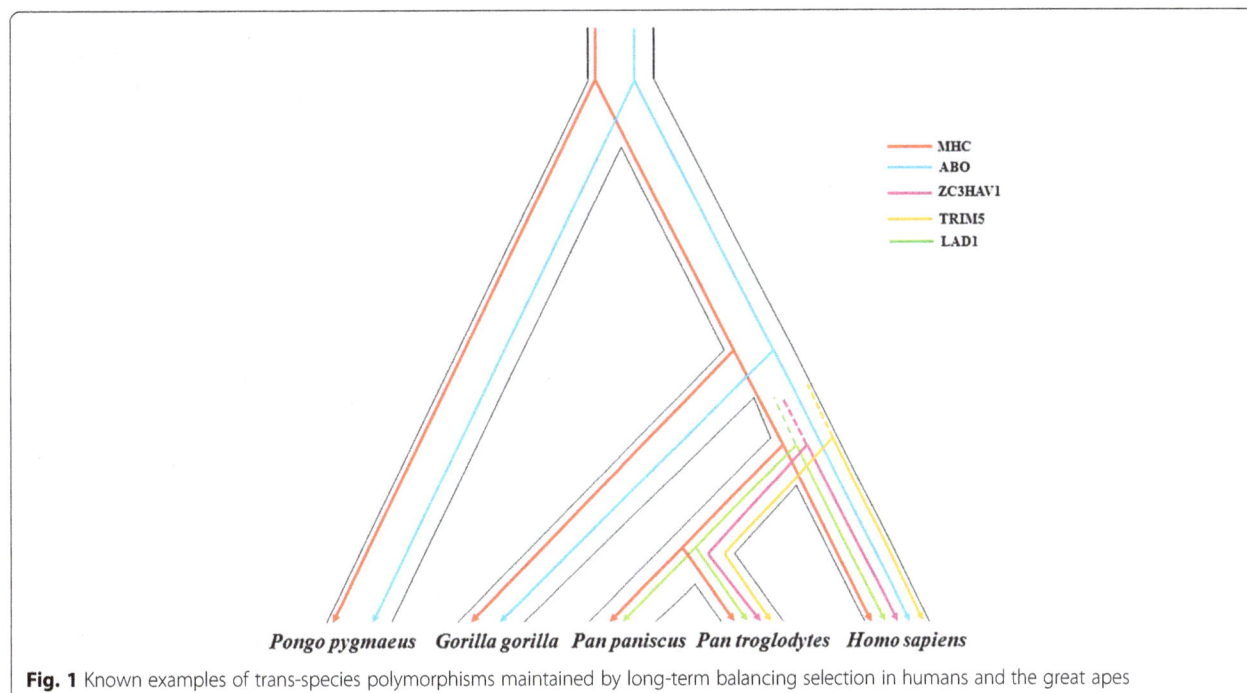

Fig. 1 Known examples of trans-species polymorphisms maintained by long-term balancing selection in humans and the great apes

alleles are functionally distinguished by the co-occurrence of two missense mutations (Leu266Met and Gly268Ala, respectively) in the encoded glycosyltransferase whereas the human O allele results from an inactivating single-nucleotide deletion (261delG) that impairs enzymatic function, resulting in failure to convert H antigen into A or B [20]. A and B alleles are shared between humans and non-human primates [21–23]. Although one of the alleles has been lost in some lineages during great ape speciation (e.g. common chimpanzees and bonobos exhibit only the A antigen, whereas the gorilla harbours the B antigen), other lineages have retained both the A and B identical-by-descent alleles, e.g. the orangutan which shared a common ancestor with humans more than 16 million years ago [24]. Although pathogen-driven selective pressure operating on the balanced A/B alleles appears less intuitive than it perhaps is for the MHC locus, one must recall that there are several instances of histo-blood group antigens being associated with differential protection against multiple infectious microbes [25–27]. For example, infection by *Helicobacter pylori* has been shown to be reduced in A and B blood types as compared with carriers of the O type [25]. Moreover, a link between pathological conditions associated with *H. pylori* colonization, such as gastric [28, 29] and pancreatic cancer [30, 31], and the ABO phenotype has also been established in humans. Assuming that the persistence of the A/B polymorphism is maintained by pathogen-driven balancing selection, one may reasonably extend these considerations to the other great ape species.

Beyond the MHC and ABO loci

Apart from the well-established examples of trans-species polymorphisms at the MHC and ABO loci which are maintained by balancing selection, few other properly substantiated examples of TSPs have been documented. One particularly interesting case is provided by the zinc-finger CCCH-type antiviral protein 1 (ZC3HAV1, also known as poly(ADP-ribose) polymerase 13-PARP13), a protein that is known to protect host cells from viral infection [32–35] and cellular stress [36]. The ZC3HAV1 polymorphic substitution Thr851Ile (rs3735007) is shared between humans and common chimpanzees and does not

occur in a hypermutable CpG site (Table 1), supporting its candidacy as a true TSP [37]; an exhaustive analysis of the genomic region adjacent to this TSP has shown that the polymorphism has been selectively maintained in both species, probably as a result of its broad protective effect against viral infection.

Another potential example of trans-species polymorphism under balancing selection is provided by the *TRIM5* gene. This gene also encodes an antiviral protein [38–40] (TRIM5, tripartite motif-containing 5), one that is known to act as a blocking factor of HIV-1 reverse transcription thereby limiting the efficiency of the infection in primates [41]. In non-primate species, TRIM5 is also active as an antiviral protein [42–45] and the reconstruction of TRIM5 evolutionary history has provided evidence for a long-term interaction with several different viruses prior to the origin of primate lentivirus [46].

A previous study has evidenced balanced TSPs in the primate *TRIM5* genes [47]. In a separate investigation, two intragenic *TRIM5* polymorphisms were found to be shared between human and common chimpanzee (Table 1): an intron 1-CTC insertion/deletion and an intron 1 transition (rs34506684), the latter occurring at an CpG dinucleotide [48]. Although the functional significance of these variants has not yet been established, the authors suggested that the intron 1 variant rs34506684 might impact transcription factor-binding sites leading to allelic differences in transcriptional activity which could underpin inter-individual differences in susceptibility to infection. However, because this shared polymorphism occurs at a hypermutable CpG site (Table 1), its candidacy as a true TSP must be in some doubt; in the absence of any convincing evidence that this variant is of direct functional significance, it may simply be a marker in linkage disequilibrium with another variant that is under balancing selection.

The discovery of these putative regulatory variants within intron 1 of the *TRIM5* gene between humans and the great apes provides evidence for the maintenance of balanced regulatory polymorphisms across species (as distinct from balanced coding sequence variants), as well as yet another example of a TSP which may impact the host pathogen response. Trans-species

Table 1 Properties and allelic frequencies of human single-nucleotide putative TSPs shared with bonobos and common chimpanzees

Gene	SNP ID	SNP flanking sequence[a]	MAF[a]	AA Change	Presence
ZC3HAV1	rs3735007	GTTTA(C/T)TGAAG	0.47 (C)	Thr851Ile	Human and common chimpanzee
TRIM5	rs34506684	TCTGG(C/T)GCCTC	0.46 (T)	Intronic	Human and common chimpanzee
LAD1	rs12088790	GGGCC(A/G)GCGAC	0.12 (G)	Leu257Pro	Human, bonobo and common chimpanzee

[a]Data extracted from dbSNP [69]

regulatory polymorphism has been functionally investigated at the MHC-DQA1 locus in eight non-human primates [49]. Numerous trans-species polymorphisms were identified within transcription factor binding sites in the MHC-DQA1 promoter region. Loisel et al. [49] assessed the functional consequences of these variants using a reporter gene assay and identified significant differences between baboon DQA1 promoter haplotypes in terms of their ability to drive transcription in vitro. Taken together with the high levels of sequence variation in this region, these findings suggest a role for balancing selection in the evolution of DQA1 transcriptional regulation in primates although the biological mechanism underlying the assumed increase in fitness remains unclear.

More recently, a scan of the human genome yielded good evidence for six non-coding genomic regions where ancestral polymorphisms shared between humans and chimpanzees have been under the influence of balancing selection [2]. The closest genes to these regions (FREM3, MTRR, PROKR2, HUS1, IGFBP7 and ST3GAL1) are all to some extent related to the innate immune response, a finding which would concur with a balancing selection mode of evolution. For three of these six regions (near HUS1, IGFBP7 and ST3GAL1), a regulatory role for the polymorphisms involved was demonstrated, indicating that physiological differences resulting from a balanced polymorphism can also be exerted at the level of gene expression. A role for regulatory variants as targets of balancing selection is not surprising since many studies have reported an important role for heterozygote advantage in the evolution of gene expression [50–52]. Although many of these variants may individually be associated with deleterious effects, they may nevertheless provide fitness advantages under certain environmental conditions [51, 53] by potentiating an optimal level of gene expression [52]. It is also quite possible that some of these regulatory balanced alleles have persisted for long periods of time in immunity-related genes thereby providing further likely examples of TSPs between humans and the great apes.

The most recently reported example of a putative TSP is that in exon 3 of the ladinin 1 (LAD1) gene (rs12088790) (Table 1) which has been claimed to be maintained by long-term selection in humans, common chimpanzees and bonobos [54]. However, once again, the shared polymorphism occurs at a CpG dinucleotide and hence may not in reality be identical-by-descent. The resulting missense change (Leu257Pro) influences the expression of LAD1: the minor allele, which occurs at a frequency of 0.12 in humans (Table 1), is associated with an increased level of LAD1 expression. Irrespective of whether this is a direct effect, or whether the missense variant is in linkage disequilibrium with another polymorphic variant with a regulatory role, it provides further evidence for the important role of shared polymorphism in modulating gene expression.

The LAD1 gene encodes a collagenous anchoring filament protein that serves to maintain dermal-epidermal cohesion and is associated with IgA bullous dermatosis, an autoimmune disease [55]. Apart from its pathogenic role in the context of IgA bullous dermatosis, there is no information as to how LAD1 might contribute to the host response against a pathogen. However, many autoimmune diseases are triggered by infectious agents in addition to environmental factors [56], and this has been specifically reported to be the case in IgA bullous dermatosis [57].

Overdominance vs. pathogen-driven frequency-dependent selection

The maintenance of TSP by long-term balancing selection has long been held to be mediated by heterozygote advantage (overdominance) or frequency-dependent selection [19]. Previous studies have claimed that the most important factor for the maintenance of MHC polymorphism is overdominant selection [58, 59]. In support of this postulate, MHC heterozygosity has been experimentally demonstrated to enhance resistance to multiple-strain infections [60]. However, a simulation-based approach has demonstrated that polymorphism at the MHC locus results not merely from overdominant selection but also from frequency-dependent host-pathogen coevolution for rare MHC alleles [61].

Host-pathogen interactions have also been proposed for the ABO TSPs whose maintenance over long periods of evolutionary time may have been due in part to coevolution with gut pathogens [62]. In accordance with this suggestion, a general model that integrates frequency-dependent selection and genetic drift would appear to account for the ABO polymorphism in humans [63]. Irrespective of the precise mechanism underlying the maintenance of a given TSP, it is important to appreciate that to be effective, the selective agent would need to exhibit the following properties: (a) it must be widespread geographically, (b) it must have been present over an extended period of evolutionary time, and (c) it must show similar tropism towards different yet evolutionarily related species. Further, heterozygosity at the targeted locus would have to mediate similar selective responses in the different species involved, and selection in favour of the heterozygote would have to be sufficiently intense to maintain the frequency of the minor allele(s) in the host species populations. The selective agent most likely to possess these properties is a pathogen. Alternative selective agents such as climate and diet would be most unlikely to remain constant over extended periods of evolutionary time [64].

Conclusion

The presence of shared polymorphisms across evolutionarily related species is a strong indicator of balancing selection. Although most of the studies to date have focussed on the classical examples of the MHC and ABO loci, a few additional examples are known between humans and the great apes. Herein, we have reviewed these cases and noted that a common feature of virtually all well-established cases of balanced TSP is their association with the host-immune response, presumably triggered by infectious agents. The diversity observed at immunity-related loci is certainly shaped by the dynamic process of host-pathogen coevolution [65], and therefore, key aspects of a species' adaptation to challenging environments are likely to have been pathogen-driven [66–68]. This raises the question of the potential relevance of TSP identification to improving our understanding of the host immune response; indeed, the study of these ancient variants could lead to new insights into immune system function with important implications for preventive medicine. Finally, it may be that the identification of additional balanced TSPs in humans and their closest relatives among the great apes might be facilitated by a guided search (e.g. by targeting immune system/autoimmune disease-associated genes specifically). The difficulty inherent in any such quest would be to prove that a newly detected shared polymorphism is identical-by-descent rather than simply identical-by-state. This would be especially true for polymorphisms residing within CpG sites where the balance of probability must lie firmly on the side of their being identical-by-state, at least until functional evidence to the contrary can be provided.

Abbreviations
LAD1: ladinin 1; HLA: human leucocyte antigen; MHC: major histocompatibility complex; TRIM5: tripartite motif-containing 5; TSP: trans-species polymorphism; ZC3HAV1: zinc-finger CCCH-type antiviral protein 1.

Competing interests
The authors declare that they have no competing interests.

Authors' contributions
LA and DNC conceived the idea, performed the literature search and wrote the manuscript. CS performed the data searches and comparative genomic analyses. AA critically revised the manuscript. All authors read and approved the final version of the manuscript.

Acknowledgements
IPATIMUP integrates the i3S Research Unit, which is partially supported by FCT, the Portuguese Foundation for Science and Technology. This work is funded by FEDER funds through the Operational Programme for Competitiveness Factors—COMPETE and National Funds through the FCT-Foundation for Science and Technology, under the projects "PEst-C/SAU/LA0003/2013". DNC gratefully acknowledges financial support from QIAGEN Inc. through a License Agreement with Cardiff University.

Author details
[1]Instituto de Investigação e Inovação em Saúde, Universidade do Porto, Porto, Portugal. [2]IPATIMUP-Institute of Molecular Pathology and Immunology, University of Porto, Rua Dr. Roberto Frias s/n, 4200-465, Porto, Portugal. [3]Department of Biology, Faculty of Sciences, University of Porto, Rua do Campo Alegre, s/n, 4169-007, Porto, Portugal. [4]Institute of Medical Genetics, School of Medicine, Cardiff University, Heath Park, Cardiff, CF14 4XN, UK.

References
1. Klein J. Origin of major histocompatibility complex polymorphism: the trans-species hypothesis. Hum Immunol. 1987;19(3):155–62.
2. Leffler EM, Gao Z, Pfeifer S, Ségurel L, Auton A, Venn O, et al. Multiple instances of ancient balancing selection shared between humans and chimpanzees. Science. 2013;339(6127):1578–82.
3. Cagliani R, Fumagalli M, Riva S, Pozzoli U, Comi GP, Menozzi G, et al. The signature of long-standing balancing selection at the human defensin β-1 promoter. Genome Biol. 2008;9(9):R143-R.
4. Gyllensten UB, Erlich HA. Ancient roots for polymorphism at the HLA-DQ alpha locus in primates. Proc Natl Acad Sci U S A. 1989;86(24):9986–90.
5. Asthana S, Schmidt S, Sunyaev S. A limited role for balancing selection. Trends Genet. 2005;21(1):30–2.
6. Mayer WE, Jonker M, Klein D, Ivanyi P, van Seventer G, Klein J. Nucleotide sequences of chimpanzee MHC class I alleles: evidence for trans-species mode of evolution. EMBO J. 1988;7(9):2765–74.
7. Spurgin LG, Richardson DS. How pathogens drive genetic diversity: MHC, mechanisms and misunderstandings. Proc R Soc B. 2010;277(1684):979–88.
8. Sommer S. The importance of immune gene variability (MHC) in evolutionary ecology and conservation. Front Zool. 2005;2(1):16.
9. Langergraber KE, Prüfer K, Rowney C, Boesch C, Crockford C, Fawcett K, et al. Generation times in wild chimpanzees and gorillas suggest earlier divergence times in great ape and human evolution. Proc Natl Acad Sci U S A. 2012;109(39):15716–21.
10. Scally A, Dutheil JY, Hillier LW, Jordan GE, Goodhead I, Herrero J, et al. Insights into hominid evolution from the gorilla genome sequence. Nature. 2012;483(7388):169–75.
11. Wroblewski EE, Norman PJ, Guethlein LA, Rudicell RS, Ramirez MA, Li Y, et al. Signature patterns of MHC diversity in three Gombe communities of wild chimpanzees reflect fitness in reproduction and immune defense against SIVcpz. PLoS Biol. 2015;13(5), e1002144.
12. Otting N, de Groot N, Doxiadis G, Bontrop R. Extensive Mhc-DQB variation in humans and non-human primate species. Immunogenetics. 2002;54(4):230–9.
13. Yao Y-F, Dai Q-X, Li J, Ni Q-Y, Zhang M-W, Xu H-L. Genetic diversity and differentiation of the rhesus macaque (Macaca mulatta) population in western Sichuan, China, based on the second exon of the major histocompatibility complex class II DQB (MhcMamu-DQB1) alleles. BMC Evol Biol. 2014;14(1):130.
14. Go Y, Satta Y, Kawamoto Y, Rakotoarisoa G, Randrianjafy A, Koyama N, et al. Mhc-DRB genes evolution in lemurs. Immunogenetics. 2002;54(6):403–17.
15. Figueroa F, Gunther E, Klein J. MHC polymorphism pre-dating speciation. Nature. 1988;335(6187):265–7.
16. Kuduk K, Babik W, Bojarska K, Śliwińska EB, Kindberg J, Taberlet P, et al. Evolution of major histocompatibility complex class I and class II genes in the brown bear. BMC Evol Biol. 2012;12:197.
17. Meyer-Lucht Y, Otten C, Püttker T, Sommer S. Selection, diversity and evolutionary patterns of the MHC class II DAB in free-ranging neotropical marsupials. BMC Genet. 2008;9:39.
18. Kamath P, Getz W. Adaptive molecular evolution of the Major Histocompatibility Complex genes, DRA and DQA, in the genus Equus. BMC Evol Biol. 2011;11(1):128.
19. Těšický M, Vinkler M. Trans-species polymorphism in immune genes: general pattern or MHC-restricted phenomenon? J Immunol Res. 2015;2015:838035.
20. Yamamoto F, Clausen H, White T, Marken J, Hakomori S. Molecular genetic basis of the histo-blood group ABO system. Nature. 1990;345(6272):229–33.
21. Martinko JM, Vincek V, Klein D, Klein J. Primate ABO glycosyltransferases: evidence for trans-species evolution. Immunogenetics. 1993;37(4):274–8.
22. Yamamoto F, Cid E, Yamamoto M, Saitou N, Bertranpetit J, Blancher A. An integrative evolution theory of histo-blood group ABO and related genes. Sci Rep. 2014;4.
23. Ségurel L, Thompson EE, Flutre T, Lovstad J, Venkat A, Margulis SW, et al. The ABO blood group is a trans-species polymorphism in primates. Proc Natl Acad Sci U S A. 2012;109(45):18493–8.

24. Steiper ME, Young NM. Primate molecular divergence dates. Mol Phylogenet Evol. 2006;41(2):384–94.

25. Boren T, Falk P, Roth KA, Larson G, Normark S. Attachment of *Helicobacter pylori* to human gastric epithelium mediated by blood group antigens. Science. 1993;262(5141):1892–5.

26. Anstee DJ. The relationship between blood groups and disease. Blood. 2010;115(23):4635–43.

27. Harris JB, Khan AI, LaRocque RC, Dorer DJ, Chowdhury F, Faruque ASG, et al. Blood group, immunity, and risk of infection with *Vibrio cholerae* in an area of endemicity. Infect Immun. 2005;73(11):7422–7.

28. Duell EJ, Bonet C, Muñoz X, Lujan-Barroso L, Weiderpass E, Boutron-Ruault M-C, et al. Variation at *ABO* histo-blood group and *FUT* loci and diffuse and intestinal gastric cancer risk in a European population. Int J Cancer. 2015;136(4):880–93.

29. Rizzato C, Kato I, Plummer M, Muñoz N, Stein A. Jan van Doorn L, et al. Risk of advanced gastric precancerous lesions in *Helicobacter pylori* infected subjects is influenced by ABO blood group and cagA status. Int J Cancer. 2013;133(2):315–22.

30. Amundadottir L, Kraft P, Stolzenberg-Solomon RZ, Fuchs CS, Petersen GM, Arslan AA, et al. Genome-wide association study identifies variants in the *ABO* locus associated with susceptibility to pancreatic cancer. Nat Genet. 2009;41(9):986–90.

31. Risch HA. Pancreatic cancer: *Helicobacter pylori* colonization, N-Nitrosamine exposures, and ABO blood group. Mol Carcinogen. 2012;51(1):109–18.

32. Gao G, Guo X, Goff SP. Inhibition of retroviral RNA production by ZAP, a CCCH-type zinc finger protein. Science. 2002;297(5587):1703–6.

33. Mao R, Nie H, Cai D, Zhang J, Liu H, Yan R, et al. Inhibition of hepatitis B virus replication by the host zinc finger antiviral protein. PLoS Pathog. 2013;9(7), e1003494.

34. Kerns JA, Emerman M, Malik HS. Positive selection and increased antiviral activity associated with the PARP-containing isoform of human zinc-finger antiviral protein. PLoS Genet. 2008;4(1), e21.

35. Guo X, Carroll J-WN, MacDonald MR, Goff SP, Gao G. The zinc finger antiviral protein directly binds to specific viral mRNAs through the CCCH zinc finger motifs. J Virol. 2004;78(23):12781–7.

36. Todorova T, Bock FJ, Chang P. Poly(ADP-ribose) polymerase-13 and RNA regulation in immunity and cancer. Trends Mol Med. 2015;21(6):373–84.

37. Cagliani R, Guerini FR, Fumagalli M, Riva S, Agliardi C, Galimberti D, et al. A trans-specific polymorphism in *ZC3HAV1* is maintained by long-standing balancing selection and may confer susceptibility to multiple sclerosis. Mol Biol Evol. 2012;29(6):1599–613.

38. Bieniasz PD. Intrinsic immunity: a front-line defense against viral attack. Nat Immunol. 2004;5(11):1109–15.

39. Kaiser SM, Malik HS, Emerman M. Restriction of an extinct retrovirus by the human TRIM5α antiviral protein. Science. 2007;316(5832):1756–8.

40. Goldschmidt V, Ciuffi A, Ortiz M, Brawand D, Muñoz M, Kaessmann H, et al. Antiretroviral activity of ancestral TRIM5α. J Virol. 2008;82(5):2089–96.

41. Stremlau M, Owens CM, Perron MJ, Kiessling M, Autissier P, Sodroski J. The cytoplasmic body component TRIM5[alpha] restricts HIV-1 infection in Old World monkeys. Nature. 2004;427(6977):848–53.

42. Schaller T, Hué S, Towers GJ. An active TRIM5 protein in rabbits indicates a common antiviral ancestor for mammalian TRIM5 proteins. J Virol. 2007;81(21):11713–21.

43. Fletcher AJ, Hué S, Schaller T, Pillay D, Towers GJ. Hare TRIM5α restricts divergent retroviruses and exhibits significant sequence variation from closely related lagomorpha *TRIM5* genes. J Virol. 2010;84(23):12463–8.

44. Ylinen LMJ, Keckesova Z, Webb BLJ, Gifford RJM, Smith TPL, Towers GJ. Isolation of an active Lv1 gene from cattle indicates that tripartite motif protein-mediated innate immunity to retroviral infection is widespread among mammals. J Virol. 2006;80(15):7332–8.

45. Jáuregui P, Crespo H, Glaria I, Luján L, Contreras A, Rosati S, et al. Ovine TRIM5α can restrict Visna/Maedi virus. J Virol. 2012.

46. Sawyer SL, Wu LI, Emerman M, Malik HS. Positive selection of primate TRIM5α identifies a critical species-specific retroviral restriction domain. Proc Natl Acad Sci U S A. 2005;102(8):2832–7.

47. Newman RM, Hall L, Connole M, Chen GL, Sato S, Yuste E, et al. Balancing selection and the evolution of functional polymorphism in Old World monkey TRIM5alpha. Proc Natl Acad Sci U S A. 2006;103(50):19134–9.

48. Cagliani R, Fumagalli M, Biasin M, Piacentini L, Riva S, Pozzoli U, et al. Long-term balancing selection maintains trans-specific polymorphisms in the human *TRIM5* gene. Hum Genet. 2010;128(6):577–88.

49. Loisel DA, Rockman MV, Wray GA, Altmann J, Alberts SC. Ancient polymorphism and functional variation in the primate MHC-DQA1 5′ *cis*-regulatory region. Proc Natl Acad Sci U S A. 2006;103(44):16331–6.

50. Arbiza L, Gronau I, Aksoy BA, Hubisz MJ, Gulko B, Keinan A, et al. Genome-wide inference of natural selection on human transcription factor binding sites. Nat Genet. 2013;45(7):723–9.

51. Schaschl H, Huber S, Schaefer K, Windhager S, Wallner B, Fieder M. Signatures of positive selection in the *cis*-regulatory sequences of the human oxytocin receptor (*OXTR*) and arginine vasopressin receptor 1a (*AVPR1A*) genes. BMC Evol Biol. 2015;15:85.

52. Sellis D, Callahan BJ, Petrov DA, Messer PW. Heterozygote advantage as a natural consequence of adaptation in diploids. Proc Natl Acad Sci U S A. 2011;108(51):20666–71.

53. Siegert S, Wolf A, Cooper DN, Krawczak M, Nothnagel M. Mutations causing complex disease may under certain circumstances be protective in an epidemiological sense. PLoS One. 2015;10(7), e0132150.

54. Teixeira JC, de Filippo C, Weihmann A, Meneu JR, Racimo F, Dannemann M, et al. Long-term balancing selection in *LAD1* maintains a missense trans-species polymorphism in humans, chimpanzees, and bonobos. Mol Biol Evol. 2015;32(5):1186–96.

55. Marinkovich MP, Taylor TB, Keene DR, Burgeson RE, Zone JJ. LAD-1, the linear IgA bullous dermatosis autoantigen, is a novel 120-kDa anchoring filament protein synthesized by epidermal cells. J Invest Dermatol. 1996;106(4):734–8.

56. Ercolini AM, Miller SD. The role of infections in autoimmune disease. Clin Exp Immunol. 2009;155(1):1–15.

57. Ahkami R, Thomas I. Linear IgA bullous dermatosis associated with vancomycin and disseminated varicella-zoster infection. Cutis. 2001;67(5):423–6.

58. Hughes AL, Nei M. Pattern of nucleotide substitution at major histocompatibility complex class I loci reveals overdominant selection. Nature. 1988;335(6186):167–70.

59. Takahata N, Nei M. Allelic genealogy under overdominant and frequency-dependent selection and polymorphism of major histocompatibility complex loci. Genetics. 1990;124(4):967–78.

60. Penn DJ, Damjanovich K, Potts WK. MHC heterozygosity confers a selective advantage against multiple-strain infections. Proc Natl Acad Sci U S A. 2002;99(17):11260–4.

61. Borghans JM, Beltman J, De Boer R. MHC polymorphism under host-pathogen coevolution. Immunogenetics. 2004;55(11):732–9.

62. Ségurel L, Gao Z, Przeworski M. Ancestry runs deeper than blood: the evolutionary history of ABO points to cryptic variation of functional importance. Bioessays. 2013;35(10):862–7.

63. Villanea FA, Safi KN, Busch JW. A general model of negative frequency dependent selection explains global patterns of human ABO polymorphism. PLoS One. 2015;10(5), e0125003.

64. Fumagalli M, Sironi M, Pozzoli U, Ferrer-Admetlla A, Pattini L, Nielsen R. Signatures of environmental genetic adaptation pinpoint pathogens as the main selective pressure through human evolution. PLoS Genet. 2011;7(11), e1002355.

65. Sironi M, Cagliani R, Forni D, Clerici M. Evolutionary insights into host-pathogen interactions from mammalian sequence data. Nat Rev Genet. 2015;16(4):224–36.

66. Siddle KJ, Quintana-Murci L. The Red Queen's long race: human adaptation to pathogen pressure. Curr Opin Genet Dev. 2014;29:31–8.

67. Andrés AM, Hubisz MJ, Indap A, Torgerson DG, Degenhardt JD, Boyko AR, et al. Targets of balancing selection in the human genome. Mol Biol Evol. 2009;26(12):2755–64.

68. Key FM, Teixeira JC, de Filippo C, Andrés AM. Advantageous diversity maintained by balancing selection in humans. Curr Opin Genet Dev. 2014;29:45–51.

69. Database of Single Nucleotide Polymorphisms (dbSNP). Available from: [http://www.ncbi.nlm.nih.gov/SNP].

Copy number variation in CEP57L1 predisposes to congenital absence of bilateral ACL and PCL ligaments

Yichuan Liu[1*], Yun Li[1], Michael E. March[1], Kenny Nguyen[1], Kexiang Xu[1], Fengxiang Wang[1], Yiran Guo[1], Brendan Keating[1], Joseph Glessner[1], Jiankang Li[2], Theodore J. Ganley[3], Jianguo Zhang[2], Matthew A. Deardorff[4], Xun Xu[2] and Hakon Hakonarson[1]

Abstract

Background: Absence of the anterior (ACL) or posterior cruciate ligament (PCL) are rare congenital malformations that result in knee joint instability, with a prevalence of 1.7 per 100,000 live births and can be associated with other lower-limb abnormalities such as ACL agnesia and absence of the menisci of the knee. While a few cases of absence of ACL/PCL are reported in the literature, a number of large familial case series of related conditions such as ACL agnesia suggest a potential underlying monogenic etiology. We performed whole exome sequencing of a family with two individuals affected by ACL/PCL.

Results: We identified copy number variation (CNV) deletion impacting the exon sequences of *CEP57L1*, present in the affected mother and her affected daughter based on the exome sequencing data. The deletion was validated using quantitative PCR (qPCR), and the gene was confirmed to be expressed in ACL ligament tissue. Interestingly, we detected reduced expression of *CEP57L1* in Epstein–Barr virus (EBV) cells from the two patients in comparison with healthy controls. Evaluation of 3D protein structure showed that the helix-binding sites of the protein remain intact with the deletion, but other functional binding sites related to microtubule attachment are missing. The specificity of the CNV deletion was confirmed by showing that it was absent in ~700 exome sequencing samples as well as in the database of genomic variations (DGV), a database containing large numbers of annotated CNVs from previous scientific reports.

Conclusions: We identified a novel CNV deletion that was inherited through an autosomal dominant transmission from an affected mother to her affected daughter, both of whom suffered from the absence of the anterior and posterior cruciate ligaments of the knees.

Keywords: Copy number variation, Rare disease, Whole exome sequencing

Introduction

Congenital absence of the anterior (ACL)/posterior cruciate ligaments (PCL) is an extremely unusual condition with a prevalence of 1.7 per 100,000 live births and was first reported in 1956 in a radiographic study of the knee [1–3]. The symptom of this disease usually associated with serious malformation or dislocation of the knees.

ACL is the most common disorder, and it can be associated with hypoplasia or total PCL.

There is little knowledge of the genetics information related to this disease, but a previous study in a large number of families suggested autosomal dominant inheritance of ACL [4]. In this study, we uncovered two individuals (mother and her daughter) in a large cohort of subjects with rare diseases, and we confirmed the underlying sequence copy number variations (CNVs) that predisposes to this disease.

* Correspondence: liuy5@email.chop.edu
[1]Center for Applied Genomics, The Children's Hospital of Philadelphia, 1014H, 3615 Civic Center Blvd, Abramson Building, Philadelphia, PA 19104, USA

Methods

Data selections

Among 662 individuals from 212 families with Mendelian disease phenotypes, there was one family with two affected patients, including a mother (ID#4779933454) and her 13-year-old daughter (ID#1993197298), suffering from an absence of bilateral ACL and PCL ligaments. Their history was notable for joint instability and difficulty with ambulation. Absence of ACL and PCL was confirmed on MRI and by arthroscopy. The proband (daughter) has an unaffected younger sister and unaffected father, for both of whom DNA was not obtainable. However, the pattern of inheritance suggests it is autosomal dominant [4]; no other extended family members have a history of joint or lower-limb disorders.

Exome capture and sequencing

Following the manufacturer's protocols, we used Agilent SureSelect Human All Exon Kit (in solution) (Agilent Technologies, Santa Clara, CA, USA) to perform exome capture for the family included in this study. We used Covaris AFA (Covaris, Woburn, MA, USA) to randomly fragment genomic DNA samples with an average size of 150–200 bp and then attached adapters to both ends. We applied AgencourtAMPure SPRI beads (Beckman Coulter, Brea CA, USA) to purify the adapter-ligated templates with an insert size of around 250 bp. We employed ligation-mediated polymerase chain reaction (LM-PCR) and SureSelect Biotinylated RNA Library (BAITS) (Agilent Technologies, Santa Clara, CA, USA) to amplify, purify, and hybridize DNA for enrichment. We then washed out non-hybridized fragments after 24 h and estimated the magnitude of enrichment through capture LM-PCR products by using an Agilent 2100 Bioanalyzer (Agilent Technologies, Santa Clara, CA, USA). The Hiseq2000 platform (Illumina, San Diego, CA, USA) started paired-end sequencing with read lengths of 90 bp. We applied Illumina base-calling software V.1.7 at its default parameters to process raw image files for base-calling.

Exome data analysis

We used two independent analysis pipelines to perform alignment, variant calling, and annotation. Pipeline (1). Sequencing reads were aligned to the human reference genome (UCSC hg19) with Burrows–Wheeler Aligner (BWA, version 0.6.2) [5]. Optical and PCR duplicates were marked and removed with Picard (version 1.73). Local realignment of reads containing indel sites and base quality score recalibration (BQSR) were performed with the Genome Analysis Tool Kit (GATK, version 2.3) [6]. Single nucleotide variation (SNV) and small indels were called with GATK UnifiedGenotyper. Variants were marked as potential sequencing artifacts if the filters on

the following annotations were evaluated to be true: (1) for SNVs, "DP < 10," "QD < 2.0," "MQ < 40.0," "FS > 60.0," "HaplotypeScore > 13.0," "MQRankSum < −12.5," "ReadPosRankSum < −8.0," (2) for small indels, "DP < 10," "QD < 2.0," "ReadPosRankSum < −20.0," "InbreedingCoeff < −0.8," "FS > 200.0." The kinship coefficient was calculated for each sample using KING [7] to confirm reported relationships and identify cryptic relationships among samples. ANNOVAR [8] and SnpEff (version 2.0.5) [9] were used for annotating variants. Human gene mutation database (HGMD) [10] was used for annotating known genes and mutations for human inherited diseases. Prediction scores from SIFT [11], Polyphen2 [12], LRT [13], and MutationTaster [14] along with conservation scores PhyloP [15] and GERP++ [16] for every potential nonsynonymous SNV in the human genome were retrieved from dbNSFP (database for nonsynonymous SNPs' functional predictions) [17].

Pipeline (2). Fastq files were aligned to the human reference genome (UCSC hg19) with the short oligonucleotide analysis package (SOAP, version 2.2.1) [18]. SOAPsnp (version 1.05) [19] was used for single nucleotide variant (SNV) detection and GATK for small insertion-deletion (indel) detection, followed by BGI's self-developed programs to perform variant functional annotation.

SNV detection, variant filtering, and prioritization

A series of common filtering criteria were applied to all candidate variants in both pipelines as described below. First we excluded variants that were (1) not within an exon, a predicted splice site, or an UTR; (2) synonymous changes, and (3) with minor allele frequency (MAF) >0.5 % in either the 1000 Genomes Project (http://www.1000genomes.org/), the Exome Sequencing Project (ESP6500; http://evs.gs.washington.edu/EVS/), or our internal exome datasets. Variants near splicing donor/recipient sites and frameshift indels were given particular attention as they could cause pathogenic changes such as exon-skipping, premature truncation, stop loss, or stop gain, as well as frameshifts.

We considered a number of possible genetic models in variant prioritization, which included filtering out of variants based on (1) evolutionary conservation, i.e., variants of PhyloP [15] value <0.95 were considered to be in non-conserved regions thus discarded; (2) prediction of pathogenicity by both PolyPhen [12] and SIFT [15]; and (3) biological and clinical relevance of identified variants with emphasis on pathways and interaction networks of known genes and/or proteins pertinent to the disease.

Given that congenital absence of ACP/PCL is thought to be a dominantly inherited syndrome, we used a de novo dominant model of Mendelian inheritance in our variant prioritization and filtering methodology [20]. Specifically, we focused on variants found to be heterozygous and strongly deleterious in the proband and affected parents. SNVs and indels were selected as potential pathogenic variants if they met all the following criteria: (1) heterozygous in proband and in the affect maternal sample; (2) not previously described or rare (MAF < 0.5 %) in a control cohort of more than 9000 control individuals (1000 genomes project, April 2012 release; 6503 exomes from NHLBI GO Exome Sequencing Project (ESP6500SI) and 1200 in-house whole-exomes; (3) nonsynonymous, or splice acceptor and donor site SNVs, or frame shift coding indels (NS/SS/I); (4) predicted to be deleterious by at least three prediction methods, i.e., SIFT, PolyPhen2, MutationTaster, and LRT; and (5) conserved PhyloP score and GERP++ score >2.0. Variants were also analyzed using the ingenuity variant analysis web-based application. As both affected were female, we also included in our analysis all variants on the *chr X*, considering both heterozygous dominant and hemizygous dominant scenarios. SNVs and indels fulfilling these criteria were retained as potential pathogenic variants if they also then met all the criteria 2–5 (above).

CNV analysis for exon sequencing

CNV analysis was performed by using the standard exome hidden Markov model (XHMM) pipeline that consisting of six steps [21]. The depth of coverage for all targets and all 662 samples used for case/control comparison for this study was performed using GATK. (2) Target regions with extreme GC content (<10 % or >90 %) and low complexity regions are filtered out from further analysis. (3) PCA normalization of read depth is performed for all samples to remove inherent biases in sample preparation and sequencing. (4) Samples with extreme variability in normalized read depth are removed. (5) Per-sample CNV Detection with a HMM is performed. (6) Quality metrics are assigned to all samples for discovered CNVs.

The raw outputs of XHMM were further processed by selecting the region with "Y" annotation, which indicates the CNV region existed in the proband. The selected regions were then compared with the clinical phenotypes; the regions were further selected if the clinical phenotype matches the existence of the CNV. In other words, if the individual in this family has the disease, CNV presence in the individual is required and CNV absence if the individual is healthy.

Computational validation for CNVs

Since CNVs are common across the human genome, it is important to determine CNV specificity during the CNV validation processes. In this study, multiple validations were applied in order to increase the confidence of the CNV hits. First, we checked whether the CNV hit is presented in 274 healthy individuals; second, we checked whether the CNV hit existed in any of the 147 families with different monogenic diseases, including 388 individuals also called at the same time for this study   (Additional file 1: Table S1); finally, we checked whether the CNV hit is present in the database of genomic variations (DGV), which contains annotated CNVs from previous scientific reports [22]. If a CNV hit presented only in the patients but not in other populations and had not been reported in other previous studies, it is more confident to conclude that the CNV may be relevant or potentially associated with the target disease.

qPCR validation for CNVs

CNV Validation by quantitative PCR (qPCR) with the Universal Probe Library (UPL): CNV validation by qPCR was performed roughly as described previously [23, 24]. UPL probes (Roche, Indianapolis, IN) and corresponding primers were selected using the ProbeFinder v2.49 software (Roche, Indianapolis, IN). Probes used and primer sequences are provided in Table 2. All primers were tested to ensure that they generated a PCR product of the correct approximate size. All primer probe combinations were also tested on a dilution series of control DNA to determine PCR efficiencies. All assays achieved PCR efficiency between 95 and 105 %. Quantitative PCR was performed on an ABI Prism™ 7900HT Sequence Detection System (Applied Biosystems, Foster City, CA). Each sample reaction was performed in triplicate, in 10 ul of reaction mixture containing 10 ng genomic DNA, 100 nM of the UPL probe, 400 nM of each PCR primer, and 1× TaqMan Gene Expression Master Mix containing UDG and ROX (Life Technologies, Carlsbad, CA). Male and female genomic DNAs (Promega, Madison, WI) were included in the analysis as positive controls for subjects with expected normal copy number. Data were evaluated using the Sequence Detection Software v2.4 (Applied Biosystems, Foster City, CA). Data was further analyzed by the $\Delta\Delta C_T$ method. The geometric mean of the C_T values for the two control sequences (GAPDH and SNCA) was calculated and used as the reference value for $\Delta\Delta C_T$ calculations. The Promega female DNA sample was considered the reference 2N sample for $\Delta\Delta C_T$ calculations. Hemizygous deletions were determined when the relative copy number value for a specific sample normalized to the reference sample was less than 0.75.

EBV cells

Epstein–Barr virus (EBV) cells were developed from the proband and affected family member (mother), together with cells from four healthy controls described in next section. The EBV-transformed cell lines from the six individuals were cultured into exponential growth, and 5 million cells were harvested for RNA isolation. Thus, two of the EBV cell lines were from the individuals harboring the hemizygous deletion in *CEP57L1*, whereas the other four cell lines were from randomly selected individuals from the CAG bio-repository representing normal *CEP57L1* genotypes and expression. Expression of *G6PD* was observed to be variable between EBV cell lines, so for the analysis of expression from the EBV cells, only *ACTB* and *B2M* were used as controls. In addition, for analysis of expression in EBV-transformed cells, a sample from an individual not carrying the predicted deletion was used as the control sample for calculations.

qPCR validation for EBV cells and ligament tissues from knee surgery

Samples of anterior cruciate ligament (ACL) tissue were isolated from four individuals undergoing ACL surgeries. Samples were stored at −80C prior to RNA isolation. RNA was isolated from homogenized tissue samples or cell lysates using the RNeasy Mini Kit (Qiagen, Venlo, Netherlands) with optional on-column DNAse digestion, following the manufacturer's instructions. ACL samples (approximately 200 mg tissue) were homogenized in buffer RLT using a PowerGen 125 Homogenizer (Fisher Scientific, Waltham, MA). EBV-transformed cell lines (5 × 10^6 cells) were homogenized in buffer RLT using QiaShredder columns (Qiagen).

Isolated RNA was converted to cDNA using the High Capacity RNA-to-cDNA Kit, following manufacturer's protocol. Reactions involving RNA from ACL tissue samples utilized 300 ng of total RNA in a 20 μl l reaction volume, while reactions with RNA from EBV cells were performed using 2 μl g of RNA in 20 μl l reactions. Additionally, control reactions were performed for each sample that excluded the reverse transcriptase. Gene expression analysis was performed by qPCR with the

reverse transcribed cDNA using the Universal Probe Library system (Roche, Indianapolis, IN). UPL probes and corresponding primers were selected using the Probe-Finder v2.49 software (Roche). Probes used and primer sequences are provided in Table 1. Quantitative PCR was performed on an ABI ViiA 7 Real-Time PCR System (Applied Biosystems, Foster City, CA). Each sample reaction was performed in triplicate, in 10 ul of reaction mixture containing cDNA, 100 nM of the UPL probe, 400 nM of each PCR primer and 1× TaqMan Gene Expression Master Mix containing UDG and ROX (Life Technologies, Carlsbad, CA). Reactions involving ACL tissue RNA contained cDNA corresponding to 5 ng of the starting RNA. Reactions with the EBV cell line RNA contained cDNA corresponding to 80 ng of starting RNA. Quantitative PCR was performed both on the converted cDNA and the corresponding control, no reverse transcriptase reactions in order to control for contamination of RNA with genomic DNA. Data was analyzed by the $\Delta\Delta C_T$ method. The geometric mean of the C_T values for the control genes (beta-actin/*ACTB*, glucose-6-phosphate dehydrogenase/*G6PD*, and beta-2-microglobulin/*B2M*) was calculated and used as the reference value for $\Delta\Delta C_T$ calculations. For analysis of ACL expression, a sample was chosen at random to use as the normalization control.

CEP57L1 modeling

Centrosomal protein 57 kDa-like protein 1 (*CEP57L1*) is a protein-coding gene that aids in (1) microtubule binding through filaments composed of tubulin monomers, (2) identical protein binding, and (3) γ-tublulin binding with the microtubule constituent protein of the same name. Each function interacts selectively and non-covalently. A three-dimensional (3D) model of *CEP57L1* was rendered using Phyre2 [25]. The structure created by Phyre2 was chosen due its high confidence, and

Table 1 Assay design for qRT-PCR analysis of gene expression in ACL tissue or EBV LCLs

Gene symbol—probe name	Gene	Amplicon position	UPL probe #	Left primer[a]	Right primer[b]
CEP57L1—exons 4-5	Centrosomal protein 57 kDa-like 1	Spans exons 4 and 5 of human CEP57L1 transcript variant 1	78	agcccaaatagccaagctc	tctctggaaagaatgttcaggtt
CEP57L1—exons 5-6	Centrosomal protein 57 kDa-like 1	Spans exons 5 and 6 of human CEP57L1 transcript variant 1	17	caaatgagagaaatctggcaca	acgagactgggctgagctt
CEP57L1—exon 1	Centrosomal protein 57 kDa-like 1	Contained within exon 1 of human CEP57L1 transcript variant 1	80	ctaagcttgcgccctgag	cgaagtctcagtgtacttctacgtct
ACTB	Beta-actin	Spans exons 4 and 5 of human ACTB transcript	11	attggcaatgagcggttc	cgtggatgccacaggact
B2M	Beta-2-microglobulin	Spans exons 1 and 2 of human B2M transcript	42	ttctggcctggaggctatc	tcaggaaatttgactttccattc
G6PD	Glucose-6-phosphate dehydrogenase	Spans exons 4 and 5 of human G6PD transcript	82	Gcaaacagagtgagcccttc	gagttgcgggcaaagaagt

[a]All primers are listed 5′ to 3′

model quality estimations were done for structural validation [26–28]. The sequence of the wild-type (wt) *CEP57L1* was obtained from the UniProt database (accession number Q7IYX8). Models and quality estimations were done through the protein model portal (PMP), which is part of the Structural Biology Knowledgebase (SBKB, sbkb.org), Protein Structure Initiative (PSI), and Nature Publishing Group (NPG). PEP-FOLD was used to determine the structure of residues 1 to 18, which resulted in an α-helical motif (Fig. 3c) [26]. A Monte Carlo simulation of CEP57L1 was performed using PELE from the Barcelona Supercomputing Center (BSC) [29]. The simulation generated 67 low-energy conformations out of 100 steps in less than 24 h using one central processing unit (CPU) as requested by the BSC for searching protein local motion.

Results

Discoveries and computational validation of rare variants underlying ACL/PCL

We used WES in search for rare causative variants explaining ACL/PCL. We ran our WES data through our discovery pipeline as previously described [27], and we did not identify any SNVs that segregated in the family that were likely to be causative of ACL/PCL. We next evaluated if there were any CNVs that could explain the disease. The major problem of CNVs discovery is the specificity of the findings due to the prevalence of CNVs across the human genome. Therefore, multiple filters have been applied to remove the potential false positive

CNVs. First, the CNV should not be present in the 274 healthy individuals called at the same time. In other words, the CNV should not be present among healthy controls to explain the association with absence of ACL/PCL. The target CNV deletions occurred within the genomic region on chr6:109466479-109485174 (Fig. 1). From the XHMM plots, the region shows clearly CNV deletions in the proband and her mother and its absence in the healthy controls.

Besides the exclusions of the CNVs in healthy controls, it is important that the CNV regions are exclusive to other rare diseases, implying that the CNV should be unique to the ACL/PCL phenotypes in this instance. In order to further validate the results, we collected additional 388 individuals who have wide ranges of rare disease (147 different diagnoses). Again, the CNV region is not identified in any of those individuals. We expanded the culprit CNV region to include 50k base pairs up/downstream at the CNV locus. The region is still exclusive and different from any CNVs identified in the healthy control or the different disease groups. We additionally evaluated the CNVs through DGV, which is a genome variation database based on published CNV regions. Many of the studies reported in DGV contain very large populations with large number of CNVs [22]. No genome variations have been reported before in DGV for the culprit region, indicating that there is a novel CNV.

The computational validation enhances the specificity of the CNV regions we identified. Based on multiple

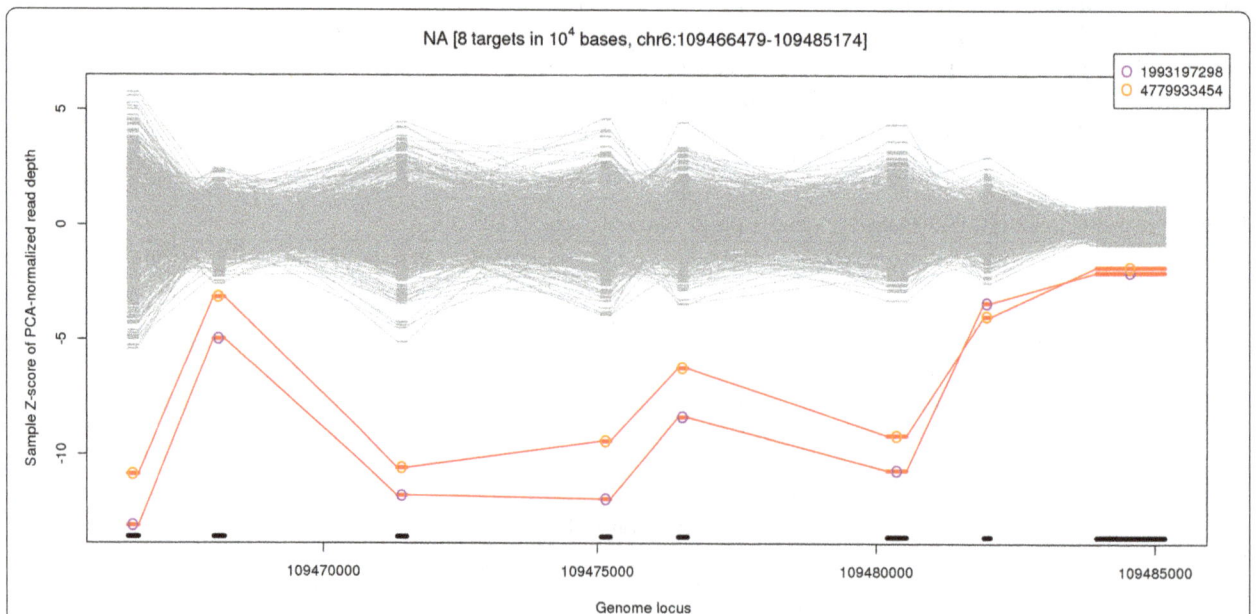

Fig. 1 XHMM plots for copy number variations (CNV): CNV deletion region for chr6:109466479-109485174. *x-axis* represents the genome locus, and y-axis is the computed Z-score of PCA normalized read depth; *positive values* indicate duplication and *negative values* indicate deletion; target individuals are highlighted with *color* while *gray lines* are the control individuals from 663 population pool

filters we applied, the results show that the CNV region could be a potential variation that is responsible for the absence of ACL/PCL phenotypes.

qPCR validation for CNVs

Quantitative PCR using UPL probes was performed to determine the copy number of *CEP57L1* genomic sequences in the proband and her mother (Fig. 2a). Primer and probe combinations were designed against *CEP57L1* genomic DNA sequence, and an assay located in an intron, approximately 400 base pairs upstream of exon 7, was selected for validation. The genomic coordinates of each targeted amplicon are listed in Table 2. The tested location was shown to exist as a hemizygous deletion (1N copy number) in the two affected individuals.

qPCR validation for CEP57L1 expression profiling in EBV cells

To determine if *CEP57L1* is expressed in anterior cruciate ligament (ACL) tissue, we extracted RNA from ACL tissue samples from four individuals who had undergone surgeries on their ACLs. These subjects were unrelated to the individuals who contained *CEP57L1* deletions. RNA from ACL tissues was converted to cDNA and subjected to qPCR using a series of assays targeting three different regions of the *CEP57L1* messenger RNA (mRNA) or one of three control mRNAs (*B2M, ACTB, G6PD*). Amplification of *CEP57L1* mRNA was observed in all four RNA samples, to variable degrees depending on the assay used. Each of the assays completely failed to amplify anything from RNA samples that had been subjected to control reverse transcriptase reactions from which the reverse transcriptase enzyme had been excluded. We conclude that *CEP57L1* is expressed in ACL tissue.

To determine if the hemizygous deletion affected expression of *CEP57L1*, we extracted RNA from cell lines generated by EBV transformation of peripheral blood mononuclear cells from the proband, her mother, and four healthy controls. RNA from the two deletion containing subjects was compared to RNA from four subjects randomly selected from the CAG bio-repository (Fig. 2b). Exon 1 of the *CEP57L1* mRNA lies outside the detected deletion so the expression of exon 1 is expected to be unaltered. Detection of *CEP57L1* mRNA with an assay targeted exon 1 showed that the t levels of expression of *CEP57L1* mRNA in the cases and controls were similar. Exons 4 through 6 are contained within the detected deletion. Using assays that detect the junction of exons 4 and 5 or the junction of exons 5 and 6 revealed relatively similar levels of expression of these more 3′ regions of the *CEP57L1* mRNA in all four of the control subjects. In comparison, both deletion subjects showed lower expression of these regions of the mRNA, at approximately 40–50 % of the control subjects. We interpret these data to indicate that the deletion subjects possess approximately half the amount of functional, full length *CEP57L1* mRNA as control subjects.

CEP57L1 structural analyses

The 3D structure of CEP57L1 was chosen from Phyre2 due to its high confidence (Fig. 3a). Residues 1 to 18 may lack a secondary structure due to the intermolecular interactions surrounding them—specifically, the 178 residues that make up the coiled-coil domain from position 51 to 228 (Fig. 3b). Residues 1 to 18 only remain after exon deletion and was modeled using PEP-FOLD, an ab initio protein structure predictor [30]. Absence of the coiled-coil domains suggest that residues 1 to 18 may result in an α-helical structure (Fig. 3c). A Monte Carlo

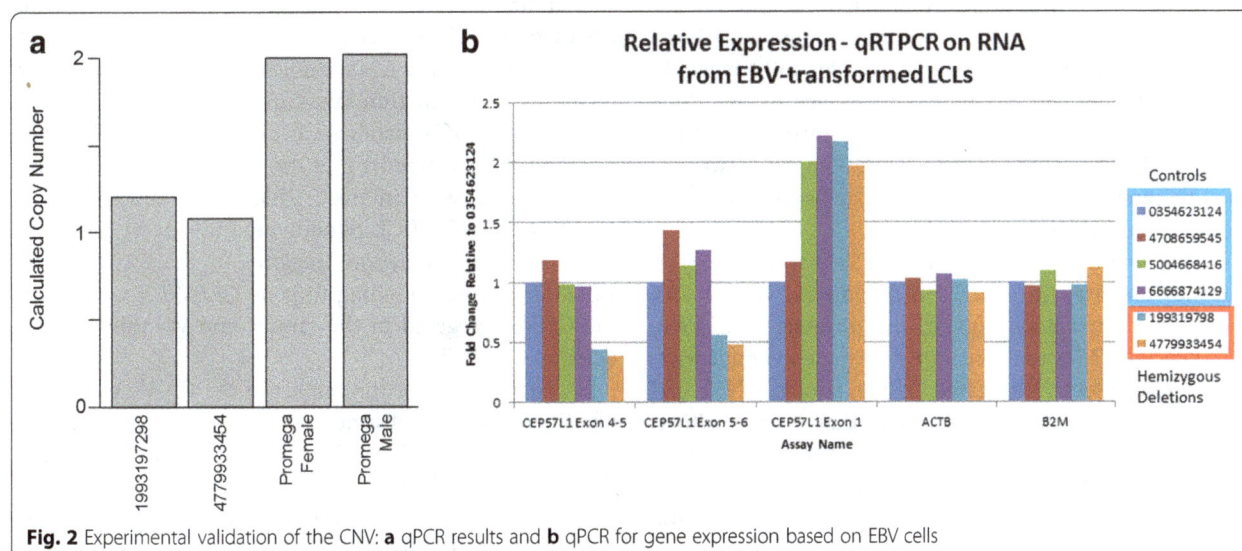

Fig. 2 Experimental validation of the CNV: **a** qPCR results and **b** qPCR for gene expression based on EBV cells

Table 2 qPCR validation primer information for EBV cells genomic DNA

Gene name	Gene	Amplicon position[a]	UPL probe #	Left primer[b]	Right primer[b]
CEP57L1	Centrosomal protein 57 kDa-like 1	Amplicon located in intronic genomic DNA between exons 6 and 7 (chr6:109474636-109474742) chr6:109474636-109474742	75	gaggggtcccgttatgttg	gcgtggtggctcatacttg
GAPDH	Glyceraldehyde 3-phosphate dehydrogenase	chr12: 6645563-6645625	10	gctgcattcgccctctta	gaggctcctccagaatatgtga
SNCA	Synuclein, alpha	chr4: 90743466-90743537	68	gctgagaagaccaaagagcaa	ctgggctactgctgtcacac

[a]Amplicon position as reported by UCSC genome browser (hg19) in silico PCR tool
[b]All primers are listed 5′ to 3′

simulation was performed (without the presence of water) to determine if a secondary structure was generated in residues 1 to 18. As a result, they did not generate a secondary structure in the presence of a coiled-coil domain and the rest of CEP57L1 [29]. Experimental evidence confirms that the C terminus is responsible for bundling and nucleating microtubules in vivo via binding [28]. As for the N terminus, the coiled-coil domain interacts with the centrosome internal to γ-tublulin and can also mutlimerize with the N terminus of other CEP57 structures [28]. Therefore, the absence of the N- and C-termini, as well as the remaining fragment of residues 1 to 18 after exon deletion, cannot support all three functions since 96 % of CEP57L1 is absent in any interaction. All images were created using pyMOL.

Discussion

Absence of the ACL/PCL constitutes a rare congenital malformation that results in knee joint instability with a prevalence of 1.7 per 100,000 live births and can be associated with other lower-limb abnormalities such as ACL agnesia and absence of the knee menisci. Previous studies of several families suggested that the potential inheritance pattern is under Mendelian etiology [4, 29, 30]. In this study, we focus on the family with both mother and daughter who are notable for joint instability and difficulty with ambulation. Absence of ACL and PCL were confirmed on MRI and by arthroscopy. Since other family member's DNA was not obtainable, we applied the analysis in different molecular levels to explore the potential causation of the disease.

We performed a thorough analysis in search for SNVs that may underlie ACL/PCL. No detrimental variant was identified that could explain the disease. We next searched for CNVs to determine if any such variant with detrimental effects segregated in the family. CNVs are structural variations in the genome that lead to either deletions or duplications of chromosome regions of a variable size. CNVs have been proving to associate with susceptibility to disease, such as cancer [31], autism [24], and schizophrenia [32]. Previous research also suggests

that inherited CNVs are associated with Mendelian diseases [33]. Due to potential underlying Mendelian etiology for the ACL/PCL phenotype, the genetic mechanism could be studied by CNV exploration. Exome sequencing has been shown to enhance Mendelian disease gene identification resulting in improved clinical diagnosis, more accurate genotype-phenotype correlations and new insights into the role of rare genomic variation in disease, both SNVs and CNVs. Therefore, we used WES to explore the potential of identifying SNVs or CNVs that may be associated with ACP/PCL with exome sequencing technology in this study.

Due to the high prevalence of CNVs in human genome [34], any potential pathogenic CNV needs to be shown to be absent in healthy individuals or patients with other unrelated diseases. We applied several filters to improve the confidences of our results. First, we collected a relatively large number of individuals who were healthy controls ($n = 274$) as well as 388 individuals that covered a wide range of 144 different rare diseases. By checking the CNVs across those individuals, we make sure the CNVs we identified are not present in healthy controls or in other rare disease group. To overcome the size limitation of the control set, we next examined the CNVs previously reported in the DGV database, which contains many large CNV studies in healthy populations, including over ten thousands individual. If the CNV is absent in the DGV database, it indicates that the CNV is novel and lends support that the CNV may be associated with the disease of interest, which was the case in our study of the ACL/PCL phenotype. Indeed, we identified a novel CNV deletion region on chr6:109466479-109485174, corresponding to gene CEP57L1, we showed the CNV segregated in the family, and we validated the deletion by qPCR.

In order to explore the impact of the CNV deletion we identified, we applied functional studies examining RNA gene expression and the 3D protein structure modeling. We performed qPCR on the corresponding EBV cells of those two individuals. The deletion occurs at region chr6:109466479-109485174, which is corresponding to

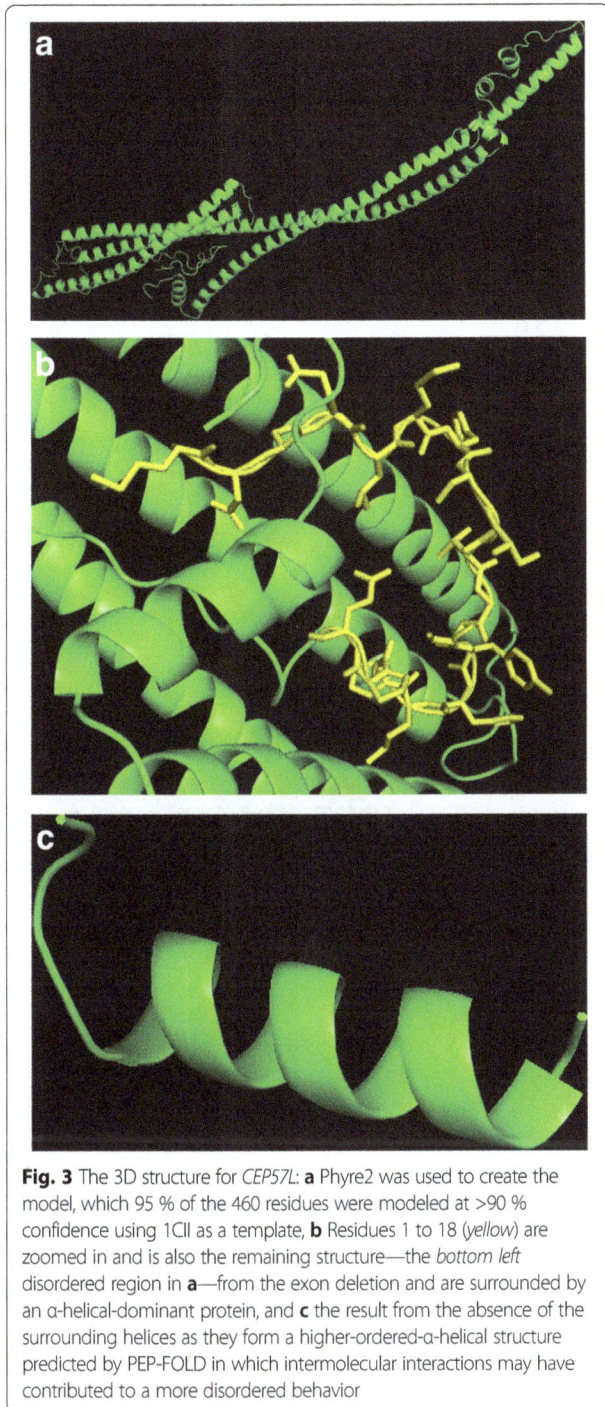

Fig. 3 The 3D structure for *CEP57L*: **a** Phyre2 was used to create the model, which 95 % of the 460 residues were modeled at >90 % confidence using 1Cll as a template, **b** Residues 1 to 18 (*yellow*) are zoomed in and is also the remaining structure—the *bottom left* disordered region in **a**—from the exon deletion and are surrounded by an α-helical-dominant protein, and **c** the result from the absence of the surrounding helices as they form a higher-ordered-α-helical structure predicted by PEP-FOLD in which intermolecular interactions may have contributed to a more disordered behavior

10 exons for gene *CEP57L1*. The rationale is if the CNVs have effects for clinical phenotype, they would cause alternation of expression at the gene level. The results from qPCR show that the CNV deletions possess approximately half the amount of functional, full length *CEP57L1* mRNA as control subjects indicates that the de novo CNV deletion would lead to decrease of gene expression. In the protein structure modeling, we prove

that the functional sites would lose while the binding site of the protein intact, indicating that with the CNV deletion, the protein could still binding to its targets without available functions.

In conclusion, we discovered a novel CNV associated with ACL/PCL, which has high potential to be the causation of the disease. The confidence is built on the robust discovery set, the novelty of the finding, and strong supports from functional studies in gene and protein levels. In other words, in this project we applied ~700 exome sequencing data for CNV discoveries, the CNVs are validated computationally through multiple filters and database, and the existences of CNV were validated through qPCR experimentally. The corresponding gene expression in tissues were measured and compared to the healthy individuals, and the 3D protein structures were simulated to enhance the confidence of the CNV impact. While interventions at the level of *CEP57L1* may not be feasible, the role of this gene in the pathogenesis of ACL/PCLagenesis helps explain disease causality.

Ethics statements

This study had been approved by the Children's Hospital of Philadelphia with IRB# 4886. All the patients who participated in this project have been consented and agree to publish the results.

Authors' contributions
Conception and study design: YL, YL, HH. Performed the experiments: MEM, KX, FW. Analyzed the data: YL, YL, NK, YG, JG, Contributed reagents/materials/data: BK, JL, TJG, JZ, MAD, XX, Drafting the manuscript: YL, YL, HH, MEM, FW. All authors read and approved the final manuscript.

Author details
[1]Center for Applied Genomics, The Children's Hospital of Philadelphia, 1014H, 3615 Civic Center Blvd, Abramson Building, Philadelphia, PA 19104, USA. [2]Beijing Genomics Institute, Shenzhen, China. [3]Center for Sports Medicine and Performance, The Children's Hospital of Philadelphia, Philadelphia, PA, USA. [4]Individualized Medical Genetics Center, The Children's Hospital of Philadelphia, Philadelphia, PA, USA.

References
1. Berruto M, Gala L, Usellini E, Duci D, Marelli B. Congenital absence of the cruciate ligaments. Knee Surg Sports Traumatol Arthrosc. 2012;20(8):1622–5. doi:10.1007/s00167-011-1816-2.
2. Thomas NP, Jackson AM, Aichroth PM. Congenital absence of the anterior cruciate ligament. A common component of knee dysplasia. J Bone Joint Surg. 1985;67(4):572–5.

3. Giorgi B. Morphologic variations of the intercondylar eminence of the knee. Clin Orthop. 1956;8:209–17.

4. Frikha R, Dahmene J, Ben Hamida R, Chaieb Z, Janhaoui N, Laziz Ben Ayeche M. [Congenital absence of the anterior cruciate ligament: eight cases in the same family. Rev Chir Orthop Reparatrice Appar Mot. 2005;91(7):642–8.

5. Li H, Durbin R. Fast and accurate short read alignment with Burrows-Wheeler transform. Bioinformatics. 2009;25(14):1754–60. doi:10.1093/bioinformatics/btp324.

6. McKenna A, Hanna M, Banks E, Sivachenko A, Cibulskis K, Kernytsky A, et al. The Genome Analysis Toolkit: a MapReduce framework for analyzing next-generation DNA sequencing data. Genome Res. 2010;20(9):1297–303. doi:10.1101/gr.107524.110.

7. Manichaikul A, Mychaleckyj JC, Rich SS, Daly K, Sale M, Chen WM. Robust relationship inference in genome-wide association studies. Bioinformatics. 2010;26(22):2867–73. doi:10.1093/bioinformatics/btq559.

8. Wang K, Li M, Hakonarson H. ANNOVAR: functional annotation of genetic variants from high-throughput sequencing data. Nucleic Acids Res. 2010;38(16):e164. doi:10.1093/nar/gkq603.

9. De Baets G, Van Durme J, Reumers J, Maurer-Stroh S, Vanhee P, Dopazo J, et al. SNPeffect 4.0: on-line prediction of molecular and structural effects of protein-coding variants. Nucleic Acids Res. 2012;40(Database issue):D935–9. doi:10.1093/nar/gkr996.

10. Stenson PD, Ball EV, Mort M, Phillips AD, Shaw K, Cooper DN. The human gene mutation database (HGMD) and its exploitation in the fields of personalized genomics and molecular evolution. Current protocols in bioinformatics/editoral board, Andreas D Baxevanis [et al]. 2012;Chapter 1:Unit1 13. doi:10.1002/0471250953.bi0113s39.

11. Kumar P, Henikoff S, Ng PC. Predicting the effects of coding non-synonymous variants on protein function using the SIFT algorithm. Nat Protoc. 2009;4(7):1073–81. doi:10.1038/nprot.2009.86.

12. Adzhubei I, Jordan DM, Sunyaev SR. Predicting functional effect of human missense mutations using PolyPhen-2. Current protocols in human genetics/editorial board, Jonathan L Haines [et al]. 2013;Chapter 7:Unit7 20. doi:10.1002/0471142905.hg0720s76.

13. Chun S, Fay JC. Identification of deleterious mutations within three human genomes. Genome Res. 2009;19(9):1553–61. doi:10.1101/gr.092619.109.

14. Schwarz JM, Rodelsperger C, Schuelke M, Seelow D. MutationTaster evaluates disease-causing potential of sequence alterations. Nat Methods. 2010;7(8):575–6. doi:10.1038/nmeth0810-575.

15. Pollard KS, Hubisz MJ, Rosenbloom KR, Siepel A. Detection of nonneutral substitution rates on mammalian phylogenies. Genome Res. 2010;20(1):110–21. doi:10.1101/gr.097857.109.

16. Davydov EV, Goode DL, Sirota M, Cooper GM, Sidow A, Batzoglou S. Identifying a high fraction of the human genome to be under selective constraint using GERP++. PLoS Comput Biol. 2010;6(12):e1001025. doi:10.1371/journal.pcbi.1001025.

17. Liu X, Jian X, Boerwinkle E. dbNSFP: a lightweight database of human nonsynonymous SNPs and their functional predictions. Hum Mutat. 2011;32(8):894–9. doi:10.1002/humu.21517.

18. Li R, Yu C, Li Y, Lam TW, Yiu SM, Kristiansen K, et al. SOAP2: an improved ultrafast tool for short read alignment. Bioinformatics. 2009;25(15):1966–7. doi:10.1093/bioinformatics/btp336.

19. Li R, Li Y, Fang X, Yang H, Wang J, Kristiansen K, et al. SNP detection for massively parallel whole-genome resequencing. Genome Res. 2009;19(6):1124–32. doi:10.1101/gr.088013.108.

20. Sirmaci A, Spiliopoulos M, Brancati F, Powell E, Duman D, Abrams A, et al. Mutations in ANKRD11 cause KBG syndrome, characterized by intellectual disability, skeletal malformations, and macrodontia. Am J Hum Genet. 2011;89(2):289–94. doi:10.1016/j.ajhg.2011.06.007.

21. Fromer M, Moran JL, Chambert K, Banks E, Bergen SE, Ruderfer DM, et al. Discovery and statistical genotyping of copy-number variation from whole-exome sequencing depth. Am J Hum Genet. 2012;91(4):597–607. doi:10.1016/j.ajhg.2012.08.005.

22. MacDonald JR, Ziman R, Yuen RK, Feuk L, Scherer SW. The database of genomic variants: a curated collection of structural variation in the human genome. Nucleic Acids Res. 2014;42(Database issue):D986–92. doi:10.1093/nar/gkt958.

23. Edelmann L, Prosnitz A, Pardo S, Bhatt J, Cohen N, Lauriat T, et al. An atypical deletion of the Williams-Beuren syndrome interval implicates genes associated with defective visuospatial processing and autism. J Med Genet. 2007;44(2):136–43. doi:10.1136/jmg.2006.044537.

24. Glessner JT, Wang K, Cai G, Korvatska O, Kim CE, Wood S, et al. Autism genome-wide copy number variation reveals ubiquitin and neuronal genes. Nature. 2009;459(7246):569–73. doi:10.1038/nature07953.

25. Kelley LA, Sternberg MJ. Protein structure prediction on the Web: a case study using the Phyre server. Nat Protoc. 2009;4(3):363–71. doi:10.1038/nprot.2009.2.

26. Thevenet P, Shen Y, Maupetit J, Guyon F, Derreumaux P, Tuffery P. PEP-FOLD: an updated de novo structure prediction server for both linear and disulfide bonded cyclic peptides. Nucleic Acids Res. 2012;40(Web Server issue):W288–93. doi:10.1093/nar/gks419.

27. Almoguera B, He S, Corton M, Fernandez-San Jose P, Blanco-Kelly F, Lopez-Molina MI, et al. Expanding the phenotype of PRPS1 syndromes in females: neuropathy, hearing loss and retinopathy. Orphanet J Rare Dis. 2014;9:190. doi:10.1186/s13023-014-0190-9.

28. Momotani K, Khromov AS, Miyake T, Stukenberg PT, Somlyo AV. Cep57, a multidomain protein with unique microtubule and centrosomal localization domains. Biochem J. 2008;412(2):265–73. doi:10.1042/BJ20071501.

29. Kwan K, Ross K. Arthrogryposis and congenital absence of the anterior cruciate ligament: a case report. Knee. 2009;16(1):81–2. doi:10.1016/j.knee.2008.08.004.

30. Coste B, Houge G, Murray MF, Stitziel N, Bandell M, Giovanni MA, et al. Gain-of-function mutations in the mechanically activated ion channel PIEZO2 cause a subtype of Distal Arthrogryposis. Proc Natl Acad Sci U S A. 2013;110(12):4667–72. doi:10.1073/pnas.1221400110.

31. Cappuzzo F, Hirsch FR, Rossi E, Bartolini S, Ceresoli GL, Bemis L, et al. Epidermal growth factor receptor gene and protein and gefitinib sensitivity in non-small-cell lung cancer. J Natl Cancer Inst. 2005;97(9):643–55. doi:10.1093/jnci/dji112.

32. Stefansson H, Meyer-Lindenberg A, Steinberg S, Magnusdottir B, Morgen K, Arnarsdottir S, et al. CNVs conferring risk of autism or schizophrenia affect cognition in controls. Nature. 2014;505(7483):361–6. doi:10.1038/nature12818.

33. McCarroll SA, Altshuler DM. Copy-number variation and association studies of human disease. Nat Genet. 2007;39(7 Suppl):S37–42. doi:10.1038/ng2080.

34. Stankiewicz P, Lupski JR. Structural variation in the human genome and its role in disease. Annu Rev Med. 2010;61:437–55. doi:10.1146/annurev-med-100708-204735.

The impact of common polymorphisms in CETP and ABCA1 genes with the risk of coronary artery disease in Saudi Arabians

Cyril Cyrus[1*], Chittibabu Vatte[1], Awatif Al-Nafie[2], Shahanas Chathoth[1], Rudaynah Al-Ali[2], Abdullah Al-Shehri[2], Mohammed Shakil Akhtar[2], Mohammed Almansori[2], Fahad Al-Muhanna[2], Brendan Keating[3] and Amein Al-Ali[1]

Abstract

Background: Coronary artery disease (CAD) is a leading cause of morbidity and mortality worldwide. Many genetic and environmental risk factors including atherogenic dyslipidemia contribute towards the development of CAD. Functionally relevant mutations in the dyslipidemia-related genes and enzymes involved in the reverse cholesterol transport system are associated with CAD and contribute to increased susceptibility of myocardial infarction (MI).

Method: Blood samples from 990 angiographically confirmed Saudi CAD patients with at least one event of myocardial infarction were collected between 2012 and 2014. A total of 618 Saudi controls with no history or family history of CAD participated in the study. Four polymorphisms, rs2230806, rs2066715 (*ABCA1*), rs5882, and rs708272 (*CETP*), were genotyped using TaqMan Assay.

Results: *CETP* rs5882 (OR = 1.45, $P < 0.005$) and *ABCA1* rs2230806 (OR = 1.42, $P = 0.017$) polymorphisms were associated with increased risk of CAD. However, rs708272 polymorphism showed protective effect (B1 vs. B2: OR = 0.80, $P = 0.003$ and B2B2 vs. B1B1: OR = 0.68, $P = 0.012$) while the *ABCA1* variant rs2066715 was not associated.

Conclusion: This study is the first to report the association of these polymorphisms with CAD in the population of the Eastern Province of Saudi Arabia. The rs5882 polymorphism (*CETP*) showed a significant association and therefore could be a promising marker for CAD risk estimation while the rs708272 polymorphism had a protective effect from CAD.

Keywords: Gene polymorphism, CAD, *CETP*, *ABCA1*, TaqMan Assay

Background

Coronary artery disease (CAD) is one of the leading causes of morbidity and disability and the most common cause of mortality worldwide equally among men and women. CAD is a disease burden in both high- and low-income countries [1, 2]. A study conducted on 17,232 people from Saudi Arabia revealed that 5.5 % had been diagnosed with CAD, with a higher prevalence in urban populations (6.2 %) compared to rural populations (4 %) [3]. Platelet aggregation and thrombus formation following the rupture of coronary atherosclerotic plaque is the major cause of myocardial infarction (MI) [4–6].

Many extrinsic and intrinsic risk factors, including hypertension, dyslipidemia, obesity, smoking, age, lack of exercise, and diabetes, are established risk factors for MI [7]. Atherogenic dyslipidemia is characterized by abnormal levels of triglycerides, low- and high-density lipoprotein (LDL-C and HDL-C) [8–10]. Functionally relevant mutations in the dyslipidemia-related genes and gene encoding enzymes involved in the reverse cholesterol transport system have been reported to be associated with high-density lipoprotein-cholesterol (HDL-C) levels [11–13]. Epidemiological and clinical studies have demonstrated a contradictory association between HDL-C concentrations and cardiovascular risk [10, 14, 15]. The anti-atherogenic effect of HDL-C may act through several mechanisms, such as anti-oxidation of low-density

* Correspondence: ccyrus@uod.edu.sa
[1]Institute for Research and Medical Consultation, University of Dammam, P.O.Box 1982, Dammam 31441, Kingdom of Saudi Arabia

lipoprotein-cholesterol (LDL-C) and anti-inflammation and inhibition of vascular endothelial cell apoptosis.

The reverse cholesterol transport system plays a vital role in these processes [16], as it is involved in the transportation of cholesterol from the peripheral tissues to the liver, where cholesterol is secreted into bile. ATP-binding-cassette A1 (ABCA1), apolipoprotein A-1 (ApoA-1), and cholesteryl ester transfer protein (CETP) play important roles in the reverse cholesterol transport system [17]. Certain ABCA1 polymorphisms have been reported to be associated with HDL-C concentrations, which in turn indicate increased cardiovascular risk [18]. The potential atherogenicity of CETP relates to its ability to transfer cholesteryl esters from the anti-atherogenic HDLs to the pro-atherogenic VLDL and LDL proteins. Mutations in the CETP gene give rise to less functional protein, which reduces the transfer of cholesteryl esters, and consequently HDL levels are elevated [19]. ABCA1 and CETP, variants including rs2230806, rs2066715, and rs5882, have been associated with increased HDL-C concentrations and rs708272 with a decreased risk for CAD [20, 21].

The objective of the present study is to evaluate the association of the two ABCA1 polymorphisms, rs2230806 [R219K: c.656G>A (p.Arg219Lys)] and rs2066715 [V825I: c.2473G>A (p.Val825Ile)], and two CETP polymorphisms, rs5882 [V422I: c.1264G>A (p.Val422Ile)] and rs708272 [TaqIB: c.118+279G>A], with the risk of CAD in the population of the Eastern Province of Saudi Arabia.

Results

Demographical and clinical data of cases and the control group, including age, sex, clinical manifestations, and biochemical parameters, are shown in Table 1. Patients were classified into subgroups based on their hypertension and diabetes status. Hypertension and diabetes were more prevalent in the patient group compared to the control group.

All genotype frequencies of the control group were consistent with Hardy-Weinberg equilibrium. The distribution

Table 1 Demographic and clinical characteristics of study subjects

Baseline characteristics	Cases ($N = 990$)	Controls ($N = 618$)	P value
Age (years)	58.37 ± 12.91	54.8 ± 8.5	<0.0001[a]
Sex M:F	708:282	423:195	0.19[b]
Total cholesterol (mg/dl)	170.84 ± 48.39	171.58 ± 43.25	0.50[a]
LDL cholesterol (mg/dl)	103.82 ± 40.58	107.25 ± 35.35	0.005[a]
HDL cholesterol (mg/dl)	36.19 ± 9.88	50.22 ± 23.89	<0.0001[a]
Clinical Characteristics			
Hypertension, n (%)	688 (69.5)	74 (12.0)	<0.05[b]
Diabetes, n (%)	577 (58.3)	87 (14.1)	<0.05[b]

[a]Student's t test
[b]Chi-square test

of analyzed genotype polymorphisms are shown in Table 2. Since all the four SNPs had a G>A transition substitution, the genotypes are denoted with the amino acid change, except for Taq1B alleles, which are designated by B1 and B2. The genotype analysis showed overall heterozygous polymorphism predominance in rs2230806 of ABCA1, rs5882, and rs708272 of CETP (Table 2).

The CETP rs708272 polymorphism showed a significantly lower risk for CAD (B1B2+B2B2 vs. B1B1: OR = 0.68, 95 % CI 0.55–0.85, P = 0.0006 and B2B2 vs. B1B1: OR = 0.68, 95 % CI 0.50–0.92, P = 0.012). There was also a significant variation of B1B2 genotypes among patients and controls (OR = 0.68, 95 % CI 0.54–0.86, P = 0.001). Genotyping for the rs5882 polymorphism in CETP exon-14 showed that the frequency of VI genotype was higher in cases than in controls (52.8 vs. 48.0 %). Our analysis revealed that CETP rs5882 polymorphism is associated with an increased risk of CAD in our Saudi population study dataset (VI+II vs. VV: OR = 1.42, 95 % CI 1.11–1.82, P = 0.005; II vs. VV: OR = 1.37, 95 % CI 1.02–1.82, P = 0.031). Allele frequency analysis of the B2 allele of rs708272 of CETP (OR = 0.80, 95 % CI 0.69–0.92, P = 0.003) and the K allele of rs2230806 of ABCA1 (OR = 1.17, 95 % CI 1.01–1.35, P = 0.029) showed a significant difference between the two tested groups (Table 2). The mutant KK genotype of rs2230806 of ABCA1 is found to be associated with an increased risk of CAD (RR vs. KK: OR = 1.42, 95 % CI 1.06–1.91, P = 0.017). There were no significant differences in allele and genotype frequencies of rs2066715 polymorphisms in ABCA1 between the patient and control groups. The power of the study observed was 100 % for protective effect at odds ratio of 0.5 and 94.9 % at 0.7 for TaqIB, and for the other three SNPs (R219K, V825I, and I405V), the results ranged from 45.48 to 96.9 % for an odds ratio of 1.2–1.5.

A joint analysis of two SNPs of both ABCA1 and CETP is shown in Table 3. All the combinations of the CETP variants exhibited no association, except B1B1+VI (OR = 1.7, 95 % CI 1.0–2.9, P = 0.048). On the other hand, for ABCA1, RK+VV and RK+VI lacked an association (Table 3). A sex-based analysis revealed a higher frequency of B1B1 genotype in men and women with CAD compared to their respective controls (Table 4). There was no statistical significance in the distribution of ABCA1 genotypes in the female cohort whereas in the male cohort genotypes KK (OR = 1.8, P = 0.001) of rs2230806 and VI (OR = 2.17, P = 0.041) of rs2066715 showed a significantly higher risk for CAD. In the CETP, rs5882 polymorphisms II (OR = 1.98, 95 % CI 1.38–2.85, P = 0.0002) revealed a significantly higher risk for CAD in the male cohort. In the male and female cohorts, rs708272 B1B2 and B2B2 genotypes showed a protective effect for CAD, respectively. The rs2230806 and rs2066715 of ABCA1 did not show any significant

Table 2 Association between *ABCA1* and *CETP* genotypes and alleles with CAD

SNP	Genotype/alleles	Cases (N = 990)	Control (N = 618)	OR (95 % CI)	P value
rs2230806 (R219K)	RR	291	195	Ref	
	RK	473	317	0.9 (0.79–1.25)	0.999
	KK	226	106	1.42 (1.06–1.91)	0.017
	RK+KK	699	423	1.10 (0.89–1.37)	0.359
	R(G)	1055	707	Ref	
	K(A)	925	529	1.17 (1.01–1.35)	0.029
rs2066715 (V825I)	VV	945	597	Ref	
	VI	45	21	1.35 (0.79–2.29)	0.26
	II	0	0	–	–
	VI+II	45	21	1.35 (0.79–2.29)	0.26
	V(G)	1935	1215	Ref	
	I(A)	45	21	1.34 (0.79–2.26)	0.27
rs5882 (V422I)	VV	178	147	Ref	
	VI	523	297	1.45 (1.12–1.88)	0.005
	II	289	174	1.37 (1.02–1.82)	0.031
	VI+II	812	471	1.42 (1.11–1.82)	0.005
	V(G)	879	591	Ref	
	I(A)	1101	645	1.14 (0.99–1.32)	0.058
rs708272 (TaqIB)	B1B1	376	183	Ref	
	B1B2	454	321	0.68 (0.54–0.86)	0.001
	B2B2	160	114	0.68 (0.50–0.92)	0.012
	B1B2+B2B2	614	435	0.68 (0.55–0.85)	0.0006
	B1(G)	1206	687	Ref	
	B2(A)	774	549	0.80 (0.69–0.92)	0.003

Table 3 Association between joint analysis of two SNPs of *ABCA1* gene and *CETP* gene with CAD

Gene	Genotype	Cases (N = 990)	Controls (N = 618)	OR	95 % CI	P value
ABCA1	RR+VV	256	186	Ref		
	RR+VI	35	9	2.82	1.32–6.02	0.007
	RK+VV	463	305	1.10	0.86–1.39	0.419
	RK+VI	10	12	0.60	0.25–1.43	0.252
	KK+VV	226	106	1.54	1.14–2.08	0.004
CETP	B1B1+VV	46	31	Ref		
	B1B1+VI	180	71	1.70	1.0–2.90	0.048
	B1B1+II	150	81	1.24	0.73–2.11	0.41
	B1B2+VV	77	71	0.73	0.41–1.27	0.27
	B1B2+VI	269	175	1.03	0.63––1.69	0.888
	B1B2+II	108	75	0.97	0.56–1.66	0.913
	B2B2+VV	55	45	0.82	0.45–1.50	0.527
	B2B2+VI	74	51	0.97	0.54–1.74	0.939
	B2B2+II	31	18	1.16	0.55–2.42	0.692

Table 4 Association between *ABCA1* and *CETP* genotypes and CAD in male and female cohorts

Genotype		Male cohort				Female cohort			
		Cases (n = 708)	Control (n = 423)	OR (95 % CI)	P value	Cases (n = 282)	Control (n = 195)	OR (95 % CI)	P value
R219K	RR	212	137	Ref		79	58	Ref	
	RK	337	229	0.95 (0.72–1.24)	0.718	136	88	1.13 (0.73–1.74)	0.566
	KK	159	57	1.80 (1.24–2.61)	*0.001*	67	49	1.0 (0.60–1.65)	0.987
V825I	VV	676	414	Ref		269	183	Ref	
	VI	32	9	2.17 (1.02–4.60)	*0.041*	13	12	0.73 (0.32–1.65)	0.458
V422I	VV	123	103	Ref		55	44	Ref	
	VI	376	232	1.35 (0.99–1.84)	0.052	147	65	1.80 (1.10–2.96)	*0.018*
	II	209	88	1.98 (1.38–2.85)	*0.0002*	80	86	0.74 (0.45–1.22)	0.246
TaqIB	B1B1	261	124	Ref		115	59	Ref	
	B1B2	340	247	0.65 (0.49–0.85)	*0.002*	114	74	0.79 (0.51–1.21)	0.282
	B2B2	107	52	0.97 (0.65–1.45)	0.91	53	62	0.43 (0.27–0.71)	*0.0008*

association with an increased risk for CAD in the female cohort compared to the control subjects.

Discussion

CAD is a multifactorial disease mediated through a complex association of environmental and genetic factors with ethnicity demonstrated to be an important determinant of disease variability. The strength of the current study is the selection of CAD patients from similar ethnic backgrounds. In the present study, two common genetic variations in each *ABCA1* and *CETP* gene were studied with reference to their effect on CAD. We tested four SNPs, namely rs2230806 (R219K), rs2066715 (V825I), rs5882 (V422I), and rs708272 (*TaqIB*) for their association with CAD. *CETP*, located on chromosome 16q21, plays a crucial role in lipid metabolism, and numerous SNPs in this gene have been reported to alter the plasma HDL-C levels and function of CETP [22, 23]. Among the *CETP* SNPs, rs708272 is the one that is most studied. Therefore, we investigated the association of two SNPs of this gene and their risk for CAD in the Saudi population.

Our overall results showed that the heterozygous and mutant of rs708272 polymorphism may confer protection against CAD (B1B2: OR = 0.68, *P* = 0.001; B2B2: OR = 0.68, *P* = 0.012) while those of rs5882 increased the risk of CAD (VI: OR = 1.45, *P* = 0.005; II: OR = 1.37, *P* = 0.031). Earlier studies had suggested that the *CETP* variant rs5882 causes low CETP and is associated with higher HDL and possibly with increased CAD among hypertriglyceridemia men [24, 25]. In contrast, in many recent studies, the rs5882 polymorphism lacked any association with CAD [26–29]. *ABCA1* encodes an important protein that facilitates the formation of HDL-C and regulates the efflux of lipids from peripheral cells into lipid-poor ApoA1 particles, stimulating reverse cholesterol transport. [30] The association between the *ABCA1* gene polymorphisms

and CAD has been the focus for many studies [31–33]. The rs2230806 is the most common polymorphism of *ABCA1*; the possible role of rs2230806 in cardiovascular diseases is still debatable as numerous studies have reported divergent results [34, 35]. The results of our study revealed that the K variant of rs2230806 (*P* = 0.029) is associated with CAD, which is in line with Zargar et al. [36]. The rs2066715 of *ABCA1* is not associated with an increased risk of CAD. The frequency of the rare allele of rs2066715 (0.02) showed a unique distribution compared to the 1000 genome database (0.113) and Han Chinese (0.44) population [37]. The K allele frequency of rs2230806 (R219K) polymorphism in our study was 0.47 compared to 0.28–0.73 in earlier studies [18, 38, 39]. The K allele frequency reported in our study is in line with other reports, namely a European ancestry study which reported a K allele frequency of 0.26–0.46 and a study on Dutch men with proven CAD, which reported a frequency of 0.46 [40, 41].

An analysis of the effect of a combination of *ABCA1* genotypes on CAD showed that the RR+VI genotype was significantly associated with a high risk of CAD (OR = 2.82, 95 % CI 1.32–6.02, *P* = 0.007). However, an analysis of the effect of the combination of CETP genotypes on CAD showed no significant pattern. All other combinations lacked a significant association with an OR ranging from 0.73 to 1.24, except for B1B1+VI (OR = 1.7, *P* = 0.048). The prevalence of CAD was higher in males (M:F 708:282) than in females in the present study. Homozygous mutant and heterozygous of rs5882 in men and women were strongly associated with an increased risk of CAD, while the other two *ABCA1* polymorphisms showed no significant association with CAD (OR = 0.73–1.13) in the female cohort. Also, heterozygousity of rs708272 alleles was strongly associated with an increased risk of CAD (OR = 0.65, *P* = 0.002), a homozygous carriage of rs5882, the rarer variants II causing amino acid substitutions showed a strong association with an increased risk of CAD (OR = 1.98, $P = 0.0002$) among the male cohort. The *CETP* rs5882 polymorphism was found to be associated with an increased risk for CAD in the overall study population and also in the male and female cohorts, which is in contrast to a recent study which reported that this polymorphism is associated with a decreased risk of CAD and MI [42].

Studies conducted in other populations have correlated the genotypes of *ABCA1* and *CETP* and risk of CHD with respect to its effect on HDL-C. A study of 119 patients in Korea showed that the B1B1 genotype of the *CETP* Taq1B polymorphism was associated with low HDL-C levels in females and non-smoking males and may be an independent genetic risk factor for CAD [43]. In the present study, the B2 allele of the *CETP* rs708272 polymorphism was associated with a reduced risk of CAD mediated by elevated HDL-C concentrations. However, Borggreve et al. in their prospective population-based study (PREVENT study) on 8141 Caucasians demonstrated that the B2 and I alleles of the rs708272 (*TaqIB*) and rs5882 (V422I) of the *CETP* gene were not associated with a decreased risk for CAD, despite their HDL-C-raising effect suggesting that the risk may be independent of the gene's influence on HDL-C levels [44]. Thus, the association of polymorphic *CETP* genotypes with a decreased cardiovascular risk seems to be independent of their effect on HDL-C levels [45–47].

Conclusion

This study is the first to report the association of these polymorphisms with the development of CAD in a Saudi population. A significant association of *CETP* rs5882 and *ABCA1* rs2230806 polymorphism with CAD was observed marking these polymorphisms as risk factors. The rs708272 showed a protective effect for CAD, and rs2066715 of *ABCA1* gene lacked any association with CAD, whereas the joint effect of the *ABCA1* gene (RR +VI and KK+VV) conferred a higher risk for CAD. A sex difference subsists with a higher prevalence of CAD among males (M:F 708:282). Female heterozygous and male homozygous for the rs5882 were shown to have an increased risk of CAD.

Methods
Study population
Patients reporting to the cardiac clinic at King Fahd Hospital of the University, Al-Khobar, and other major hospitals in the Eastern Province of Saudi Arabia were screened, and angiographically confirmed CAD cases with at least one event of MI ($N = 990$) were enrolled in this study. A total of 618 age-matched normal Saudi controls with no history or family history of CAD were recruited from the blood banks of the same hospitals. This study was approved by the Ethical Committee of the University of Dammam. Signed written informed consent was obtained from all participants.

Genotyping
Blood samples (5 ml) were obtained from 1608 subjects in EDTA-coated tubes and DNA was extracted using QIAamp DNA isolation kit (Qiagen, Germany) as per the manufacturer's instructions. Allele-Specific TaqMan® PCR procedures were used to detect the genetic variants rs2230806, rs2066715, rs708272, and rs5882.

Statistical analysis
The difference between cases and controls was evaluated using *t* test for continuous variables and Chi-square test for discrete variables. Allele frequencies were estimated by direct counting of the test allele divided by the total number of alleles. To assess the risk for CAD, odds ratio was determined by univariate analysis. All statistical analyses were performed using SPSS software (version19). The power of the study was calculated using online software sampsize.sourceforge.net.

Competing interests
The authors declare that they have no competing interests.

Authors' contributions
CC, CV, and SC designed the study, performed the assay, and drafted the manuscript. AAS and MA provided the CAD patient samples, AAN and FAM provided the age- and sex-matched controls. MSA and RAA collected all medical data of the individual participant from the hospital records. CV performed the statistical analyses. BK was involved in drafting the manuscript for important intellectual content. AAA and FAM provided critical review of the manuscript. All authors made significant intellectual contributions and have read and reviewed the manuscript. All authors read and approved the final manuscript.

Acknowledgements
The authors acknowledge the King Abdulaziz City for Science and Technology (KACST) for funding the for the current research grant (LGP 32-44). We thank Mr. Geoffrey James Tam Moro and Mr. Florentino Jr Mata for the technical support.

Author details
[1]Institute for Research and Medical Consultation, University of Dammam, P.O.Box 1982, Dammam 31441, Kingdom of Saudi Arabia. [2]King Fahd Hospital of the University, University of Dammam, P.O.Box 4001, Al-Khobar 31952, Kingdom of Saudi Arabia. [3]Department of Pediatrics, Perelman School of Medicine, University of Pennsylvania, Philadelphia, PA, USA.

References
1. Kubo M, Kiyohara Y, Kato I, et al. Trends in the incidence, mortality, and survival rate of cardiovascular disease in a Japanese community: the Hisayama study. Stroke. 2003;34(10):2349–54.
2. Gaziano TA, Bitton A, Anand S, et al. Growing epidemic of coronary heart disease in low-and middle-income countries. Curr Probl Cardiol. 2010;35(2): 72–115.
3. Al-Nozha MM, Arafah MR, Al-Mazrou YY, et al. Coronary artery disease in Saudi Arabia. Saudi Med J. 2004;25(9):1165–71.
4. Yamada Y, Izawa H, Ichihara S, et al. Prediction of the risk of myocardial infarction from polymorphisms in candidate genes. N Engl J Med. 2002; 347(24):1916–23.

5. Licastro F, Chiappelli M, Porcellini E, et al. Gene-gene and gene-clinical factors interaction in acute myocardial infarction: a new detailed risk chart. Curr Pharm Des. 2010;16(7):783–8.

6. Ohki T, Itabashi Y, Kohno T, et al. Detection of periodontal bacteria in thrombi of patients with acute myocardial infarction by polymerase chain reaction. Am Heart J. 2012;163(2):164–7.

7. Romero-Corral A, Somers VK, Sierra-Johnson J, et al. Normal weight obesity: a risk factor for cardio metabolic dysregulation and cardiovascular mortality. Eur Heart J. 2010;31(6):737–46.

8. Vinueza R, Boissonnet CP, Acevedo M, et al. Dyslipidemia in seven Latin American cities: CARMELA study. Prev Med. 2010;50(3):106–11.

9. Pöss J, Custodis F, Werner C, et al. Cardiovascular disease and dyslipidemia: beyond LDL. Curr Pharm Des. 2011;17(9):861–70.

10. Voight BF, Peloso GM, Orho-Melander M, et al. Plasma HDL cholesterol and risk of myocardial infarction: a Mendelian randomization study. Lancet. 2012; 380(9841):572–80.

11. Rejeb J, Omezzine A, Rebhi L, et al. Association of the cholesteryl ester transferprotein Taq1 B2B2 genotype with higher high-density lipoprotein-cholesterol concentrations and lower risk of coronary artery disease in a Tunisian population. Arch Cardiovasc Dis. 2008;101(10):629–36.

12. van Acker BA, Botma GJ, Zwinderman AH, et al. High HDL cholesterol does not protect against coronary artery disease when associated with combined cholesteryl ester transfer protein and hepatic lipase gene variants. Atherosclerosis. 2008;200(1):161–7.

13. Hiura Y, Shen CS, Kokubo Y, et al. Identification of genetic markers associated with high-density lipoprotein-cholesterol by genome-wide screening in a Japanese population: the Suita study. Circ J. 2009;73(6):1119–26.

14. Holzmann MJ, Jungner I, Walldius G, et al. Dyslipidemia is a strong predictor of myocardial infarction in subjects with chronic kidney disease. Ann Med. 2012;44(3):262–70.

15. Holmes MV, Asselbergs FW, Palmer TM, et al. Mendelian randomization of blood lipids for coronary heart disease. Eur Heart J. 2015;36(9):539–50.

16. Tsompanidi EM, Brinkmeier MS, Fotiadou EH, et al. HDL biogenesis and functions: role of HDL quality and quantity in atherosclerosis. Atherosclerosis. 2010;208:3–9.

17. Nakamura A, Niimura H, Kuwabara K, et al. Gene-gene combination effect and interactions among ABCA1, APOA1, SR-B1, and CETP polymorphisms for serum high-density lipoprotein-cholesterol in the Japanese population. PLoS One. 2013;8(12):e82046.

18. Porchay I, Péan F, Bellili N, et al. ABCA1 single nucleotide polymorphisms on high-density lipoprotein-cholesterol and overweight: the D.E.S.I.R.E study. Obesity. 2006;14(11):1874–9.

19. Barter PJ, Brewer Jr HB, Chapman MJ, et al. Cholesteryl ester transfer protein, a novel target for raising HDL and inhibiting atherosclerosis. Arterioscler Thromb Vasc Biol. 2003;23(2):160–7.

20. Brousseau ME, O'Connor Jr JJ, Ordovas JM, et al. Cholesteryl ester transfer protein TaqIB2B2 genotype is associated with higher HDL cholesterol level and lower risk of coronary heart disease endpoints in men with HDL deficiency: veterans affairs HDL cholesterol intervention trial. Arterioscler Thromb Vasc Biol. 2002;22(7):1148–54.

21. Boekholdt SM, Kuivenhoven JA, Hovingh GK, et al. CETP gene variation: relation to lipid parameters and cardiovascular risk. Curr Opin Lipidol. 2004;15:393–8.

22. Thompson A, Di Angelantonio E, Sarwar N, et al. Association of cholesteryl ester transfer protein genotypes with CETP mass and activity, lipid levels, and coronary risk. JAMA. 2008;299(23):2777–88.

23. Ridker PM, Paré G, Parker AN, et al. Polymorphism in the CETP gene region, HDL cholesterol, and risk of future myocardial infarction: genome wide analysis among 18245 initially healthy women from the Women's genome health study. Circ Cardiovasc Genet. 2009;2(1):26–33.

24. Bruce C, Sharp DS, Tall AR. Relationship of HDL and coronary heart disease to a common amino acid polymorphism in the cholesteryl ester transfer protein in men with and without hypertriglyceridemia. J Lipid Res. 1998;39:1071–8.

25. Gudnason V, Kakko S, Nicaud V, et al. Cholesteryl ester transfer protein gene effect on CETP activity and plasma high-density lipoprotein in European populations. The EARS Group. Eur J Clin Invest. 1999;29(2):116–28.

26. Padmaja N, Ravindra KM, Soya SS, et al. Common variants of Cholesteryl ester transfer protein gene and their association with lipid parameters in healthy volunteers of Tamilian population. Clin Chim Acta. 2007;375:140–6.

27. Parra ES, Panzoldo NB, Kaplan D, et al. The I405V and Taq1B polymorphisms of the CETP gene differentially affect sub- clinical carotid atherosclerosis. Lipids Health Dis. 2012;11:130.

28. Wang J, Wang LJ, Zhong Y, et al. CETP gene polymorphisms and risk of coronary atherosclerosis in a Chinese population. Lipids Health Dis. 2013;12:176.

29. Wang Q, Zhou SB, Wang LJ, et al. Seven functional polymorphisms in the CETP gene and myocardial infarction risk: a meta-analysis and meta-regression. PLoS One. 2014;9(2):e88118.

30. Hayden MR, Clee SM, Brooks-Wilson A, et al. Cholesterol efflux regulatory protein, Tangier disease and familial high-density lipoprotein deficiency. Curr Opin Lipidol. 2000;11:117–22.

31. Abd El-Aziz TA, Mohamed RH, Hagrass HA. Increased risk of premature coronary artery disease in Egyptians with ABCA1 (R219K), CETP (TaqIB), and LCAT (4886C/T) genes polymorphism. J Clin Lipidol. 2014;8(4):381–9.

32. Marvaki A, Kolovou V, Katsiki N, et al. Impact of 3 common ABCA1 gene polymorphisms on optimal vs non-optimal lipid profile in Greek young nurses. Open Cardiovasc Med J. 2014;8:83–7.

33. Liu N, Hou M, Ren W, et al. The R219K polymorphism on ATP-binding cassette transporter A1 gene is associated with coronary heart disease risk in Asia population: evidence from a meta-analysis. Cell Biochem Biophys. 2015;71(1):49–55.

34. Hou R, Zhu X, Pan X, et al. ATP-binding cassette transporter A1 R219K polymorphism and ischemic stroke risk in the Chinese population: a meta-analysis. Neurological Sci. 2014;336(1):57–61.

35. Yin YW, Li JC, Gao D, et al. Influence of ATP-binding cassette transporter R219K and M883I polymorphisms on development of atherosclerosis: a meta-analysis of 58 studies. PLoS One. 2014;9(1):e86480.

36. Zargar S, Wakil S, Mobeirek AF, et al. Involvement of ATP-binding cassette, subfamily A polymorphism with susceptibility to coronary artery disease. Biomed Rep. 2013;1(6):883–8.

37. Cao XL, Yin RX, Wu DF, et al. Genetic variant of V825I in the ATP-binding cassette transporter A1 gene and serum lipid levels in the Guangxi Bai Ku Yao and Han populations. Lipids Health Dis. 2011;10:14.

38. Pasdar A, Yadegarfar G, Cumming A, et al. The effect of ABCA1 gene polymorphisms on ischaemic stroke and relationship with lipid profile. BMC Med Genet. 2007;8:30.

39. Kolovou V, Kolovou G, Marvaki A, et al. ATP-binding cassette transporter A1 gene polymorphisms and serum lipid levels in young Greek nurses. Lipids Health Dis. 2011;10:56.

40. Koren-Morag N, Tanne D, Graff E, et al. Low and high density lipoprotein cholesterol and ischemic cerebrovascular disease: the bezafibrate infarction prevention registry. Arch Intern Med. 2002;162:993–9.

41. Clee SM, Zwinderman AH, Engert JC, et al. Common genetic variation in ABCA1 is associated with altered lipoprotein levels and a modified risk for coronary artery disease. Circulation. 2001;103(9):1198–205.

42. Isaacs A, Sayed-Tabatabaei FA, Hofman A, et al. The cholesteryl ester transfer protein I405V polymorphism is associated with increased high-density lipoprotein levels and decreased risk of myocardial infarction: the Rotterdam Study. Eur J Cardiovasc Prev Rehabil. 2007;14(3):419–21.

43. Park KW, Choi JH, Kim HK, et al. The association of cholesteryl ester transfer protein polymorphism with high-density lipoprotein cholesterol and coronary artery disease in Koreans. Clin Genet. 2003;63(1):31–8.

44. Borggreve SE, Hillege HL, Wolffenbuttel BH, et al. An increased coronary risk is paradoxically associated with common cholesteryl ester transfer protein gene variations that relate to higher HDL cholesterol: a population based study. J Clin Endocrinol Metab. 2006;91(9):3382–8.

45. Freeman DJ, Griffin BA, Holmes AP, et al. Regulation of plasma HDL cholesterol and subfraction distribution by genetic and environmental factors: associations between the TaqIB RFLP in the CETP gene and smoking and obesity. Arterioscler Thromb. 1994;14(3):336–44.

46. Fumeron F, Betoulle D, Luc G, et al. Alcohol intake modulates the effect of a polymorphism of the cholesteryl ester transfer protein gene on plasma high density lipoprotein and the risk of myocardial infarction. J Clin Invest. 1995; 96(3):1664–71.

47. Dullaart RP, Hoogenberg K, Riemens SC, et al. Cholesteryl ester transfer protein gene polymorphism is a determinant of HDL cholesterol and of the lipoprotein response to a lipid-lowering diet in type 1diabetes. Diabetes. 1997;46(12):2082–7.

Copy number variation of human AMY1 is a minor contributor to variation in salivary amylase expression and activity

Danielle Carpenter, Laura M. Mitchell and John A. L. Armour*

Abstract

Background: Salivary amylase in humans is encoded by the copy variable gene AMY1 in the amylase gene cluster on chromosome 1. Although the role of salivary amylase is well established, the consequences of the copy number variation (CNV) at AMY1 on salivary amylase protein production are less well understood. The amylase gene cluster is highly structured with a fundamental difference between odd and even AMY1 copy number haplotypes. In this study, we aimed to explore, in samples from 119 unrelated individuals, not only the effects of AMY1 CNV on salivary amylase protein expression and amylase enzyme activity but also whether there is any evidence for underlying difference between the common haplotypes containing odd numbers of AMY1 and even copy number haplotypes.

Results: AMY1 copy number was significantly correlated with the variation observed in salivary amylase production (11.7% of variance, $P < 0.0005$) and enzyme activity (13.6% of variance, $P < 0.0005$) but did not explain the majority of observed variation between individuals. AMY1-odd and AMY1-even haplotypes showed a different relationship between copy number and expression levels, but the difference was not statistically significant ($P = 0.052$).

Conclusions: Production of salivary amylase is correlated with AMY1 CNV, but the majority of interindividual variation comes from other sources. Long-range haplotype structure may affect expression, but this was not significant in our data.

Keywords: Genome instability, Amylase, CNV, Gene expression

Introduction

The enzyme amylase plays a major role in starch hydrolysis, which begins in the oral cavity and continues into the stomach and then small intestine. Amylase is the most abundant protein in saliva, accounting for at least 50% of salivary protein [1], but the quantity and enzyme activity of salivary amylase varies greatly among individuals. This variation in amylase production could be attributable to a number of factors including environmental factors, such as stress [2] and circadian rhythms [3], oral health [4] and the genetic background of an individual's amylase gene cluster. Whilst it has been suggested that quantitative variation in amylase protein patterns does not always reflect variation in the amylase gene cluster [5], some studies have shown a relationship between the observed copy number variation (CNV) at the salivary amylase gene

(AMY1) and an increased level of amylase protein expression [6, 7]. Perry et al. [7], using immunoblotting to investigate amylase protein levels, identified a significant positive correlation ($R = 0.59$) between CNV at AMY1 and levels of amylase protein in saliva. Mandel and colleagues [6], also using immunoblotting, observed a similar correlation ($R = 0.50$) with copy number at AMY1 and amylase protein levels as well as a correlation ($R = 0.52$) between CNV at AMY1 and salivary enzyme activity. These results suggest that approximately 20–35% of the variance in salivary amylase expression can be attributed to variation in AMY1 copy number.

The human amylase genes form a cluster on chromosome 1 which contains both the salivary (AMY1) and pancreatic (AMY2) amylase genes, both of which vary in copy number [5, 8, 9]. The CNV at AMY1 has an observed range of 2–18 copies per person [7, 10–12] and an average of 6 copies per person, whilst the CNV at AMY2 has an observed range of 2–12 copies per person and an average

* Correspondence: john.armour@nottingham.ac.uk
School of Life Sciences, University of Nottingham, Nottingham NG7 2UH, UK

of 4 copies per person. The amylase gene cluster is highly structured [13–15], with a correlation between the CNV at *AMY1* and the CNV at *AMY2* [10, 12]. Recent observations have identified a fundamental difference in the underlying genomic structure across the amylase gene cluster between majority haplotypes containing an odd number of copies of *AMY1* and one copy each of *AMY2A* and *AMY2B*, and less common variant haplotypes containing even copy number haplotypes of *AMY1* and deletions or duplications of *AMY2* genes. Consequently, the majority of individuals (60–70%), with two *AMY1*-odd haplotypes, have an even copy number of *AMY1* and no CNV of *AMY2*, whereas those individuals with an odd copy number of *AMY1* (usually heterozygous for an *AMY1*-even haplotype) also display CNV of *AMY2* [10, 12].

Previous studies investigating the relationship between CNV at *AMY1* and salivary amylase protein expression have used qPCR to measure copy number. However, qPCR measurement of *AMY1* has since been shown to be subject to systematic error and, in one study, consistently underestimated the *AMY1* copy number [7, 10]. We aimed to re-evaluate the relationship between CNV at *AMY1* and salivary amylase protein expression using alternative copy number measurement methods that have been shown to be precise and reproducible [10]. Our experimental plan was designed to measure both the expression of salivary amylase total protein and amylase enzyme activity in saliva from a larger cohort of individuals than has been previously studied, in parallel with determination of copy number at *AMY1*. Knowledge of the haplotype structures also allows us to test whether all copies of *AMY1* are functionally equivalent or whether there is any evidence for context dependence of gene expression. Therefore, our aim is to explore the functional consequences of the multi-allelic copy variable gene *AMY1* on more (N = 119) samples than previously investigated and using novel methods of *AMY1* copy number measurement, capable of resolving single integer copy numbers.

Results

Variation in AMY1 and AMY2 copy numbers

Copy number measurement of both *AMY1* and *AMY2* was performed on all 119 independent UK samples (see "Methods"). The *AMY1* copy number distribution is shown in Fig. 1 and shows a predominance of even copy numbers (75%), with a range of 2–15 and a modal copy number of 6, consistent with prior studies of *AMY1* copy number [7, 10–12]. Variation in *AMY2* copy number was also observed with *AMY2A* copy variable in 24% of samples and *AMY2B* showing CNV in about 10% of samples (Table 1).

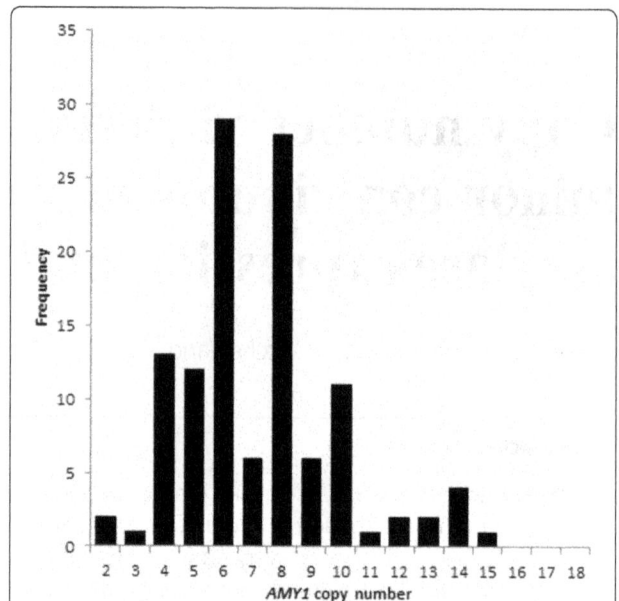

Fig. 1 Distribution histogram of *AMY1* copy number in 119 unrelated UK samples; a clear majority have even copy numbers (89 out of 119)

Correlation of AMY1 copy number with protein production and enzyme activity

We investigated both amylase protein levels and salivary amylase enzyme activity. Our data are consistent with previous studies in exhibiting considerable variation in protein expression [6, 7] and include some samples (across all copy numbers) with very low amounts of amylase protein (lowest value 0.48 mg/mL), as also detected by Mandel et al. [6].

The raw data for total protein (antigen) concentration (Fig. 2) and for enzyme activity (Fig. 3) did not show a strong relationship with copy number, and the residuals of the regression are far from normally distributed (Additional file 1: Figure S2). A linear regression was performed using $\log_{10}$ of protein and of enzyme activity giving residuals that follow a normal distribution (Additional file 1: Figure S2) and satisfy other assumptions of linear regression modelling, and therefore, all further analyses were performed with the transformed data.

Table 1 CNV of *AMY2* in 119 UK samples studied

Copy number	AMY2A	AMY2B
0	0	0
1	14	0
2	91	107
3	13	12
4	1	0
TOTAL	119	119

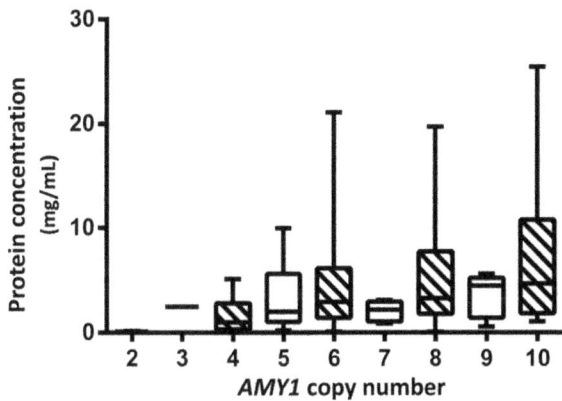

Fig. 2 A *box* and *whiskers plot* of amylase protein concentration by *AMY1* copy number, with mean for all samples at a given copy number shown as a *black bar*, the standard deviation as the *box*, and *whiskers* showing the observed full range of data. Each contributing data point is the mean of three experimental replicates, and further details of biological replicates from the same subjects can be found in the "Methods"

A significant correlation was observed between *AMY1* copy number and amylase protein ($R = 0.342$) ($P < 0.0005$). Our data suggests that *AMY1* copy number accounts for 11.7% of the variation observed in salivary amylase protein levels, much less than previous reports of 35% from a study of 50 European American individuals [7], and 25% in a study of 62 individuals of unspecified ancestry [6]. Our observation suggests that the CNV at *AMY1* plays much less of a role in the variation of salivary amylase protein levels than previously proposed. Furthermore, the great spread of observed levels of amylase protein

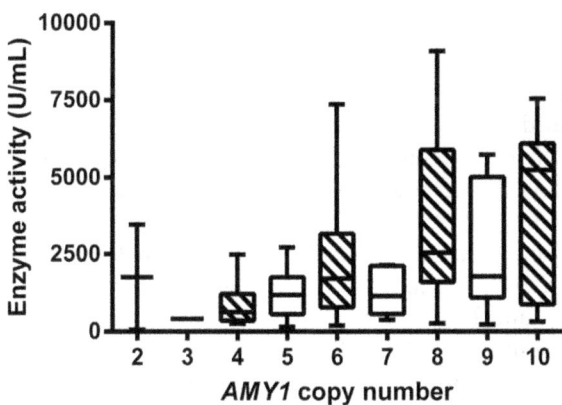

Fig. 3 A *box* and *whiskers plot* of amylase enzyme activity by *AMY1* copy number, with mean of all samples at each copy number shown as a *black bar*, and *whiskers* showing the observed range of data. The *boxes* indicate the standard deviation at each copy number. Each data point used in the analysis is the mean of two experimental replicates, and further details of biological replicates from the same subjects can be found in the "Methods"

production suggests that the genetic contribution of the CNV is not simply proportional to protein production.

A similar relationship was observed with salivary amylase enzyme activity. A significant correlation was observed between *AMY1* copy number and enzyme activity ($R = 0.369$) ($P < 0.0005$). Again, our data suggest that copy number plays less of a role than previously reported, with copy number accounting for 13.6% of the variation in enzyme activity observed in our study, rather than 27% as previously suggested [6]. A similar observation was reported in a Chinese population ($n = 92$) which found the gene copy number provided 12.2% of the observed salivary enzyme activity variation [16]. However, direct comparison between these studies is not straightforward, as Chinese populations do have a distinct *AMY1* distribution to Europeans [10, 12].

As our copy number measurement system is accurate enough to assign single integers, we were able to investigate whether there are differences in gene expression between those individuals that have even *AMY1* copy numbers and those with odd. We fitted a logistic regression model to protein and copy number data and used the model to predict protein expression from copy number to examine whether the odd or even number status was associated with a systematic difference in the relationship. This analysis gave marginally non-significant evidence ($P = 0.052$) for a difference between the regressions of log protein with even ($R = 0.168$) and odd ($R = 0.017$) copy numbers (Additional file 1: Figure S3), suggesting that the relationship between copy number and protein may be different for the odd and even copy numbers.

Discussion

This is only the second study to investigate both protein levels and enzyme activity in the same samples. There is a highly significant correlation between the two measures ($R = 0.66$; $P < 0.0001$) (Additional file 1: Figure S4); whilst this is not a strong relationship, it is consistent with previous observations between these two measures ($R = 0.61$) [6]. This observation does suggest that for a particular quantity of amylase protein, there are variations in measurable enzyme activity, and supports the proposal that the enzymatic functions of amylase may be affected by protein modifications or the formation of complexes [6, 17].

With the complex underlying structure at the amylase gene cluster, it is possible that longer-range structure, including the CNV at *AMY2*, may influence *AMY1* expression, but there was no significant correlation observed between CNV at either *AMY2A* or *AMY2B* with either salivary amylase protein production or enzyme activity.

Conclusions

To re-assess the relationship between copy number and salivary amylase protein expression and activity, our work used a more accurate and precise *AMY1* copy number measurement method than the qPCR methods previously employed. Our previous work demonstrated the reliability of PRT-based methods and the susceptibility of qPCR methods to measurement error [10]. Our data clearly show that copy number plays much less of a role in salivary amylase expression and activity than has been previously suggested, and that it is not possible to predict an individual's salivary amylase concentration or enzyme activity solely from their copy number, with implications for studies of the effects of *AMY1* copy number variation, such as with diet. It is interesting to speculate on possible reasons for the differences between our results and those of other researchers. In addition to the improved methodology for copy number measurement, there may also have been differences in the sampling regime for saliva; in our work, we specified the method and time of collection but did not examine or standardise other factors, such as the timing relative to meals. Because of the fundamental structural difference between haplotypes containing odd or even numbers of copies, we wanted to test the possibility that odd or even number haplotypes might have different relationships between copy number and gene expression. Our data neither confirm nor exclude functional differences that arise from the underlying genomic structure across the amylase region between odd and even copy number haplotypes. Further studies would be needed to support the idea, but it does remain possible that the longer-range genomic structure, in addition to *AMY1* copy number itself, may have a role in determining variation in gene expression.

Methods
Study population
Our analysis utilised 120 independent volunteers from the University of Nottingham staff and student body, with 10 randomly selected to provide repeat samples. The blood, for DNA extraction, and saliva, for salivary amylase analysis, were taken with full consent from individuals and under local ethical approval (University of Nottingham Medical School Ethics Committee approval reference number BT10/02/2010). All samples were of the UK origin with no known clinical phenotype. DNA was extracted using isolated lymphocytes and a standard 'salting out' method for protein removal followed by phenol-chloroform extraction. DNA concentration was measured using a NanoDrop spectrophotometer, and DNA purity was assessed from the 260:280 nm absorbance ratio. All samples were diluted to a working concentration of 10 ng/mL. DNA was successfully extracted for all samples, except one, giving a total sample size of 119 independent individuals.

Sample preparation
The saliva samples were collected from each volunteer at approximately 9.30 am (+/−10 min). The volunteers chewed on a 4 cm piece of parafilm for 30 s to allow saliva to be produced and then collected into a 15 ml sterile polypropylene container. The tubes were centrifuged at 13 rpm for 5 min to remove any solids from the suspension, and the remaining saliva was stored at −80 °C. For genotyping, 20 mL of the whole blood was taken from each volunteer from which genomic DNA was isolated and stored at −80 °C.

Measurement of AMY1 and AMY2 copy numbers
The copy number of *AMY1* was measured from genomic DNA using a paralogue ratio test (PRT) in combination with a TATC microsatellite assay, as previously described [10]. *AMY2* copy number was measured using an *AMY2A:AMY2B* ratio assay, an *AMY2A:AMY2A* pseudogene ratio assay and an *AMY2A/2B* duplication junction assay, as previously described [10].

PRT PCR reactions were performed using previously described primers PRT_ref12 [10] that amplify from each copy of *AMY1* and from a reference locus at hg19 chr12:9,867,565–9,867,813. PCR products were mixed with 10 µl HiDi formamide with ROX-500 marker (Applied Biosystems, Warrington, UK), and subsequent fragment analysis was carried out by electrophoresis on an ABI3130xl 36 cm capillary using POP-7 polymer with an injection time of 30 s at 1 kV. GeneMapper software (Applied Biosystems, Warrington, UK) was used to extract the peak areas for the PRT and calculate the ratio of test (244 bp) to reference (249 bp) products. Copy number values were calculated by calibrating the ratios using HapMap CEU samples [NA11930 with *AMY1* copy number (CN) = 2; NA06993 with CN = 6; NA10852 with CN = 6; NA10835 with CN = 8; NA12248 with CN = 8; NA11931 with CN = 8; NA11993 with CN = 10 and NA07347 with CN = 11], which were included in every experiment in duplicate.

For further confirmation of *AMY1* gene copy number, a TATC microsatellite PCR was performed for each sample [10]. A single PCR reaction was performed and the products were mixed with 10 µl HiDi formamide with ROX-500 marker (Applied Biosystems, Warrington, UK), and fragment analysis was carried out by electrophoresis on an ABI3130xl 36 cm capillary using POP-7 polymer with an injection time of 30 s at 1 kV. GeneMapper software (Applied Biosystems, Warrington, UK) was used to extract the peak areas.

The ratio of *AMY2A* copy number to *AMY2B* copy number and the ratio of *AMY2A* copy number to *AMY2A* pseudogene copy number were measured as previously described [10]. One microliter of PCR products from both assays were mixed and added to 10 µl

HiDi formamide with ROX-500 marker (Applied Biosystems, Warrington, UK), and fragment analysis was carried out by electrophoresis on an ABI3130xl 36 cm capillary using POP-7 polymer, injecting at 1 kV for 10 s. GeneMapper software (Applied Biosystems, Warrington, UK) was used to extract the peak areas and calculate the ratio of *AMY2A* (163 bp) to *AMY2B* (167 bp) and the ratio of *AMY2A* (197 bp) to *AMY2A* pseudogene (232 bp).

The *AMY2A/2B* duplication junction assay is a three-primer assay producing PCR amplicons of 424 bp in all samples and a specific 323 bp only from the duplication junction. The products were visualised on a 2% (*w/v*) agarose gel, as previously described [10].

Measurement of amylase protein

The concentration of amylase protein antigen present in the saliva was measured using a sandwich amylase ELISA with 1 µg/mL of anti-salivary amylase antibody (Abcam, Cambridge, UK) and 100 µg/mL of biotinylated detection antibody (Biorbyt, Cambridge, UK). Assays were performed using serial dilutions (1:5) of natural human salivary amylase protein of known concentration (200 µg/mL) (Sigma-Aldrich, Gillingham, Dorset, UK) to generate a standard curve. Assays were performed in duplicate for each unknown sample, and the standard curve was measured in triplicate.

Measurement of amylase enzyme activity

The amylase enzyme activity within saliva was measured using the EnzCheck® *Ultra* Amylase Assay Kit (Invitrogen, ThermoFisher, Paisley, UK) according to manufacturer's instructions. The saliva samples were added to the substrate solution, vortexed, and the fluorescence of the samples was measured after 10 min incubation at room temperature. Assays were performed using serial dilutions (1:5) of natural salivary amylase protein of known concentration (200 µg/mL) (Sigma-Aldrich, Gillingham, Dorset, UK) to generate a standard curve. Assays were performed in duplicate for each unknown sample, and the standard curve was measured in triplicate.

In order to deduce the variation in concentration for protein expression and enzyme activity within each individual, the repeat samples from an initial cohort of 10 individuals were investigated on four separate occasions (Additional file 1: Figure S1). These 10 repeat samples comprise samples with 2 (×1), 4 (×1), 6 (×4), 8 (×2) and 9 (×2) copies of *AMY1*. Analysis of repeat measures found that neither of the within-subject factors (time and measurement) was significant for measurement of salivary amylase protein expression or enzyme activity, and therefore, a single time point is suitable for comparing measurements of salivary amylase protein and enzyme activity between individuals.

Bradford assay for measurement of total protein concentration

The total protein concentration in the saliva samples was measured using Bradford Reagent (Sigma-Aldrich, Gillingham, Dorset, UK) according to manufacturer's instructions. Assays were performed using serial dilutions (1:2) of bovine serum albumin (BSA) protein of known concentration (1.4 mg/mL) (Sigma-Aldrich, Gillingham, Dorset, UK) to generate a standard curve. Assays were performed in duplicate for each unknown sample, and the standard curve measured in triplicate.

Statistical analysis

Correlations between groups of copy number data and either protein expression or enzyme activity were assessed using logistic regression in SPSS V22 (IBM, Armonk, New York, USA), and figures were drawn with the software GraphPad Prism (GraphPad Software, Inc., La Jolla, CA, USA). The repeat measures were analysed in SPSS using a general linear model with repeated measures, with the within-subject factors defined as time and measurement and the between-subject factor defined as copy number.

Additional files

Additional file 1: Additional Figures for Carpenter et al., "Copy number variation of human AMY1 is a minor contributor to variation in salivary amylase expression and activity". **Figure S1.** Graphs for the time trial samples illustrating the overall variation observed for each sample from the four separate time points. The samples are sorted by copy number for amylase protein concentration (A) and enzyme activity (B). **Figure S2.** Normal P-P plots of the residuals from logistic regression analysis for (A) amylase protein concentration, (B) Log10 protein concentration, (C) amylase enzyme activity and (D) Log10 enzyme activity. **Figure S3.** Graph of the residuals from logistic regression between log10 protein (LogProtein) with odd (blue circles) and even (green circles) AMY1 copy numbers shown separately. **Figure S4.** Correlation between salivary amylase protein levels (mg/mL) and amylase enzyme activity (U/mL).

Additional file 2: Dataset

Abbreviations

BSA: Bovine serum albumin; CN: Copy number; CNV: Copy number variation/variant; ELISA: Enzyme-linked immunosorbent assay; HiDi: Highly deionised (formamide)

Acknowledgements

The authors would like to thank Dr David Sirl (University of Nottingham) for statistical guidance.

Funding

This work was supported by the Biotechnology and Biological Sciences Research Council [grant BB/1006370/1] (awarded to JALA). The funding body had no role in the design of the study, collection, analysis, interpretation of data or in writing the manuscript.

Authors' contributions

DC participated in the study design, coordinated the study, performed copy number typing, optimised the ELISA assay, performed statistical analysis and wrote the manuscript. LM recruited volunteers, performed copy number typing, optimised and performed ELISA assays, optimised and performed enzyme activity assays, performed Bradford assay and generated the DNA from the samples. JALA participated in the study design and wrote the manuscript. All authors read and approved the final manuscript.

Competing interests

The authors declare that they have no competing interests.

Consent for publication

Not applicable.

References

1. Noble RE. Salivary alpha-amylase and lysozyme levels: a non-invasive technique for measuring parotid vs submandibular/sublingual gland activity. J Oral Science. 2000;42:83–6.
2. Granger DA, Kivlighan KT, El-Sheikh M, Gordis EB, Stroud LR. Salivary α-amylase in biobehavioral research. Ann N Y Acad Sci. 2007;1098:122–44.
3. Nater UM, Hoppmann CA, Scott SB. Diurnal profiles of salivary cortisol and alpha-amylase change across the adult lifespan: evidence from repeated daily life assessments. Psychoneuroendocrinology. 2013;38:3167–71.
4. Lawrence HP. Salivary markers of systemic disease: noninvasive diagnosis of disease and monitoring of general health. J Canadian Dental Association. 2002;68:170–4.
5. Bank RA, Hettema EH, Muijs MA, Pals G, Arwert F, Boomsma DI, Pronk JC. Variation in gene copy number and polymorphism of the human salivary amylase isoenzyme system in Caucasians. Hum Genet. 1992;89:213–22.
6. Mandel AL, Peyrot des Gachons C, Plank KL, Alarcon S, Breslin PAS. Individual differences in *AMY1* gene copy number, salivary α-amylase levels, and the perception of oral starch. PLoS One. 2010;5:e13352.
7. Perry GH, Dominy NJ, Claw KG, Lee AS, Fiegler H, Redon R, et al. Diet and the evolution of human amylase gene copy number variation. Nat Genet. 2007;39:1256–60.
8. Pronk JC, Frants RR, Jansen W, Eriksson AW, Tonino GJM. Evidence for duplication of the human salivary amylase gene. Hum Genet. 1982;60:32–5.
9. Sudmant PH, Kitzman JO, Antonacci F, Alkan C, Malig M, Tsalenko A, et al. Diversity of human copy number variation and multicopy genes. Science. 2010;330:641–6.
10. Carpenter D, Dhar S, Mitchell L, Fu B, Tyson J, Shwan N, et al. Obesity, starch digestion and amylase: association between copy number variants at human salivary (AMY1) and pancreatic (AMY2) amylase genes. Hum Mol Genet. 2015;24:3472–80.
11. Falchi M, El-Sayed Moustafa JS, Takousis P, Pesce F, Bonnefond A, Andersson-Assarsson JC, et al. Low copy number of the salivary amylase gene predisposes to obesity. Nat Genet. 2014;46:492–7.
12. Usher CL, Handsaker RE, Esko T, Tuke MA, Weedon MN, Hastie AR, et al. Structural forms of the human amylase locus and their relationships to SNPs, haplotypes and obesity. Nat Genet. 2015;47:921–5.
13. Groot PC, Bleeker MJ, Pronk JC, Arwert F, Mager WH, Planta RJ, et al. The human α-amylase multigene family consists of haplotypes with variable numbers of genes. Genomics. 1989;5:29–42.
14. Groot PC, Mager WH, Frants RR. Interpretation of polymorphic DNA patterns in the human α-amylase multigene family. Genomics. 1991;10:779–85.
15. Groot PC, Mager WH, Henriquez NV, Pronk JC, Arwert F, Planta RJ, et al. Evolution of the human alpha-amylase multigene family through unequal, homologous, and interchromosomal and intrachromosomal crossovers. Genomics. 1990;8:97–105.
16. Yang Z-M, Lin J, Chen L-H, Zhang M, Chen W-W, Yang X-R. The roles of AMY1 copies and protein expression in human salivary α-amylase activity. Physiol Behav. 2015;138:173–8.
17. Iontcheva I, Oppenheim FG, Troxler RF. Human salivary mucin MG1 selectively forms heterotypic complexes with amylase, proline-rich proteins, statherin, and histatins. J Dent Res. 1997;76:734–43.

Variants in congenital hypogonadotrophic hypogonadism genes identified in an Indonesian cohort of 46,XY under-virilised boys

Katie L. Ayers[1,2], Aurore Bouty[1,3], Gorjana Robevska[1], Jocelyn A. van den Bergen[1], Achmad Zulfa Juniarto[4], Nurin Aisyiyah Listyasari[4], Andrew H. Sinclair[1,2†] and Sultana M. H. Faradz[4*†]

Abstract

Background: Congenital hypogonadotrophic hypogonadism (CHH) and Kallmann syndrome (KS) are caused by disruption to the hypothalamic-pituitary-gonadal (H-P-G) axis. In particular, reduced production, secretion or action of gonadotrophin-releasing hormone (GnRH) is often responsible. Various genes, many of which play a role in the development and function of the GnRH neurons, have been implicated in these disorders. Clinically, CHH and KS are heterogeneous; however, in 46,XY patients, they can be characterised by under-virilisation phenotypes such as cryptorchidism and micropenis or delayed puberty. In rare cases, hypospadias may also be present.

Results: Here, we describe genetic mutational analysis of CHH genes in Indonesian 46,XY disorder of sex development patients with under-virilisation. We present 11 male patients with varying degrees of under-virilisation who have rare variants in known CHH genes. Interestingly, many of these patients had hypospadias.

Conclusions: We postulate that variants in CHH genes, in particular PROKR2, PROK2, WDR11 and FGFR1 with CHD7, may contribute to under-virilisation phenotypes including hypospadias in Indonesia.

Keywords: Congenital hypogonadotrophic hypogonadism, Under-virilisation, Hypospadias, Targeted gene sequencing, Disorder of sex development

Background

Proper function of the hypothalamic-pituitary-gonadal (H-P-G) axis is essential for the development of the reproductive system. Gonadotrophin-releasing hormone (GnRH), secreted by the hypothalamus, stimulates the biosynthesis and the release of gonadotrophins from the anterior pituitary gland. These gonadotrophins (luteinising hormone (LH) and follicle-stimulating hormone (FSH)) both play distinct roles in the gonads during embryonic development. In males, FSH stimulates the proliferation of immature Sertoli cells and spermatogonia [1]. FSH also stimulates the secretion of inhibin, which acts in a negative feedback loop directly to the anterior pituitary. LH stimulates the production and secretion of testosterone from the Leydig cells, which is thought to occur through the LH receptor after 10 weeks post conception [2]. Disruption to the H-P-G axis (through deficient production, secretion or action of the gonadotrophins) can result in hypogonadotrophic hypogonadism (HH). While this can be associated with additional anomalies or syndromes such as Dandy-Walker syndrome, Gorden Holmes syndrome and CHARGE [3], when observed alone, it is termed congenital or idiopathic HH (CHH) (OMIM 146110). CHH can be coupled with a decreased or absent sense of smell due to the abnormal migration of the GnRH neurons [4, 5]. The co-occurrence of CHH with anosmia is termed Kallmann syndrome (KS (OMIM 308700, 147950, 244200, 610628, 612370 and 612702)).

* Correspondence: sultanafaradz@gmail.com
†Equal contributors
[4]Division of Human Genetics, Centre for Biomedical Research, Faculty of Medicine, Diponegoro University (FMDU), JL. Prof. H. Soedarto, SH, Tembalang, Semarang 50275, Central Java, Indonesia

Estimates of the prevalence of CHH range between 1 and 10 in 100,000 live births, with approximately two thirds of cases arising from KS [6]. CHH in 46,XY males can cause a reduced level of circulating androgens due to hypogonadism. Isolated or apparently isolated CHH (i.e. in a patient with KS who does not complain of an absent or diminished sense of smell) is most commonly diagnosed in teenagers or young men who present with pubertal failure. During foetal development, testosterone is responsible for virilisation of the reproductive tract and dihydrotestosterone (DHT), a highly potent derivative of testosterone, drives differentiation of the external genitalia. The appearance of clinical characteristics depends on when HH begins. When GnRH deficiency occurs in the late foetal or early neonatal periods, a significant decrease in androgens can lead to some CHH patients being diagnosed postnatally with under-virilisation phenotypes such as cryptorchidism, micropenis [7] and, in some rare cases, hypospadias [8]. Patients also typically showed delayed or absent puberty including minimal virilisation, low libido, lack of sexual function and a reduced or absent growth spurt [3]. In addition to the physical anomalies, physiological impairments have also been reported such as low self-esteem, distorted body image and, in some cases, problems in sexual identity [9, 10]. Finally, CHH may be diagnosed following adolescence, later in life when infertility is a concern. Given these complex and significant physical and psychological implications, early diagnosis and treatment of CHH is essential. Clinically, CHH can phenocopy partial androgen insensitivity syndrome (PAIS) or other disorders of sex development (DSDs) in which a reduction of testosterone during development can cause reduced virilisation. If blood hormone testing is not routinely carried out, these patients may be misdiagnosed and clinical management may differ.

Genetically, CHH is highly complex. More than 30 genes have been implicated in CHH and/or KS including nine genes that cause an overlapping syndrome [3]. To complicate matters, a large degree of variability in inheritance, penetrance and expressivity is seen in CHH and an increasing body of evidence suggests that this disorder can be caused by variants in more than one gene (oligogenicity) [11, 12]. Variants in known CHH genes currently account for only 50% of CHH cases [13] meaning that more genes are yet to be found. Here, we present genetic mutational analysis of CHH genes in Indonesian 46,XY patients presenting with under-virilisation phenotypes.

Materials and methods
Clinical data
Patients with 46,XY DSD were referred to the Center for Biomedical Research, Faculty of Medicine, Diponegoro University (FMDU), Semarang, Indonesia. The medical ethics committee of the Dr. Kariadi Hospital/FMDU

approved this study, and informed consent was obtained from all participants, as well as their parents or guardians, prior to their participation in this study. Following informed consent, a detailed interview was performed at recruitment and data concerning medical history, age of initial presentation, sex of rearing, family history (relatives with a genital disorder) and consanguinity were collected. Patients were clinically evaluated by a trained andrologist; a detailed description of the external genitalia was obtained and, in many cases, images taken. A blood sample was obtained for karyotyping, hormonal analysis and DNA extraction. Referral and data collection took place between 2004 and 2010. Eighty-eight of these patients have been described previously [14]. A total of 47 males with 46,XY under-virilisation phenotypes (including uni- or bilateral cryptorchidism, hypospadias, bifid scrotum, micropenis and, in some cases, severe hypospadias) were included in this study. Hormone analysis was carried out for some patients including base level LH and FSH and testosterone (T). Reference levels for FSH and LH are based on paediatric measurements depending on age [15]. In some cases, T levels were also measured following Leydig cell stimulation by human chorionic gonadotrophin (hCG). For more details on blood hormone analysis, see [14].

Gene panel sequencing
Genomic DNA was obtained from peripheral EDTA-blood samples using the salting out method [16]. The DNA underwent quality control at the Murdoch Childrens Research Institute (MCRI), Melbourne, Australia. Total genomic DNA was sequenced using a targeted panel (Haloplex, Agilent) that covers 64 diagnostic DSD genes [17]. This included 19 genes implicated in CHH (CHD7, GNRH1, GNRHR, HESX1, LEP, PROKR2, PROP1, TAC3, FGFR1, KAL1, LHX3, FGF8, PROK2, KISS1R, WDR11, SPRY4, FSHB, CGA, SOX10). Library preparation and sequencing were carried out as detailed in [17]. Raw data was analysed using a modified pipeline created at MCRI—C-pipe, which calls variants and provides data on frequency and pathogenicity [18].

Following C-pipe analysis, variants were checked for quality and depth and were filtered for those less than 1% minor allele frequency (MAF) in both the ESP6500 and 1000 genome project. As non-affected controls from Indonesia were not included, variants that were found very frequently in our screen (greater than 5% of total samples run) were also discounted. We manually check variant frequency in EVS and extracted ExAC data on frequency in Asia (South Asia and East Asia). Variants were checked for previous implication in human disease via ClinVAR and HMGD. Predicted pathogenicity of each variant was analysed using a range of up-to-date in silico prediction tools (SIFT, PolyPhen-2, LRT and

MutationTaster). Effects on protein structure and function were predicted using the HOPE tool [19].

Results

Patient cohort

All forty-seven 46,XY DSD patients from Indonesia were first analysed for mutations in DSD genes that cause androgen insensitivity or reduced testosterone production (e.g. *androgen receptor* (*AR*), *SRD5A2*, *HSD17B3*). Rare and damaging mutations in these genes were found in 19 patients [17] of the 47. The other 28 did not have a causative variant identified. These patients ranged in age from newborn to 14 years old, with a variety of 46,XY DSD phenotypes including hypospadias, bifid scrotum, cryptorchidism/undescended testis, microtestis and micropenis. All patients identify as male. Many have undergone hypospadias repair. The phenotypes of the eleven patients with a CHH variant are shown in Table 1, and representative images are shown in Fig. 1.

The hallmarks of CHH can include low levels of testosterone (due to hypogonadism), which can often be increased by hCG stimulation. Indeed, we found that all of the patients tested showed moderate to high increases in testosterone after hCG stimulation (Table 1). In addition, low levels of LH and FSH are often indicative of CHH; however, the natural levels of these hormones are low during childhood. Indeed, most patients were between mini-puberty and puberty when FSH/LH levels are expected to be low (<0.1–4 IU/l for LH and <0.1–8 IU/l for FSH). For all patients of this age, assayed LH and FSH levels were within the normal range. Patient 169, who was 14 at the time, had an LH measurement of 2.7 IU/l, which is within normal range, but an FSH of 9.24 IU/l, which is considered slightly elevated. Patient 147 was within mini-puberty at the time of measurement and subsequently had an elevated LH level of 10.8 and FSH of 6.23. This may suggest that secretion of these gonadotrophins is not inhibited in this patient. Patient 143 did not have hormonal analysis.

CHH genetic variants

The remaining 28 patients were then analysed for mutations in the exonic regions of CHH genes as previously detailed in [17]. Eleven patients had one or more rare variants (<1% MAF in g1000 and ESP6500) in a CHH gene (Table 2). In total, we found 14 variants in CHH genes in these patients. The variants are described below.

PROKR2

Four patients had variants in the *PROKR2* gene. Two patients (173 and 143) had the same variant—*PROKR2*:c.C563T:pS188L (Table 2). This variant has not been found in our DSD panel previously (in over 300

DSD patients; see [17]) but has a total allele frequency of 1.65e–05 in ExAC (although it has not been recorded in SA or EA). This change has been recorded to be likely pathogenic (ClinVar) [20, 21] (Table 2). Previous functional analysis has shown this variant has a strong defect in G-protein coupling [21]. The two patients with this variant (patients 173, 143) had under-virilisation phenotypes including micropenis, scrotal hypospadias and cryptorchidism (Fig. 1, Table 1). Interestingly, one of these patients also had additional anomalies. Patient 173 had spina bifida, incontinence and suspected intellectual disability—suggesting additional genetic or environmental contributors (Table 1). Indeed, the mother of this patient had a suspected folic acid deficiency during pregnancy.

Two other patients had heterozygous missense variants in the *PROKR2* gene (Table 2). c.G991A:p.V331M was found in patient 159 who has perineal hypospadias and unilateral cryptorchidism (Fig. 1, Table 1). This variant (rs117106081) has a total frequency in ExAC of 0.0065 (and was greater than 0.01 in both SA and EA). It is not predicted to be damaging in any of the in silico prediction tools and was not highly conserved. Nevertheless, this variant has been previously reported in CHH/KS patients, and functional analysis in both publications suggested a reduction in function (in particular a mild G-protein coupling defect) [21–23]. In contrast, another variant c.T1054G:p.W352G was not found in ExAC or EVS and was predicted to be damaging and highly conserved among different species (Table 2 and Fig. 2a). This was found in a patient 171, who has bilateral cryptorchidism and scrotal hypospadias (Fig. 1 and Table 1). In this case, the mutated residue is located on the surface of a domain with unknown function (Fig. 2b). The mutant residue (glycine) is smaller than the wild-type residue and differs in hydrophobicity to the wild-type residue (tryptophan). This may cause a loss of external interactions in particular a loss of hydrophobic interactions with other molecules on the surface of the protein.

From this, we hypothesise that *PROKR2* variants, in particular the variant p.S188L, represent a significant cause of under-virilisation including cryptorchidism, micropenis and, in some cases, hypospadias in Indonesian 46,XY DSD patients.

PROK2

One patient (47) was found to harbour a variant in this gene (*PROK2*:c.G68A:p.R23H). This missense heterozygous variant was not found in any of the online databases; however, it was not predicted to be pathogenic (Table 2). This patient has a micropenis, scrotal hypospadias and unilateral cryptorchidism (Table 1). The first 27 amino acids of PROK2 are a signal peptide, important

Table 1 Patient clinical details

Patient ID	Age at initial appointment	Gender		Clinical description				Associated malformations	Anosmia reported?	hCG stimulation test	Image provided?
		Genetic	Sex of rearing	Testes	Scrotum	Micropenis	Urethral meatus (type of hypospadias)			Increased T?	
173	12	46,XY	Male	Bilaterally non palpable	Bifid	Yes	Scrotal	Spina bifida	Unknown	Moderate	
143	6	46,XY	Male	R, not palpable L, 1 ml, scrotal	Bifid	Yes	Scrotal		No		Figure 1b
159	2	46,XY	Male	R, 1 ml, scrotal L, fetractile	Bifid	No	Perineal		Unknown		
171	4	46,XY	Male	R, 1–2 ml, scrotal L, 2 ml, scrotal	Bifid	Yes	Scrotal		No	Yes	Figure 1a
47	3	46,XY	Female, changed to male at 3 years	R, 2 ml, scrotal L, not palpable	Bifid	Yes	Scrotal		No	Yes	
174	3	46,XY	Male	R, not palpable L, 1 ml, scrotal	Fused	Yes	Penoscrotal		No	Yes	
164	3	46,XY	Male	Bilaterally 2 ml, scrotal	Bifid	No	Penoscrotal		No	Yes	
163	10	46,XY	Male	Bilaterally 3 ml, scrotal	Bifid	Yes	Penile		No	Yes	Figure 1c
147	1 m	46,XY	Male	R, inguinal L, not palpable	Bifid	No	Scrotal		No	Yes	
101	3	46,XY	Male	Bilaterally 2 ml, scrotal	Bifid	Yes	Scrotal		Unknown	Yes	Figure 1d
169	14	46,XY	Male	R, 4 ml, scrotal L, 6 ml, scrotal	Bifid	no	Penoscrotal		No	Yes	

Patient identification number and age at first consultation are shown, as well as sex chromosome complement and gender. A description of anomalies is also included. Response to hCG stimulation is shown. Testosterone reference levels were considered as 0.3–0.5 nmol/l except for patients 169, 163 and 8 (where reference was considered 3–6.5 nmol/l)

for its secretion. The affected amino acid (arginine at position 23) lies within this region.

WDR11

Three patients had heterozygous missense variants in *WDR11* (Table 2, patients 174, 164, 163). The first of these was one of a pair of twins, who have concordant phenotypes (patient 174, twin not analysed). This variant, *WDR11*:c.G2409T:p.W803C, was not found in online databases and is predicted to be pathogenic with strong conservation—even down to zebrafish (Table 2,

Fig. 2c). Patient 174 has a micropenis with penoscrotal hypospadias and chordee (Fig. 1, Table 1). The second variant found was *WDR11*:c.A1352G:p.H451R (Table 2). This variant (rs199920020) has a total frequency of 0.0001 in ExAC but is rare in Asia and was found in a patient with penoscrotal hypospadias and bifid scrotum but no micropenis or cryptorchidism (patient 164, Fig. 1, Table 1). Like the previous variant, this amino acid is highly conserved (Fig. 2c). The third variant (*WDR11*:c.T1279A:p.L427I) was not previously found in any online databases and was predicted to be

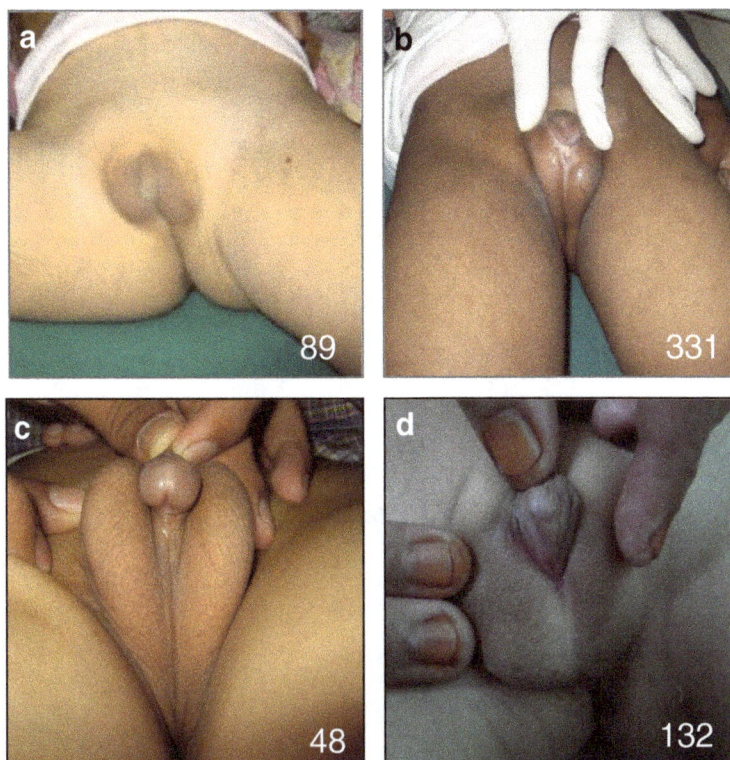

Fig. 1 Under-virilisation in patients with CHH gene variants. **a–d** Representative images of external genitalia for four patients presenting with 46,XY DSD (see Table 1 for details)

damaging and highly conserved (Fig. 2c). This was found in patient 163, who has a micropenis and penile hypospadias (Fig. 1, Table 1).

FGFR1

Three patients had the *FGFR1* variant rs140382957 (*FGFR1*:c.C320T:p.S107L) (Table 2). This variant has a MAF in EVS of 0.0077 and in ExAC of 0.0023 (and while it is predicted to be pathogenic in two prediction tools, one record in ClinVar has it logged as being benign [24].)

CHD7

Curiously, two of the three patients who had a *FGFR1* variant also had a variant in *CHD7*. One of these, *CHD7*:c.G1565T:p.G522V has an ExAC MAF of 0.002318 and was predicted damaging (Table 2); however, it has been reported as benign for CHARGE on ClinVar. This was found in patient 101. Another novel variant was rare (ExAC MAF = 9.2e–05) (*CHD7*:c.C2347T:p.P783S). No other *CHD7* variants were found.

Discussion

In this study, we have investigated variants in CHH-related genes in a cohort of Indonesian 46,XY DSD patients who had an under-virilisation phenotype. After excluding patients with mutations in known DSD genes, we found rare and damaging variants in CHH genes in 11 of the remaining 28 patients. CHH and KS can present at birth with under-virilisation phenotypes in males such as micropenis and cryptorchidism [25]. Our study suggests that CHH may be a cause of under-virilisation in Indonesia. While forty-seven 46,XY DSD patients were initially recruited, 19 of these were found to harbour mutation in a known DSD genes such as *AR* (data not shown). Of the remaining 28, we found a likely CHH variant(s) in 11 patients, making this a total of 25% of the total original cohort. In addition, while our targeted DSD panel has a comprehensive list of diagnostic DSD genes, it only covers 19 of approximately 24 genes that cause CHH/KS without an associated syndrome. Sequencing of the entire list of known CHH genes, including those that cause CHH in association with additional anomalies, may increase the diagnostic yield of a genetic screen like this. This will be important for future studies of this cohort.

Penile and urethral morphology is established before 14 weeks gestation meaning that the foetal pituitary-hypothalamic axis is typically thought to be unnecessary for normal penile development (instead relying on maternal hCG). However, after week 14, continued increase in penile length is dependent upon the hypothalamic-

Table 2 Rare variants found in CHH genes in 46,XY DSD patients

Patient ID	CHH gene	Variant location	Change	Variant details	dbSNP	EVS MAF	ExAC total freq.	ExAC SA/EA	ClinVar/ HGMD	In silico predictions	GERP+ + RS score	Previous functional studies
173	PROKR2	chr20:5283278-5283279	G/A	PROKR2:NM_144773:c.C563T:p.S188L	rs376239580	0.0077	0.00002	0/0	Yes—likely pathogenic for CHH	3 of 4	5.31	Cole et al. (2008); Zhu et al. (2015)
143	PROKR2	chr20:5283278-5283279	G/A	PROKR2:NM_144773:c.C563T:p.S188L	rs376239580	0.0077	0.00002	0/0	Yes—likely pathogenic for CHH	3 of 4	5.31	Cole et al. (2008); Zhu et al. (2015)
159	PROKR2	chr20:5282850-5282851	C/T	PROKR2:NM_144773:c.G991A:p.V331M	rs117106081	0.0154	0.00652	0.03119/0.02901	Yes—CHH	0 of 4	2.01	Dodé (2006); Monnier et al. (2009); Cole et al. (2008)
171	PROKR2	chr20:5282787-5282788	A/C	PROKR2:NM_144773:c.T1054G:p.W352G	Not found	0	0.00000	0.00	Not found	4 of 4	5.05	
47	PROK2	chr3:71834136-7184137	C/T	PROK2:NM_001126128:c.G68A:p.R23H	Not found	0	0.00000	0/0	Not found	0 of 4	2.47	
174	WDR11	chr10:122650293-122650294	G/T	WDR11:NM_018117:c.G2409T:p.W803C	Not found	0	0.00000	0/0	Not found	4 of 4	5.79	
164	WDR11	chr10 :122630739-122630740	A/G	WDR11:NM_018117:c.A1352G:p.H451R	rs199920020	0	0.00007	0/0.00104	Not found	2 of 4	3.575	
163	WDR11	chr10:122626666-122626667	T/A	WDR11:NM_018117:c.T1279A:p.L427I	Not found	0	0.00000	0/0	Not found	3 of 4	3.11	
147	FGFR1	chr8:38287238-38287239	G/A	FGFR1:NM_001174063:c.C320T:p.S107L	rs140382957	0.0077	0.00253	0.0002393/0.0454	1 record—benign	2 of 4	3.6	Sato (2004); Sykiotis (2010); Fukami et al. (2013)
101	CHD7	chr8:61655556-61655557	G/T	CHD7: NM_017780:c.G1565T:p.G522V	rs142962579	0	0.00232	0.0003717/0.03098	Not found	3 of 4	5.67	
	FGFR1	chr8:38287238-38287239	G/A	FGFR1:NM_001174063:c.C320T:p.S107L	rs140382957	0.0077	0.00253	0.0002393/0.0454	1 record—benign	2 of 4	3.6	Sato (2004); Sykiotis (2010); Fukami et al (2013)
169	FGFR1	chr8:38287238-38287239	G/A	FGFR1:NM_001174063:c.C320T:p.S107L	rs140382957	0.0077	0.00253	0.0002393/0.0454	1 record—benign	2 of 4	3.6	Sato (2004); Sykiotis (2010); Fukami et al. (2013)
	CHD7	chr8:61713055-61713056	C/T	CHD7:NM_017780:c.C2347T:p.P783S	rs373873996	0	0.00009	6.152e-05/0.00117	1 record—benign for CHARGE	2 of 4	5.81	
	LEP	chr7:127892124-127892125	A/G	LEP:NM_000230:c.A53G:p.Y18C	rs148407750	0.0461	0.00041	6.056e-05/0.003004	Not found	0 of 4	1.13	

Patient number is shown and the gene, variant location and DNA change. The allele frequency (from ExAC) is shown for all populations (MAF) and also specifically for both South Asia (AS) and East Asia (EA). Details are shown in the variant in found in Clinvar or in HMGD, and if reported previously, the reference is shown. Four in silico prediction programs were used for each variant, and the number of these showing a likely pathogenic/damaging score is shown. GERP++ scores are also shown

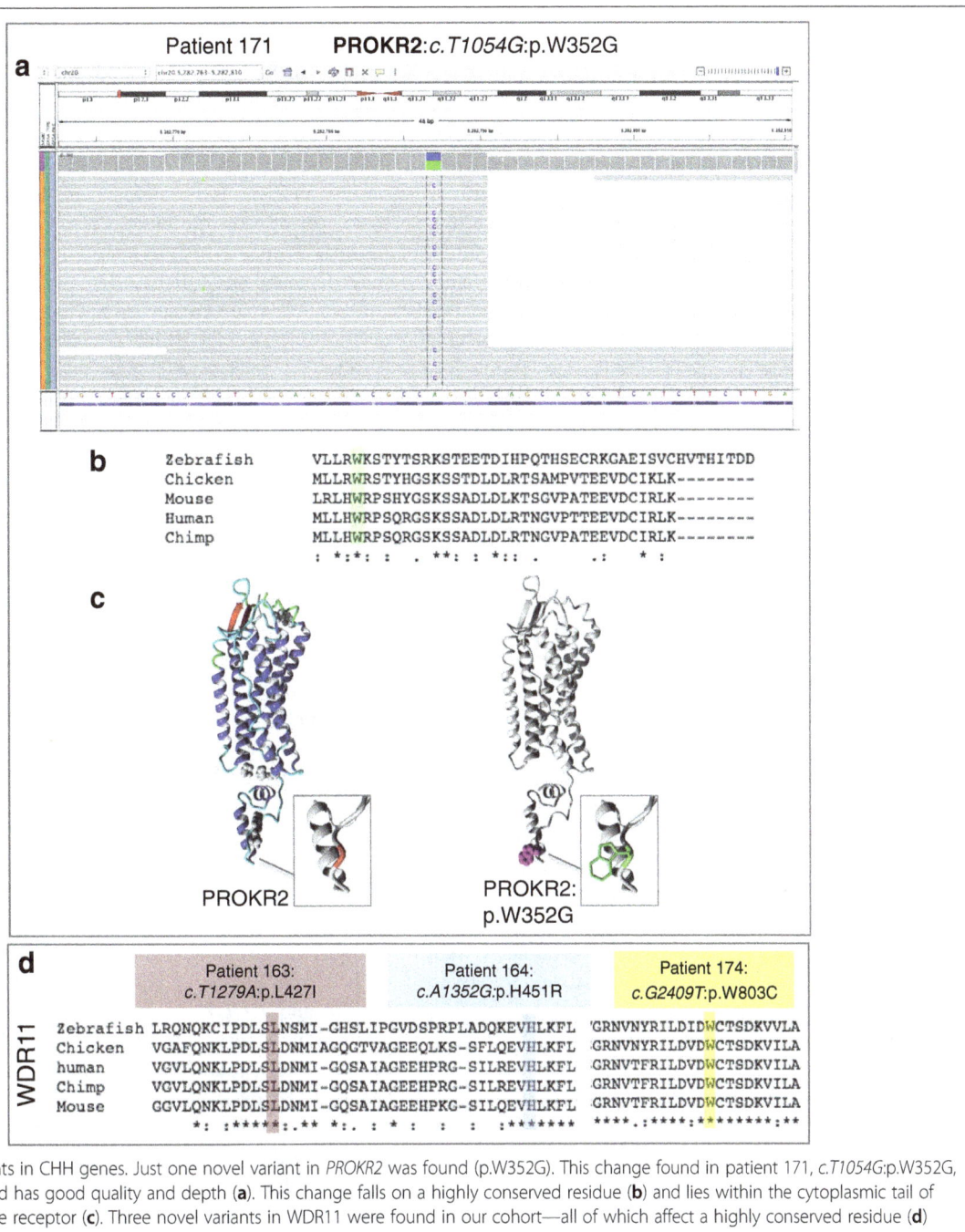

Fig. 2 Novel variants in CHH genes. Just one novel variant in *PROKR2* was found (p.W352G). This change found in patient 171, *c.T1054G*:p.W352G, is heterozygous and has good quality and depth (**a**). This change falls on a highly conserved residue (**b**) and lies within the cytoplasmic tail of this transmembrane receptor (**c**). Three novel variants in WDR11 were found in our cohort—all of which affect a highly conserved residue (**d**)

pituitary axis. Therefore, boys with hypogonadism will often have micropenis but normal phallic morphology. However, we found that many of individuals in our cohort had varying degrees of hypospadias. Given this, it is interesting to note that while rare, hypospadias in patients with CHH or KS has been described. A large study found two patients with CHH and hypospadias [26], and several other studies have described patients with KS or CHH and hypospadias of varying degrees [8, 27–29]. Nevertheless, hypospadias in CHH is a rare combination,

and it is interesting to speculate why our cohort has an over-representation of variants in CHH genes in patients with hypospadias. It is possible that in our cohort of Indonesian patients, CHH and KS manifest in a unique way, as we have not found this association in patients with 46,XY DSD of other nationalities (data not shown). Or, it may be that these variants simply contribute to a phenotype in these patients that could involve additional undetected variants in genes controlling either gonadal or penile development. It is also possible that these genes/variants have an interaction with environmental cues in this population, resulting in more common under-virilisation in CHH than in other populations. Indeed, many of the described patients come from low socio-economic communities, and many of them are involved in agriculture. Both genetic and environmental factors are thought to contribute to isolated hypospadias (reviewed in [30]), and numerous studies in different populations have shown agriculture and pesticides to be a risk factor for reproductive development and health, e.g. [31–34]. Finally, it is possible that these variants detected in CHH genes are non-damaging variants over-represented in the Indonesian population. However, three patients had a *PROKR2* variant previously shown to be deleterious in functional studies, and we have sequenced more than 100 individuals from Indonesia (include severe DSD patients, parents and siblings) who were not enriched for these or other rare variants in CHH genes (data not shown).

Mutations in *PROK2* and *PROKR2* are thought to contribute to around 9% of patients with KS [23]. In our cohort, a total of four patients had a variant in *PROKR2* and one with *PROK2*. We also had three patients with *WDR11* variants, meaning this gene may also play a significant role in Indonesian 46,XY DSD patients. Overall, we have found eight variants in CHH genes that have not been previously described in this disorder. Of these, four are not present in online variant databases ExAC or EVS. The PROKR2 variant p.W352G lies within the cytoplasmic tail of this transmembrane receptor. Other variants have been described in this region (such as p.V331M—which we also found, and p.R357W) [21]. In this case, the mutant residue (glycine) is smaller than the wild-type residue and differs in hydrophobicity to the wild-type residue (tryptophan). This may cause a loss of external interactions in particular a loss of hydrophobic interactions with other molecules on the surface of the protein. One patient had a variant in *PROK2* (p.R23H) that has not been previously described. The affected amino acid lies within the signal peptide region, and the mutant residue (histidine) is smaller than the wild-type residue and has a different charge (neutral rather than positive). This may change the activity of the signal peptide, and a patient with a variant affecting the neighboring amino acid (p.A24P) has been described previously in CHH [21].

The PROKR2 p.V331M variant that we and others have found has been shown to have reduced functional activity (albeit weaker than other variants) [21–23]; however, it is not predicted to be damaging by any of the four prediction tools used. This is likely due to the fact that several orthologous proteins in other species have a methionine in this protein position. It has been suggested that filtering variants based on currently available

pathogenicity tools may lead to under-reporting of such compensated variants [35]. Therefore, while we have included the *in silico* predictions of pathogenicity in our pipeline, we have chosen to report all rare variants in this manuscript regardless of these predictions.

Finally, two novel *WDR11* variants were found in our screen. WDR11 is predicted to exhibit two β propellers made up of WD domains. Protein structure modelling has predicted that WDR11 has 12 WD domains and that nine of them (second through tenth) participate in the genesis of two consecutive β propellers [36]. The p.W803C variant in which a tryptophan is replaced by a cysteine at position 803 falls within the 12th WD domain and is a highly conserved amino acid [36]. Cysteine is a smaller residue than the wild-type residue, which could interrupt with the WD function. The p.L427I change is predicted to fall adjacent to WD domain 6, where at least two other human variants have been described [36].

Interestingly, we also found two patients with both *FGFR1* and *CHD7* variants. Indeed, oligogenicity has been described to be a feature of CHH (for a summary, see [3]). Specifically, oligogenic inheritance has been previously reported for *FGFR1*, while no reports for *CHD7* oligogenicity have yet been published. While *CHD7* has most frequently been associated with CHARGE syndrome, of which hypospadias can be a feature, a recent paper has detailed patients in which *CHD7* single-nucleotide variants (SNVs) were not associated with classical CHARGE syndromic features. Indeed, they show that rare deleterious SNVs in this gene contribute to the mutational burden of patients with both KS and CHH in the absence of full CHARGE syndromic features [37]. It may be that a combination of variant alleles in *FGFR1* and *CHD7* can cause hypospadias and under-virilisation. However, several of the *FGFR1* and *CHD7* variants had a total MAF of around 0.2%, with a prevalence of 0.3 or 0.4% in East Asia indicating that they may be over-represented in the Indonesian population. Further studies to address the pathogenicity of these variants and the interaction between *FGFR1* and *CHD7* are required and are beyond the scope of this study.

Hormonal analysis at the right age can be highly informative in a clinical diagnosis of CHH. This includes assays of the levels of the gonadotrophins FSH and LH, as well as testosterone levels before and after hCG stimulation. Diagnosis of KS and CHH in many of these patients has been limited by access to detailed blood hormone analysis (in particular as many are pre-pubescent children meaning that measuring LH and FSH is not informative). Most of our patients showed low levels of testosterone (consistent with their age), but these levels were stimulated by hCG. Nevertheless, the genetic results of this study suggest that boys presenting with under-virilisation phenotypes in Indonesian clinics should be tested for CHH or KS. The patients presented here will be monitored as they develop, and we recommend they have their gonadotrophin levels retested at a later date when reduced levels can be detected.

A genetic diagnosis can inform family planning and fertility investigations, as well as direct clinical management. Treatments exist for many of the features of CHH. In early life, this can include low-dose testosterone or gonadotrophins for micropenis and stimulation of gonadal development. Later, during adolescence or adulthood, testosterone therapy can also induce puberty including psychosocial development [3]. CHH-associated infertility can also be treated, for example, by administering GnRH or gonadotrophins [3]. Thus, given the therapeutic options, having a genetic diagnosis may allow earlier or tailored intervention. Gene panel testing is a viable option to deliver this genetic diagnosis.

Conclusion

We conclude that variants in CHH genes, in particular *PROKR2*, *PROK2*, *WDR11* and *FGFR1* with *CHD7*, may contribute to under-virilisation phenotypes including hypospadias in Indonesian boys. We suggest that in this population, 46,XY DSD patients should be monitored for signs of CHH including hormonal and genetic analysis.

Abbreviations

CHH: Congenital hypogonadotrophic hypogonadism; DHT: Dihydrotestosterone; DSD: Disorder of sex development; FSH: Follicle-stimulating hormone; GnRH: Gonadotrophin-releasing hormone; hCG: Human chorionic gonadotrophin; KS: Kallmann syndrome; LH: Luteinising hormone

Acknowledgements
The authors would like to thank the patients who kindly consented to be part of this study.

Funding
KA, GR and JvdB are funded by a National Health and Medical Research Council (NHMRC) program grant (number APP1074258). AS is funded by a NHMRC research fellowship.

Authors' contributions
KLA analysed and interpreted the data and wrote the manuscript. AB interpreted the data and critically revised the manuscript. GR and JvdB handled the DNA, carried out sequencing, and analysed data. NAL and AZJ collected and analysed the patient clinical data. AHS and SMHF supervised the work, contributed to the study design and interpretation, and critically reviewed the manuscript. All authors have read and approved the manuscript for publication.

Competing interests
The authors declare that they have no competing interests.

Consent for publication
All participants included in this study have given their informed consent for the publication of this manuscript.

Author details
[1]Murdoch Childrens Research Institute, Melbourne, Victoria, Australia. [2]Department of Paediatrics, University of Melbourne, Melbourne, Victoria, Australia. [3]The Royal Children's Hospital, Melbourne, Victoria, Australia. [4]Division of Human Genetics, Centre for Biomedical Research, Faculty of Medicine, Diponegoro University (FMDU), JL. Prof. H. Soedarto, SH, Tembalang, Semarang 50275, Central Java, Indonesia.

References

1. Walker WH, Cheng J. FSH and testosterone signaling in Sertoli cells. Reproduction. 2005;130:15–28. Society for Reproduction and Fertility.
2. Svechnikov K, Landreh L, Weisser J, Izzo G, Colón E, Svechnikova I, et al. Origin, development and regulation of human Leydig cells. Horm Res Paediatr. 2010;73:93–101.
3. Boehm U, Bouloux P-M, Dattani MT, de Roux N, Dodé C, Dunkel L, et al. Expert consensus document: European Consensus Statement on congenital hypogonadotropic hypogonadism—pathogenesis, diagnosis and treatment. Nat Rev Endocrinol. 2015;11:547–64. Nature Publishing Group.
4. Teixeira L, Guimiot F, Dodé C, Fallet-Bianco C, Millar RP, Delezoide A-L, et al. Defective migration of neuroendocrine GnRH cells in human arrhinencephalic conditions. The Journal of clinical investigation. Am Soc Clin Invest. 2010;120:3668–72.
5. Schwanzel-Fukuda M, Pfaff DW. Origin of luteinizing hormone-releasing hormone neurons. Nature. 1989;338:161–4. Nature Publishing Group.
6. Bianco SDC, Kaiser UB. The genetic and molecular basis of idiopathic hypogonadotropic hypogonadism. Nat Rev Endocrinol. 2009;5:569–76. Nature Publishing Group.
7. Fraietta R, Zylberstejn DS, Esteves SC. Hypogonadotropic hypogonadism revisited. Clinics (Sao Paulo). Hospital das Clinicas da Faculdade de Medicina da Universidade de Sao Paulo. 2013;68 Suppl 1:81–8.
8. Moriya K, Mitsui T, Tanaka H, Nakamura M, Nonomura K. Long-term outcome of pituitary-gonadal axis and gonadal growth in patients with hypospadias at puberty. J Urol. 2010;184:1610–4.
9. Ediati A, Juniarto AZ, Birnie E, Drop SLS, Faradz SMH, Dessens AB. Body image and sexuality in Indonesian adults with a disorder of sex development (DSD). J Sex Res. 2013;52:15–29.
10. Ediati A, Faradz SMH, Juniarto AZ, van der Ende J, Drop SLS, Dessens AB. Emotional and behavioral problems in late-identified Indonesian patients with disorders of sex development. J Psychosom Res. 2015;79:76–84. Elsevier Inc.
11. Izumi Y, Suzuki E, Kanzaki S, Yatsuga S, Kinjo S, Igarashi M, et al. Genome-wide copy number analysis and systematic mutation screening in 58 patients with hypogonadotropic hypogonadism. Fertil Steril. 2014;102: 1130–3.
12. Raivio T, Sidis Y, Plummer L, Chen H, Ma J, Mukherjee A, et al. Impaired fibroblast growth factor receptor 1 signaling as a cause of normosmic idiopathic hypogonadotropic hypogonadism. J Clin Endocrinol Metab. 2009;94:4380–90.
13. Miraoui H, Dwyer AA, Sykiotis GP, Plummer L, Chung W, Feng B, et al. Mutations in FGF17, IL17RD, DUSP6, SPRY4, and FLRT3 are identified in individuals with congenital hypogonadotropic hypogonadism. Am J Hum Genet. 2013;92:725–43.
14. Juniarto Z, van der Zwan YG, Santosa A, Ariani MD, Eggers S, Hersmus R, et al. Hormonal evaluation in relation to phenotype and genotype in 286 patients with a disorder of sex development from Indonesia. Clin Endocrinol (Oxf). 2016;n/a–n/a.
15. Soldin OP, Hoffman EG, Waring MA, Soldin SJ. Pediatric reference intervals for FSH, LH, estradiol, T3, free T3, cortisol, and growth hormone on the DPC IMMULITE 1000. Clin Chim Acta. 2005;355:205–10.
16. Miller SA, Dykes DD, Polesky HF. A simple salting out procedure for extracting DNA from human nucleated cells. Nucleic Acids Res. 1988;16: 1215. Oxford University Press.
17. Eggers S, Sadedin S, van den Bergen JA, Robevska G, Ohnesorg T, Hewitt J, et al. Disorders of sex development: insights from targeted gene sequencing of a large international patient cohort. Genome Biol. 2016;17: 243. BioMed Central.
18. Sadedin SP, Dashnow H, James PA, Bahlo M, Bauer DC, Lonie A, et al. Cpipe: a shared variant detection pipeline designed for diagnostic settings. Genome Med. 2015;7:68.
19. Venselaar H, Beek Te TAH, Kuipers RKP, Hekkelman ML, Vriend G. Protein structure analysis of mutations causing inheritable diseases. An e-Science

approach with life scientist friendly interfaces. BMC Bioinformatics. 2010;11: 548. BioMed Central.

20. Zhu J, Choa RE-Y, Guo MH, Plummer L, Buck C, Palmert MR, et al. A shared genetic basis for self-limited delayed puberty and idiopathic hypogonadotropic hypogonadism. J Clin Endocrinol Metab. 2015;100: E646–54. Endocrine Society Chevy Chase.

21. Cole LW, Sidis Y, Zhang C, Quinton R, Plummer L, Pignatelli D, et al. Mutations in prokineticin 2 and prokineticin receptor 2 genes in human gonadotrophin-releasing hormone deficiency: molecular genetics and clinical spectrum. J Clin Endocrinol Metab. 2008;93:3551–9.

22. Monnier C, Dodé C, Fabre L, Teixeira L, Labesse G, Pin J-P, et al. PROKR2 missense mutations associated with Kallmann syndrome impair receptor signalling activity. Hum Mol Genet. 2009;18:75–81. Oxford University Press.

23. Dodé C, Rondard P. PROK2/PROKR2 signaling and Kallmann syndrome. Front Endocrinol (Lausanne). 2013;4:19. Frontiers.

24. Fukami M, Iso M, Sato N, Igarashi M, Seo M, Kazukawa I, et al. Submicroscopic deletion involving the fibroblast growth factor receptor 1 gene in a patient with combined pituitary hormone deficiency. Endocr J. 2013;60:1013–20.

25. Costa-Barbosa FA, Balasubramanian R, Keefe KW, Shaw ND, Tassan Al N, Plummer L, et al. Prioritizing genetic testing in patients with Kallmann syndrome using clinical phenotypes. J Clin Endocrinol Metab. 2013;98:E943–53.

26. Vizeneux A, Hilfiger A, Bouligand J, Pouillot M, Brailly-Tabard S, Bashamboo A, et al. Congenital hypogonadotropic hypogonadism during childhood: presentation and genetic analyses in 46 boys. Veitia RA, editor. PLoS ONE. 2013;8:e77827. Public Library of Science.

27. Kurzrock EA, Delair S. Hypospadias and Kallmann's syndrome: distinction between morphogenesis and growth of the male phallus. J Pediatr Urol. 2006;2:515–7.

28. Knorr JR, Ragland RL, Brown RS, Gelber N. Kallmann syndrome: MR findings. AJNR Am J Neuroradiol. 1993;14:845–51.

29. Ponticelli C, Frosini P, Masi L. Kallmann's syndrome. Apropos of 2 personal cases. Acta Otorhinolaryngol Ital. 1991;11:603–8.

30. Bouty A, Ayers KL, Pask A, Heloury Y, Sinclair AH. The genetic and environmental factors underlying hypospadias. Sex Dev. 2015;9:239–59. Karger Publishers.

31. Strazzullo M, Matarazzo MR. Epigenetic effects of environmental chemicals on reproductive biology. Curr Drug Targets. 2016.

32. Bianca S, Li Volti G, Caruso-Nicoletti M, Ettore G, Barone P, Lupo L, et al. Elevated incidence of hypospadias in two sicilian towns where exposure to industrial and agricultural pollutants is high. Reprod Toxicol. 2003;17:539–45.

33. Xu L-F, Liang C-Z, Lipianskaya J, Chen X-G, Fan S, Zhang L, et al. Risk factors for hypospadias in China. Asian J Androl. 2014;16:778–81.

34. Kristensen P, Irgens LM, Andersen A, Bye AS, Sundheim L. Birth defects among offspring of Norwegian farmers, 1967-1991. Epidemiology. 1997;8:537–44.

35. Azevedo L, Mort M, Costa AC, Silva RM, Quelhas D, Amorim A, et al. Improving the in silico assessment of pathogenicity for compensated variants. Eur J Hum Genet. 2016;25:2–7.

36. Kim H-G, Ahn J-W, Kurth I, Ullmann R, Kim H-T, Kulharya A, et al. WDR11, a WD protein that interacts with transcription factor EMX1, is mutated in idiopathic hypogonadotropic hypogonadism and Kallmann syndrome. Am J Hum Genet. 2010;87:465–79.

37. Balasubramanian R, Choi J-H, Francescatto L, Willer J, Horton ER, Asimacopoulos EP, et al. Functionally compromised CHD7 alleles in patients with isolated GnRH deficiency. Proceedings of the National Academy of Sciences. National Acad Sci. 2014;111:17953–8.

In silico prioritization and further functional characterization of SPINK1 intronic variants

Wen-Bin Zou[1,2,3,4†], Hao Wu[1,2,3,4†], Arnaud Boulling[2,3], David N. Cooper[5], Zhao-Shen Li[1,4*], Zhuan Liao[1,4*], Jian-Min Chen[2,3,6*] and Claude Férec[2,3,6,7]

Abstract

Background: *SPINK1* (serine protease inhibitor, kazal-type, 1), which encodes human pancreatic secretory trypsin inhibitor, is one of the most extensively studied genes underlying chronic pancreatitis. Recently, based upon data from qualitative reverse transcription-PCR (RT-PCR) analyses of transfected HEK293T cells, we concluded that 24 studied *SPINK1* intronic variants were not of pathological significance, the sole exceptions being two canonical splice site variants (i.e., c.87 + 1G > A and c.194 + 2T > C). Herein, we employed the splicing prediction tools included within the Alamut software suite to prioritize the 'non-pathological' *SPINK1* intronic variants for further quantitative RT-PCR analysis.

Results: Although our results demonstrated the utility of in silico prediction in classifying and prioritizing intronic variants, we made two observations worth noting. First, we established that most of the prediction tools employed ignored the general rule that GC is a weaker donor splice site than the canonical GT site. This finding is potentially important because for a given disease gene, a GC variant donor splice site may be associated with a milder clinical manifestation. Second, the non-pathological c.194 + 13T > G variant was consistently predicted by different programs to generate a new and viable donor splice site, the prediction scores being comparable to those for the physiological c.194 + 2T donor splice site and even higher than those for the physiological c.87 + 1G donor splice site. We do however provide convincing in vitro evidence that the predicted donor splice site was not entirely spurious.

Conclusions: Our findings, taken together, serve to emphasize the importance of functional analysis in helping to establish or refute the pathogenicity of specific intronic variants.

Keywords: Aberrant mRNA transcripts, Chronic pancreatitis, In silico, Intronic variants, Non-canonical splice sites, Quantitative RT-PCR analysis, *SPINK1*, Splicing phenotype prediction

Background

SPINK1 (serine protease inhibitor, kazal-type, 1; OMIM #167790), which encodes pancreatic secretory inhibitor, is one of the most extensively studied genes underlying chronic pancreatitis [1]. Of the some 90 different nucleotide sequence variants listed in the *Chronic Pancreatitis Genetic Risk Factors Database* (http://www.pancreasgenetics.org/index.php; accessed 2 Jan 2017), 31 (34%) are intronic, a difficult category of sequence variant to ascertain in terms of their potential pathological relevance. Recently, using a 'maxigene' expression assay for which the full-length *SPINK1* genomic sequence (approximately 7 kb stretching from the translational initiation codon to the stop codon of the four-exon gene) was cloned into the pcDNA3.1/V5-His-TOPO vector, we analyzed the functional consequences of 24 *SPINK1* intronic variants for the mRNA splicing phenotype in transfected HEK293T cells by means of reverse transcription-PCR (RT-PCR) analysis. Based upon the observed splicing patterns, we concluded that none of the studied variants, apart from the two canonical splice site variants (i.e., c.87 + 1G > A and c.194 + 2T > C), were of pathological significance [2, 3].

However, upon reflection, we felt that whereas our conclusions regarding the two canonical splice site

---

* Correspondence: zhaoshenli@hotmail.com; liaozhuan@smmu.edu.cn; Jian-Min.Chen@univ-brest.fr
†Equal contributors
[1]Department of Gastroenterology, Changhai Hospital, Second Military Medical University, Shanghai, China
[2]Institut National de la Santé et de la Recherche Médicale (INSERM), U1078 Brest, France
Full list of author information is available at the end of the article

variants were solid, those relating to the other 22 intronic variants could have been too hasty. For example, some of these 22 intronic variants might have caused aberrant splicing albeit to a limited extent. However, such aberrantly spliced transcripts may have been rapidly degraded by the cellular mRNA quality control system as compared with the correctly spliced transcripts, resulting in a quantitative decrease in terms of the correctly spliced transcripts. To explore this possibility, we employed the commonly used in silico splicing prediction programs to prioritize these *SPINK1* intronic variants for further quantitative RT-PCR analysis.

Materials and methods
In silico splicing prediction
All 24 of the *SPINK1* intronic variants previously analyzed by the maxigene assay [2, 3] were re-examined in the context of in silico splicing prediction by means of Alamut® Visual version 2.7.1 (Interactive Biosoftware, Rouen, France) that included five prediction algorithms viz. SpliceSiteFinder-like, MaxEntScan, NNSPLICE, GeneSplicer, and Human Splicing Finder. We focused exclusively on the potential impact of the *SPINK1* intronic variants in terms of the disruption of known splice sites or the creation of new potential splice sites. We firstly used data derived from the two canonical splice site variants (i.e., c.87 + 1G > A and c.194 + 2T > C) as a first means to assess the performance of each of the five prediction programs. We then used the selected programs to prioritize variants for quantitative RT-PCR analysis.

Quantitative RT-PCR analysis of four prioritized variants in the context of a maxigene assay
Four variants (i.e., c.87 + 363A > G, c.194 + 13T > G, c.194 + 1504A > G, and c.195-323C > T) were prioritized for quantitative RT-PCR analysis. The wild-type and variant expression vectors harboring the corresponding full-length genomic *SPINK1* genes have been previously described [2, 3]. HEK293T cell culture, transfection, reverse transcription, and real-time quantitative RT-PCR analyses were performed as described [4], except that the primer pair used for amplifying the full-length target gene transcripts was changed to 5′-GGAGACC CAAGCTGGCTAGT-3′ (forward) and 5′-AGACC GAGGAGAGGGTTAGG-3′ (reverse); the forward and reverse primers are located within the pcDNA3.1 5′- and 3′-untranslated regions, respectively (i.e., the primer pair Q1 as described in [5]).

Further analyses of the c.194 + 13T > G variant in the context of a maxigene assay
Analysis of nonsense-mediated mRNA decay (NMD)
This analysis was performed as described in [4], with the *CEL-HYB1* and *CEL-HYB2a* expression vectors being

replaced by the aforementioned full-length *SPINK1* wild-type and c.194 + 13T > G variant gene expression vectors, respectively.

Identification of the in silico predicted aberrant transcript
The c.194 + 13T > G variant was consistently predicted by the selected four programs to create a putative splice donor site (see Results and discussion). An allele-specific forward primer, 5′-ATGTTTTGAAAATCGGTGAG TAC-3′, was used together with the reverse primer of the aforementioned primer pair Q1 [5] to amplify this predicted aberrant transcript. The PCR program comprised an initial denaturation at 95 °C for 15 min, followed by 35 cycles of denaturation at 94 °C for 45 s, annealing at 58 °C for 45 s, and extension at 72 °C for 1 min.

Estimation of the frequency of the aberrant transcripts relative to the wild-type transcripts
Using the aforementioned primer pair Q1, we performed RT-PCR to amplify the full-length transcripts prepared from the c.194 + 13T > G variant maxigene-transfected HEK293T cells treated with cycloheximide. After addition of 3′-A overhangs, the purified products were cloned into the pcDNA3.1/V5-His-TOPO vector (Invitrogen, Carlsbad, CA, USA). Transformation was performed using XL10-Gold Ultracompetent Cells (Stratagene, La Jolla, CA, USA), and transformed cells were spread onto LB agar plates with 50 mg/mL ampicillin and were incubated at 37 °C overnight [4]. Bacterial colonies were picked to be added into a 25 μl PCR mixture, which contained 12.5 μl HotStarTaq Master Mix (QIAGEN), 0.4 μM each primer (i.e., primer pair C1 in [5]; the forward and reverse primer sequences are located within the beginning and the end regions of the *SPINK1* coding sequence, respectively). The PCR had an initial denaturation step at 95 °C for 15 min, followed by 30 cycles of denaturation at 94 °C for 45 s, annealing at 58 °C for 45 s, and extension at 72 °C for 45 s. The samples showing the expected band on the gel were cleaned and sequenced using the forward primer.

Further analyses of the c.194 + 13T > G variant in the context of a minigene assay
Minigene construction and mutagenesis
A 567-bp fragment spanning the exon 3 of *SPINK1* as well as 230 bp flanking intronic sequences on both sides (Additional file 1) was amplified using primers 5′-CGGGCCCCCCCTCGAGTTTCAGAAGGGCCATAG GAC-3′ (forward) and 5′-TAGAACTAGTGGATCCC CAAGCTATCGACTATTTTGCTG-3′ (reverse). PCR was performed in a 25 μl reaction mixture containing 0.5 U KAPA HiFi HotStart DNA Polymerase, 5 μl 5× KAPA HiFi Buffer, 0.75 μl dNTP Mix, 20 ng expression vector containing the full-length wild-type *SPINK1*

genomic sequence [5], and 0.3 μM each primer. The PCR program comprised an initial denaturation at 95 °C for 5 min followed by 30 cycles of denaturation at 98 °C for 20 s, annealing at 66 °C for 15 s, extension at 72 °C for 15 s, and a final extension at 72 °C for 5 min. The PCR products were cloned into the Exontrap vector pET01 (MoBiTec) that was linearized by restriction enzymes *BamH*I and *Xho*I, using the In-Fusion® HD Cloning kit (Clontech) in accordance with the manufacturer's instructions. The resulting expression vector was termed the wild-type *SPINK1* exon 3 minigene.

The c.194 + 13T > G variant was introduced into the wild-type *SPINK1* exon 3 minigene as previously described for introducing the same variant into the wild-type *SPINK1* maxigene vector [2], except that the extension time was reduced from 13 to 5 min. The successful introduction of the variant was confirmed by DNA sequencing using primers 5′-GTAGCTGCCAGGAAG GAGTG-3′ (forward) and 5′-GGCCTCCAAAACCTA CACAT-3′ (reverse).

Qualitative and quantitative RT-PCR analyses

HEK293T cell culture, transfection, and reverse transcription were performed as previously described [2]. Qualitative RT-PCR was performed in a 25 μl mixture containing 12.5 μl HotStarTaq Master Mix (QIAGEN), 0.4 μM each primer (5′-GAGGGATCCGCTTCCTGGCCC-3′ (forward) and 5′-CTCCCGGGCCACCTCCAGTGCC-3′ (reverse)), and 1 μl cDNA. The program had an initial denaturation at 95 °C for 15 min followed by 30 cycles of denaturation at 94 °C for 45 s, annealing at 58 °C for 45 s, and extension at 72 °C for 2 min. PCR products were cleaned by ExoSAP-IT (Affymetrix) and were sequenced using a BigDye Terminator v1.1 Cycle Sequencing Kit (Applied Biosystems).

Quantitative RT-PCR analysis was done as previously described [4], except that the primers used for target gene amplification were changed to those used for the aforementioned qualitative RT-PCR analysis.

Results and discussion

The increasingly routine use of whole genome sequencing has led to the discovery of an increasing number of rare variants, particularly intronic variants, owing to the relatively large size of intronic sequences as compared to the exonic sequences of protein-coding genes. It is, in practice, generally unrealistic to functionally analyze the large number of intronic variants detected in protein-coding genes. In silico prediction is therefore commonly used both to classify and prioritize intronic variants for further functional analysis [6, 7]. Herein, we adopted this approach to prioritize *SPINK1* intronic variants, which had been previously classified as 'non-pathological' in accordance with data obtained from qualitative RT-PCR

analyses of transfected HEK293T cells [2, 3], for further quantitative analysis. To this end, we predicted the impact on mRNA splicing of all 24 previously functionally analyzed *SPINK1* intronic variants by means of the widely used Alamut software suite, focusing on their potential disruption or creation of splice sites in accordance with previous studies [7, 8]. The corresponding score changes for each variant, predicted by each of the five prediction programs included within Alamut, are summarized in Additional file **2**.

Using data from two canonical splice site variants as a first means to assess the relative performance of the five splicing prediction programs

There are currently no consensus guidelines as to which prediction tools should be used or how the results of predictions should be interpreted [9]. Importantly, of the 24 *SPINK1* intronic variants studied here, c.87 + 1G > A and c.194 + 2T > C affected canonical splice donor splice sites and were the only tested variants that had been previously shown to result in aberrant splicing using a maxigene assay [2, 3]. We therefore used data from these two variants as a means to assess the relative performance of the five splicing prediction programs included within the Alamut software suite.

GeneSplicer was excluded from consideration owing to its poor performance in relation to the two corresponding wild-type alleles

We first assessed the relative performance of the five splicing prediction programs by evaluating their prediction scores with respect to the corresponding wild-type alleles of the two canonical splice site variants. SpliceSiteFinder-like, MaxEntScan, NNSPLICE, and Human Splicing Finder all yielded scores that were ≥79.8% of their respective maxima. However, GeneSplicer only yielded a score of <27% of its maximum (Additional file 3) and was therefore excluded from further consideration.

A particular observation with respect to the c.194 + 2T > C variant

We then correlated the prediction scores from the four selected programs and the known splicing phenotypes in relation to the two canonical splice site variant alleles. In the context of the c.87 + 1G > A variant allele, SpliceSiteFinder-like, MaxEntScan, NNSPLICE, and Human Splicing Finder all predicted a score of zero (Table 1). This prediction was consistent with the complete exon 2 skipping that was observed in the maxigene assay [2].

A particular observation was made in the context of the c.194 + 2T > C variant. This variant was shown to lead to two distinct transcripts, one wild-type, the other

Table 1 In vitro observed and in silico predicted mRNA splicing phenotypes associated with the two canonical splice site variants and four intronic variants prioritized for quantitative RT-PCR analysis

Intron	SPINK1 variant	SpliceSiteFinder-like (0–100)	MaxEntScan (0–12)	NNSPLICE (0–1)	Human Splicing Finder (0–100)	In vitro observed mRNA splicing phenotype[a]
Canonical splice donor site variants						
2	c.87 + 1G > A	dss 79.8 → 0	dss 8.3 → 0	dss 0.9 → 0	dss 84.1 → 0	Complete exon 2 skipping
3	c.194 + 2T > C	dss 82.6 → 72.3	dss 11.1 → 0	dss 1.0 → 0	dss 92.1 → 0	Partial exon 3 skipping
Variants prioritized for quantitative RT-PCR analysis						
2	c.87 + 363A > G	–	–	–	dss 0 → 65.5	Normal
					ass 0 → 83.3	
3	c.194 + 13T > G	dss 0 → 82.0	dss 0 → 9.5	dss 0 → 0.9	dss 0 → 86.9	Normal
3	c.194 + 1504A > G	dss 0 → 77.2	–	dss 0 → 0.7	dss 0 → 83.2	Normal
3	c.195-323C > T	–	dss 0 → 6.3	dss 0 → 0.7	dss 0 → 75.1	Normal

Abbreviations: dss donor splice site, *ass* acceptor splice site
[a]In accordance with Zou et al. [2, 3]

lacking exon 3, the aberrant transcript being expressed at a much higher level than the wild-type transcript in our maxigene assay [2]. These results not only concur with those obtained from the analysis of mRNA derived from stomach tissue (in which SPINK1 is also abundantly expressed) from a c.194 + 2T > C homozygote [10] but are also consistent with the general rule that GC is a lesser frequently used and weaker splice donor site as compared to the canonical GT site [11, 12]. However, only SpliceSiteFinder-like predicted a consistent reduced score (from 82.6 to 72.3) whilst MaxEntScan, NNSPLICE, and Human Splicing Finder all predicted a score of zero (Table 1).

Taken together, in the context of the two canonical splice site variants, a good correlation was noted between the predictions of SpliceSiteFinder-like, MaxEntScan, NNSPLICE, and Human Splicing Finder and the in vitro maxigene-obtained functional results, although only SpliceSiteFinder-like yielded a perfect correlation in both cases.

In silico prioritization of the 22 'non-pathological' variants for further quantitative analysis
We classified the 22 empirically tested 'non-pathological' SPINK1 intronic variants into three categories in accordance with the predictions by SpliceSiteFinder-like, MaxEntScan, NNSPLICE, and Human Splicing Finder (Additional file 2).

Category 1 comprised 13 variants, none of which were predicted to affect splicing by any of the four programs. These variants were excluded from further functional analysis.

Category 2 comprised three variants that were only predicted (and only by Human Splicing Finder) to disrupt a putative splice acceptor or donor site (i.e., c.88-352A > G, c.194 + 90A > T, and c.194 + 184 T > A). These predictions were clearly inappropriate because (i)

the SPINK1 gene is not known to have alternative transcripts and (ii) we did not observe any alternative transcripts from the wild-type SPINK1 gene in the maxigene assay. They were therefore also excluded from further functional analysis.

Category 3 comprised 6 variants, each being predicted by at least one program to create a new donor or acceptor splice site. Of these, c.194 + 13T > G was predicted by all four programs to create a potential donor splice site and both c.194 + 1504A > G and c.195-323C > T were predicted by three of the four programs to create a potential donor splice site. These three variants were selected for further functional analysis (Table 1).

The remaining three category 3 variants were all predicted by one and the same program (i.e., Human Splicing Finder) to create a new donor or acceptor splice site (i.e., c.87 + 363A > G, c.195-1570C > A, and c.195-1399G > A). Of these, we selected that which had the highest predicted score (i.e., 83.3 of the maximum score of 100; c.87 + 363A > G), for further analysis (Table 1).

All four prioritized variants were not found to significantly reduce mRNA expression by quantitative RT-PCR analysis
We performed quantitative RT-PCR analysis of the HEK293T cells co-transfected with each of the expression vectors harboring the respective full-length variant SPINK1 genes and the reference PGL3-GP2 plasmid [4, 5]. We did not however observe statistically significant differences in mRNA expression level between any of the four variants and the wild-type (Fig. 1).

Further analyses of the c.194 + 13T > G variant
Of the four variants subjected to quantitative RT-PCR analysis, c.194 + 13T > G was the only one that was

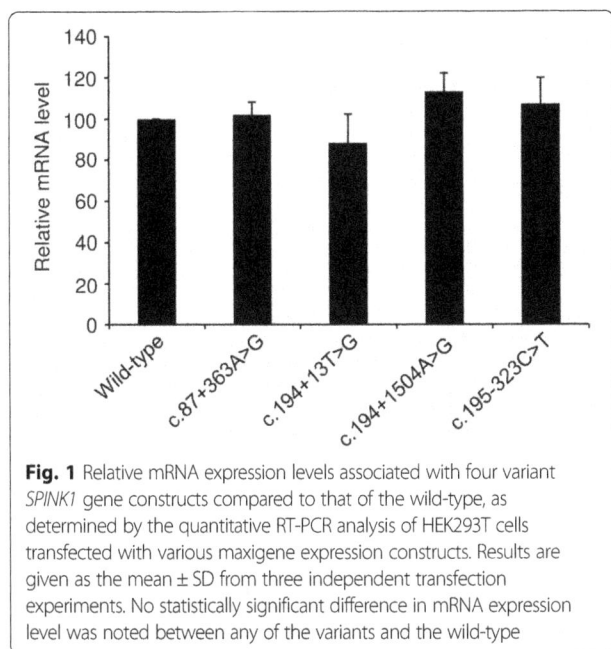

Fig. 1 Relative mRNA expression levels associated with four variant *SPINK1* gene constructs compared to that of the wild-type, as determined by the quantitative RT-PCR analysis of HEK293T cells transfected with various maxigene expression constructs. Results are given as the mean ± SD from three independent transfection experiments. No statistically significant difference in mRNA expression level was noted between any of the variants and the wild-type

consistently predicted by the SpliceSiteFinder-like, MaxEntScan, NNSPLICE, and Human Splicing Finder programs to create a potential donor splice site, with each program predicting a relatively high score (Table 1). Indeed, the predicted scores for this variant were even higher than, or at least equal to, the corresponding ones for the physiological c.87 + 1G donor splice site (Table 1). Moreover, the predicted scores for this variant were comparable to those for the physiological c.194 + 2T donor splice site (Table 1), the two sites being separated by only 10 bp.

The use of the predicted splice donor site would lead to the generation of an aberrant transcript containing a premature stop codon (Fig. 2a). Such a transcript might be subject to significant degradation by NMD [13], leading to a reduced mRNA expression level. In this regard, the aforementioned quantitative RT-PCR analyses hinted at a possible reduced level of expression of the correctly spliced transcripts from the c.194 + 13T > G variant maxigene (Fig. 1). To clarify this issue, we performed further quantitative RT-PCR analysis of the c.194 + 13T > G variant maxigene-transfected HEK293T cells with or without treatment with cycloheximide (a known NMD inhibitor [14]) as previously described [4], but no statistically significant changes were observed (Additional file 4). Additionally, we analyzed the impact on splicing of the c.194 + 13T > G variant in the context of a minigene assay. We found only a single band whose size was similar to that of the wild-type (Additional file 5a) and whose identity to the wild-type sequence was confirmed by sequencing the RT-PCR product. We further performed quantitative RT-PCR analysis of the HEK293T cells co-

transfected with the c.194 + 13T > G variant-containing minigene expression vector and the PGL3-GP2 plasmid. We did not observe statistically significant differences in mRNA expression level between the c.194 + 13T > G variant and the wild-type (Additional file 5b).

We then speculated that such an aberrant transcript might exist but at such an extremely low level as compared to the correctly spliced transcript that it would be beyond the detection limit of quantitative RT-PCR analysis. We therefore designed an allele-specific forward primer (rightward blue arrow in Fig. 2a) in an attempt to detect this aberrant transcript if it were to exist. Use of this forward primer and a reverse primer located within the pcDNA3.1 3'-untranslated region succeeded in detecting a specific RT-PCR product from the c.194 + 13T > G variant maxigene-transfected HEK293T cells (both with and without cycloheximide treatment) but not in the wild-type *SPINK1* maxigene-transfected HEK293T cells (Fig. 2b). Sequencing of this specific RT-PCR product confirmed the use of predicted novel splice donor site (Fig. 2c).

To obtain an approximate estimate of the expression level of the aberrant transcript relative to the correctly spliced transcript, we then performed colony PCRs followed by sequencing of the full-length transcripts amplified from the c.194 + 13T > G variant maxigene-transfected HEK293T cells treated with cycloheximide. Sequencing of 100 PCR products of expected size revealed only wild-type transcript. [note that the wild-type and aberrant transcripts are indistinguishable by gel analysis owing to their length difference of only 12 bp; see Fig. 2a.] This suggested that the c.194 + 13T > G variant resulted in less than 1% of aberrantly spliced transcripts relative to the amount of wild-type transcript.

Conclusions

In silico prioritization and subsequent quantitative RT-PCR analyses of selected *SPINK1* intronic variants for further functional characterization in a maxigene assay supported our previous classification of 24 *SPINK1* intronic variants as having pathological relevance (or not) in chronic pancreatitis [2, 3]. As in many studies, our results demonstrated the utility of in silico prediction in classifying and prioritizing intronic variants. However, we made two observations worth noting during this study. First, we found that most of the prediction programs included within the commonly used Alamut software suite ignore the general rule that GC is a weaker donor splice site as compared with the canonical GT donor splice site. This finding serves to remind us of a key point in medical genetics: in a given disease gene, a C introduced into the second position of a canonical GT donor splice site may have a milder clinical manifestation than a G or A.

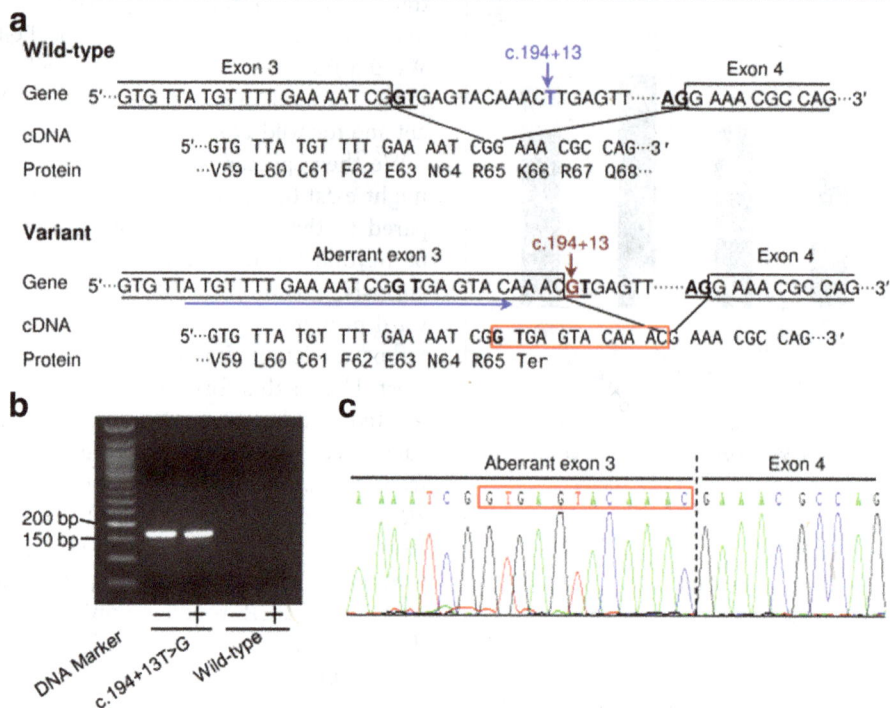

Fig. 2 Confirmation of the Alamut-predicted creation of a new splice donor site by the *SPINK1* c.194 + 13T > G variant. **a** Schemas for the splicing of intron 3 with respect to the wild-type and variant alleles of c.194 + 13T > G. The splice donor (GT) and splice acceptor (AG) signals potentiating the normal and aberrant splicing of intron 3 are highlighted in *bold* and *underlined*. The *rightward pointing blue arrow* indicates the forward allele-specific primer designed to amplify the predicted aberrant transcript (as shown in **b**). The 12 bp intronic sequence inappropriately included within the predicted aberrant transcript is indicated by a *red box*. The amino acid sequences of the wild-type and predicted mutant proteins are also shown. **b** PCR identification of the aberrant transcripts expressed from the HEK293T cells transfected with the c.194 + 13T > G variant-containing maxigene expression construct. The primers used for amplification were the forward allele-specific primer as illustrated in (**a**) and a reverse primer located within the 3′ untranslated region of the expression vector. No PCR products were identified in cells transfected with the wild-type maxigene expression vector. *Plus* and *minus* symbols refer to cells treated with and without cycloheximide, respectively. **c** Sequence of the c.194 + 13T > G/+ PCR products as illustrated in (**b**). The 12 bp intronic sequence included within the aberrant transcript is indicated by a *red box*

Second, the non-pathological c.194 + 13T > G variant was consistently predicted by the selected four programs to generate a potential donor splice site; the prediction scores being even higher than the physiological c.87 + 1G splice site. However, by means of allele-specific PCR, we provided convincing in vitro evidence that the predicted donor splice site was not entirely spurious (Fig. 2). These findings, taken together, serve to emphasize the importance of functional analysis in helping to establish or refute the pathogenicity of certain intronic variants, an issue of increasing importance in the new age of precision medicine [15, 16]. This notwithstanding, it should be pointed out that in the context of the c.194 + 13T > G variant, it would be desirable to investigate its effect in the corresponding carrier's pancreatic tissue. However, obtaining such a tissue sample would be extremely difficult particularly given that the c.194 + 13T > G variant has so far been reported only once [17].

Additional files

Additional file 1: Figure S1. The *SPINK1* sequence cloned into the Exontrap vector pET01.

Additional file 2: Table S1. In vitro observed and in silico predicted mRNA splicing phenotypes associated with the 24 *SPINK1* intronic variants under study.

Additional file 3: Figure S2. Alamut-predicted impact of the *SPINK1* c.87 + 1G > 1, c.194 + 2T > C, and c.194 + 13T > G variants on the disruption or creation of splice sites.

Additional file 4: Figure S3. Relative mRNA expression levels of the *SPINK1* c.194 + 13T > G variant-containing maxigene in transfected HEK293T cells in the presence (gray) and absence (black) of cycloheximide as determined by quantitative RT-PCR analysis.

Additional file 5: Figure S4. Further analyses of the *SPINK1* c.194 + 13T > G variant in a minigene assay.

Abbreviations
ass: Acceptor splice site; dss: Donor splice site; NMD: Nonsense-mediated mRNA decay; RT-PCR: Reverse transcription-PCR

Acknowledgements
Not applicable.

Funding
WBZ was a joint PhD student between Changhai Hospital and INSERM U1078 and received a 1 year scholarship (2015) from the China Scholarship Council (No. 201403170271). HW is a joint PhD student between the Changhai Hospital and INSERM U1078 who was in receipt of a 1 year scholarship from the China Scholarship Council (No. 201503170355). Support for this study came from the Youth Foundation of Shanghai Changhai Hospital (Grant No. CH201707; to WBZ), the National Natural Science Foundation of China (Grant Nos. 81470884 and 81422010; to ZL), the Shuguang Program of Shanghai Education Development Foundation and Shanghai Municipal Education Commission (Grant No. 15SG33; to ZL), and the Chang Jiang Scholars Program of Ministry of Education, People's Republic of China (Grant No. Q2015190; to ZL). Work in the Brest Genetics Laboratory was supported by the Conseil Régional de Bretagne, the Association des Pancréatites Chroniques Héréditaires, the Association de Transfusion Sanguine et de Biogénétique Gaetan Saleun, and the Institut National de la Santé et de la Recherche Médicale (INSERM), France.

Authors' contributions
JMC, ZL, ZSL, and CF conceived and designed the experiments. WBZ, HW, and AB performed the experiments. All authors analyzed the data. DNC significantly revised the manuscript. ZSL and CF contributed reagents/materials/analysis tools. JMC wrote the paper. All authors read and approved the final manuscript.

Competing interests
The authors declare that they have no competing interests.

Consent for publication
Not applicable.

Author details
[1]Department of Gastroenterology, Changhai Hospital, Second Military Medical University, Shanghai, China. [2]Institut National de la Santé et de la Recherche Médicale (INSERM), U1078 Brest, France. [3]Etablissement Français du Sang (EFS)–Bretagne, Brest, France. [4]Shanghai Institute of Pancreatic Diseases, Shanghai, China. [5]Institute of Medical Genetics, School of Medicine, Cardiff University, Cardiff, UK. [6]Faculté de Médecine et des Sciences de la Santé, Université de Bretagne Occidentale (UBO), Brest, France. [7]Laboratoire de Génétique Moléculaire et d'Histocompatibilité, Centre Hospitalier Universitaire (CHU) Brest, Hôpital Morvan, Brest, France.

References
1. Witt H, Luck W, Hennies HC, Classen M, Kage A, Lass U, et al. Mutations in the gene encoding the serine protease inhibitor, Kazal type 1 are associated with chronic pancreatitis. Nat Genet. 2000;25:213–6.
2. Zou WB, Boulling A, Masson E, Cooper DN, Liao Z, Li ZS, et al. Clarifying the clinical relevance of SPINK1 intronic variants in chronic pancreatitis. Gut. 2016;65:884–6.
3. Zou WB, Masson E, Boulling A, Cooper DN, Li ZS, Liao Z, et al. Digging deeper into the intronic sequences of the SPINK1 gene. Gut. 2016;65:1055–6.
4. Zou WB, Boulling A, Masamune A, Issarapu P, Masson E, Wu H, et al. No association between CEL-HYB hybrid allele and chronic pancreatitis in Asian populations. Gastroenterology. 2016;150:1558–60.e5.
5. Boulling A, Chen JM, Callebaut I, Férec C. Is the SPINK1 p.Asn34Ser missense mutation per se the true culprit within its associated haplotype? WebmedCentral GENETICS, vol. 3. 2012. p. WMC003084. Available at: https://www.webmedcentral.com/wmcpdf/Article_WMC003084.pdf.
6. Shirley BC, Mucaki EJ, Whitehead T, Costea PI, Akan P, Rogan PK. Interpretation, stratification and evidence for sequence variants affecting mRNA splicing in complete human genome sequences. Genomics Proteomics Bioinformatics. 2013;11:77–85.
7. Jian X, Boerwinkle E, Liu X. In silico prediction of splice-altering single nucleotide variants in the human genome. Nucleic Acids Res. 2014;42:13534–44.
8. Frisso G, Detta N, Coppola P, Mazzaccara C, Pricolo MR, D'Onofrio A, et al. Functional studies and in silico analyses to evaluate non-coding variants in inherited cardiomyopathies. Int J Mol Sci. 2016;17:1883.
9. Tang R, Prosser DO, Love DR. Evaluation of bioinformatic programmes for the analysis of variants within splice site consensus regions. Adv Bioinformatics. 2016;2016:5614058.
10. Kume K, Masamune A, Kikuta K, Shimosegawa T. [-215G>A; IVS3+2T>C] mutation in the SPINK1 gene causes exon 3 skipping and loss of the trypsin binding site. Gut. 2006;55:1214.
11. Churbanov A, Winters-Hilt S, Koonin EV, Rogozin IB. Accumulation of GC donor splice signals in mammals. Biol Direct. 2008;3:30.
12. Thanaraj TA, Clark F. Human GC-AG alternative intron isoforms with weak donor sites show enhanced consensus at acceptor exon positions. Nucleic Acids Res. 2001;29:2581–93.
13. Karam R, Wengrod J, Gardner LB, Wilkinson MF. Regulation of nonsense-mediated mRNA decay: implications for physiology and disease. Biochim Biophys Acta. 1829;2013:624–33.
14. Pereverzev AP, Gurskaya NG, Ermakova GV, Kudryavtseva EI, Markina NM, Kotlobay AA, et al. Method for quantitative analysis of nonsense-mediated mRNA decay at the single cell level. Sci Rep. 2015;5:7729.
15. MacArthur DG, Manolio TA, Dimmock DP, Rehm HL, Shendure J, Abecasis GR, et al. Guidelines for investigating causality of sequence variants in human disease. Nature. 2014;508:469–76.
16. Aronson SJ, Rehm HL. Building the foundation for genomics in precision medicine. Nature. 2015;526:336–42.
17. Gomez-Lira M, Bonamini D, Castellani C, Unis L, Cavallini G, Assael BM, et al. Mutations in the SPINK1 gene in idiopathic pancreatitis Italian patients. Eur J Hum Genet. 2003;11:543–6.

A pipeline combining multiple strategies for prioritizing heterozygous variants for the identification of candidate genes in exome datasets

Teresa Requena[1*], Alvaro Gallego-Martinez[1] and Jose A. Lopez-Escamez[1,2]

Abstract

Background: The identification of disease-causing variants in autosomal dominant diseases using exome-sequencing data remains a difficult task in small pedigrees. We combined several strategies to improve filtering and prioritizing of heterozygous variants using exome-sequencing datasets in familial Meniere disease: an in-house Pathogenic Variant (PAVAR) score, the Variant Annotation Analysis and Search Tool (VAAST-Phevor), Exomiser-v2, CADD, and FATHMM. We also validated the method by a benchmarking procedure including causal mutations in synthetic exome datasets.

Results: PAVAR and VAAST were able to select the same sets of candidate variants independently of the studied disease. In contrast, Exomiser V2 and VAAST-Phevor had a variable correlation depending on the phenotypic information available for the disease on each family. Nevertheless, all the selected diseases ranked a limited number of concordant variants in the top 10 ranking, using the three systems or other combined algorithm such as CADD or FATHMM. Benchmarking analyses confirmed that the combination of systems with different approaches improves the prediction of candidate variants compared with the use of a single method. The overall efficiency of combined tools ranges between 68 and 71% in the top 10 ranked variants.

Conclusions: Our pipeline prioritizes a short list of heterozygous variants in exome datasets based on the top 10 concordant variants combining multiple systems.

Keywords: Exome sequencing, Variants filtering, Phenotype, Autosomal dominant diseases, Human phenotype ontology, Hearing loss, Meniere disease

Background

Whole-exome sequencing (WES) has become the preferred tool to discover new variants for the diagnosis of genetic diseases, since the protein-coding regions and their boundaries represent only 1.5–2% of the human genome and they accumulate most of the disease-causing mutations: missense and protein-truncating variants (frameshift, splice-acceptor, splice-donor, and nonsense variants) [1, 2]. On average, 45,000 single-nucleotide variants (SNVs) are obtained by WES, 39% are located in coding regions, while 4% are in untranslated regions (UTR),

and 56% are in intronic regions near to UTR. In addition, ~90% of SNVs obtained by WES are described in the dbSNP138 based in reference genome (GRCh37 hg19) [3]. However, novel and rare variants (minor allelic frequency (MAF) ≤0.01) identified by WES cannot be interpreted as pathogenic only with this information, and causality must be validated by replication in different individuals with the same phenotype and by functional studies in an appropriate cellular or animal model for each disease. Nevertheless, WES has already shown the efficiency to identify potential disease-causing variants in monogenic diseases [4, 5]. Particularly, WES has been successfully used in rare Mendelian disorders, since most of the disease-causing variants are located in protein-coding regions [5]. Recently, WES studies have been also extended for diagnosis

* Correspondence: mariateresa.requena@genyo.es
[1]Otology & Neurotology Group CTS495, Department of Genomic Medicine, GENYO - Centre for Genomics and Oncological Research – Pfizer/University of Granada/Junta de Andalucía, PTS, 18016 Granada, Spain

in oligogenic and complex genetic disorders [6–10] and for predicting disease progression [11, 12]. However, when the disease is poorly characterized at the molecular level, the filtering and prioritizing of WES datasets requires a more elaborated search strategy based not only in single variant effects on protein structure or evolutionary conservation but also upon the phenotype description and mathematical interaction models.

The high efficiency of WES data in Mendelian disorders is explained because most of the causal variants in recessive disorders are rare homozygous variants or compound heterozygous variants observed in familiar cases, which are not found in healthy relatives or individuals in the same population [13]. However, the situation is more complex with autosomal dominant (AD) disorders, where a single heterozygous de novo variant can affect the gene function and hundreds of candidate variants need to be filtered. So, an improved workflow to identify potential candidate variants involved in the disease is needed. Software package as MendelScan try to solve this providing a composite score improved with tissue expression data [14]. However, systemic disease or disease involving tissues with multiple cells types and low-quality gene expression data as the cochlea are not easy to analyze with this approach.

Hearing and vestibular disorders are the most common sensory deficits in humans. Hearing loss affect around 5.3% of the world population according to the World Health Organization. Non-syndromic autosomal dominant sensorineural hearing loss (AD-SNHL) remains a challenge for genetic diagnosis, and 33 genes and 60 loci have been involved according to Hereditary Hearing loss Homepage [15], with a considerable overlap in the phenotype and pleiotropy [16].

Meniere's disease (MD) is clinically defined by episodes of vertigo, tinnitus, and SNHL (MD, [MIM 156000]) [17], and it has a prevalence about 0.5–1/1000 individuals. Most of the patients are considered sporadic, although around 8–10% are familial cases in European descendent population [18–20]. Previous linkage studies in familial MD (FMD) have found candidate loci at 12p12.3 in a large Swedish family [21] and 5q14-15 in another German family [22], but the involved genes were not identified. Recently, WES analyses have identified *DTNA*, *FAM136A*, and *PRKCB* as potential causal genes in FMD [9, 10]. MD is a clinical syndrome, and its phenotype may overlap with different conditions including vestibular migraine or autoimmune inner ear disease [16]. In contrast, other AD diseases with a more precise phenotype, such as Centro Nuclear Myopathy (CNM), an inherited neuromuscular disorder characterized by congenital myopathy with a histopathological diagnosis (centrally placed nuclei on muscle biopsy), have a reduced number of causal variants.

The aim of this study is to develop a workflow to improve the filtering and prioritizing of candidate variants and genes in AD disorders by using WES data. We focus mainly in AD familial MD, a complex clinical scenario with clinical and genetic heterogeneity, few cases per family, incomplete penetrance, and variable expressivity [23, 24]. The pipeline proposed is based on (1) the combination of several tools to score variants according to its effect on protein structure and phylogenetic conservation, (2) the ranking according to available information on phenotype databases, (3) the comparison with two integrated systems (CADD and FATHMM), and (4) the use of un-affected relatives as control to filter candidate variants. The pipeline is summarized in Fig. 1.

Results

Six prioritizing systems were selected and combined in the pipeline to filter and rank rare variants in exome sequencing data. Two of them were based upon protein structure and sequence conservation across species: (a) an in-house Pathogenic Variant (PAVAR) score and (b) the Variant Annotation Analysis and Search Tool (VAAST) [25], and the other two prioritize according to the Phenotype Ontology information: (c) Exomiser v2 [26] and (d) VAAST-Phevor [27]. And finally two integrated tools were compared and added to the system CADD [28] and FATHMM [29].

Comparison of prioritizing strategies with FMD exome datasets

Table 1 shows the number of variants obtained for each FMD dataset with the six systems after filtering by several control datasets. We included the number of ranked variants with enough score to be prioritized, according to each of the six systems (thresholds are described in the "Material and methods" section). Mean values obtained for each family dataset were highly variable for each system, and they were dependent on the number of cases and controls available for each family.

We selected the top 10, 20, and 50 ranked variants from each prioritizing system and filtered them using the different control datasets (F, T-F, and T) to analyze the concordance between methods. Figure 2 shows the concordance between all systems. Although PAVAR score and VAAST use a different methodology, both systems show the highest concordance rate to filter and prioritize the candidate variants. Between 20 and 55% of ranked variants were matched in top 10, top 20, and top 50. However, the observed variability in the ranked variants between the different systems is caused by the control datasets (F, T-F, or T) used to filter the variants. In contrast, Exomiser v2 and VAAST-Phevor prioritized according to the Phenotype Ontology information (HPO term) [30], but the maximum correlation between

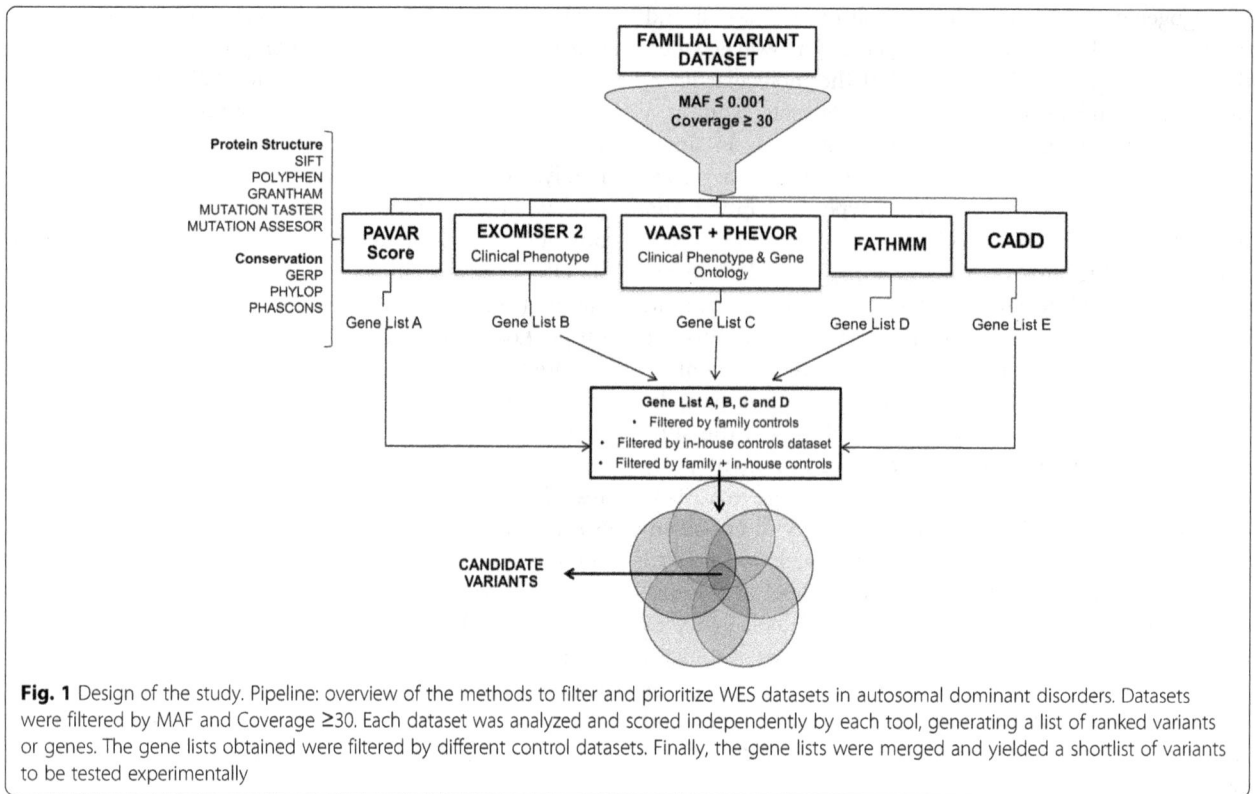

Fig. 1 Design of the study. Pipeline: overview of the methods to filter and prioritize WES datasets in autosomal dominant disorders. Datasets were filtered by MAF and Coverage ≥30. Each dataset was analyzed and scored independently by each tool, generating a list of ranked variants or genes. The gene lists obtained were filtered by different control datasets. Finally, the gene lists were merged and yielded a shortlist of variants to be tested experimentally

Table 1 Number of remaining variants per family dataset according to the filtering strategy

Family dataset	FMD exomes (N)	Control dataset (N)	PAVAR score ≥5 (N)	Exomiser score ≥1.46 × 10⁻⁵ (N)	VAAST (p value ≤1)	VAAST-Phevor (p value ≤1)	CADD score ≥15 (N)	FATHMM score ≤−1.5 (N)
1	3	F (1)	17 (134)	308 (1437)	40	39	15 (38)	7 (35)
		T-F (29)	15 (106)	78 (296)	48	44	18 (36)	7 (34)
		T (30)	10 (68)	42 (175)	27	27	12 (25)	5 (23)
2	2	F (3)	4 (58)	60 (270)	53	22	9 (18)	1 (14)
		T-F (27)	9 (73)	89 (369)	146	135	12 (28)	1 (25)
		T (30)	2 (34)	9 (39)	19	16	5 (13)	0 (11)
3	3	F (2)	9 (68)	151 (862)	23	23	9 (20)	1 (14)
		T-F (28)	13 (92)	67 (309)	38	38	17 (25)	5 (20)
		T (30)	6 (32)	24 (104)	16	16	7 (10)	1 (7)
4	3	F (0)	31 (283)	394 (2198)	54	46	34 (90)	4 (86)
		T (30)	4 (34)	20 (72)	19	17	5 (14)	1 (14)
5	3	F (3)	16 (83)	93 (391)	68	22	7 (20)	1 (15)
		T-F (27)	14 (113)	89 (430)	52	45	14 (35)	7 (28)
		T (30)	5 (36)	18 (67)	11	9	4 (9)	1 (6)
Mean (1–5)	21	F	15.4 ± 10.21 (125)	251.5 ± 143.83 (1032)	47 ± 16.95	30.4 ± 11.33	14.8 ± 9.96	2.8 ± 2.4
		T-F	12.75 ± 2.63 (96)	85 ± 28.66 (351)	71 ± 50.35	65.5 ± 46.44	13.5 ± 4.38	5.0 ± 2.44
		T	5.2 ± 2.97 (51)	31 ± 13.94 (155)	28.2 ± 5.81	5.81 ± 6.44	6.60 ± 2.87	1.6 ± 1.74

All variants with a MAF >0.001 were discarded. Setting for each software threshold is described in the "Material and methods" section

p values for VAAST and Phevor were not corrected since they were used as thresholds according to the user's guide

F family controls exome dataset, T-F in-house controls exome dataset without family control dataset, T in-house and family control datasets

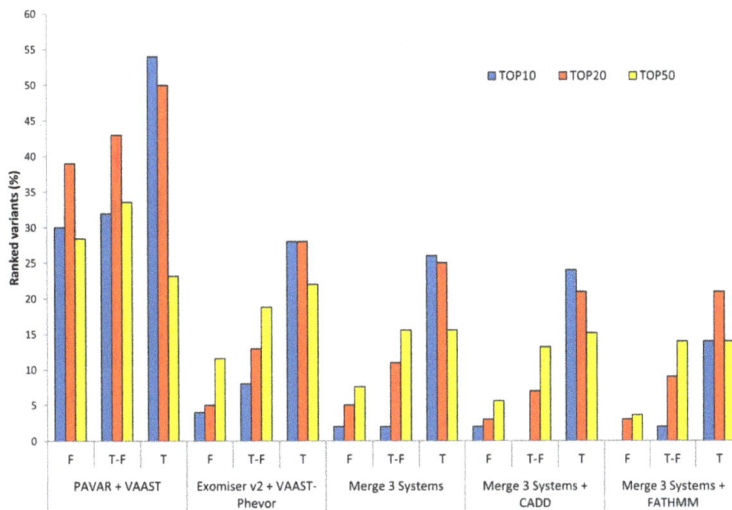

Fig. 2 Prioritized variants in FMD datasets. Percentage of the variants ranked and shared in top 10 (*blue*), 20 (*red*), and 50 (*yellow*) ranked variants by (a) PAVAR score and VAAST; (b) Exomiser v2 score and Phevor; (c) the combination of the three systems (PAVAR, Exomiser v2, VAAST, VAAST-Phevor); (d) the combination of the three systems (PAVAR, Exomiser v2, VAAST, VAAST-Phevor) and CADD; and (e) the combination of the three systems (PAVAR, Exomiser v2, VAAST, VAAST-Phevor) and FATHMM. F = family controls exome dataset, T-F = in-house controls exome dataset without family control datasets, and T = in-house and family control datasets

systems was 28% when the largest control dataset (T) was used to filter. Therefore, only the variants located in genes previously associated with the phenotype were matched by different systems. Consequently, the combinations of PAVAR, VAAST-Phevor, and Exomiser v2 only matched in few variants (2–26%), which were top ranked and highly related with MD HPO terms. A similar concordance was obtained between the combination of that three and other combined systems as CADD or FATHMM.

The maximum correlation between CADD and the merge of three systems was 24% in top 10, whereas for FATHMM was 21% in top 20. In both cases, this correlation was obtained after using the largest controls' dataset (T) to filter the variants.

Benchmark in exome datasets containing variants described in AD-SNHL and CNM genes

We compared the ability of these variant prioritizing tools to identify AD variants in small familial exome data files by a benchmarking procedure. Since the structure of the families as well as the number of cases and controls available for each pedigree could generate a bias in the benchmarking analyses, multiple families were tested.

Figure 3 shows the percentage of ranked variants in top 10, 20, and 50 by the six systems for both, hearing loss variants (Fig. 3a) and CNM variants (Fig. 3b). In top 10 and 20, the observed percentages were highly variable between each system, particularly depending on the control dataset used.

Next, we selected the top 10, 20, and 50 ranked variants from each prioritizing system and filtered them for

the different datasets (F, T-F, and T) to analyze the concordance between the different methods. Figure 4 illustrates a progressive increase of concordance between systems in the top 10, 20, and 50 ranked variants for both disorders. Exomiser v2 and VAAST-Phevor yielded higher correlations in the top 10 and 20, highlighting that both tools identify similar genes associated with the HPO term for a given phenotype. This pattern was more prominent in top 10 ranked variants for AD-SNHL datasets in the benchmarking, reaching a 50% of concordance (Fig. 4a), whereas in CNM datasets, only 34% of concordance was found (Fig. 4b). In contrast, low correlations were obtained between PAVAR score and VAAST (9–33%), mainly in the top 10 ranked, means that few variants are considered as candidates by both systems as real pathogenic variants. As a result, potentially pathogenic variants located in genes with HPO terms associated with the disease were shared by PAVAR, Exomiser v2, and VAAST-Phevor and tending to be ranked in the top 10.

A similar percentage was obtained when we add CADD to the combined system. However, the combination of multiple systems with CADD did not reduce the list of candidate variants in the top 10 ranking.

Next, 200 variants were randomly selected for each disease to build synthetic datasets. So, 42% for AD-SNHL and 25.5% CNM were previously described in HGDB as pathogenic (Additional file 1: Table S1 and S2). So, multiple logit regression models were performed to assess the accuracy to predict correctly candidate variants associated with each phenotype. The area under the curve

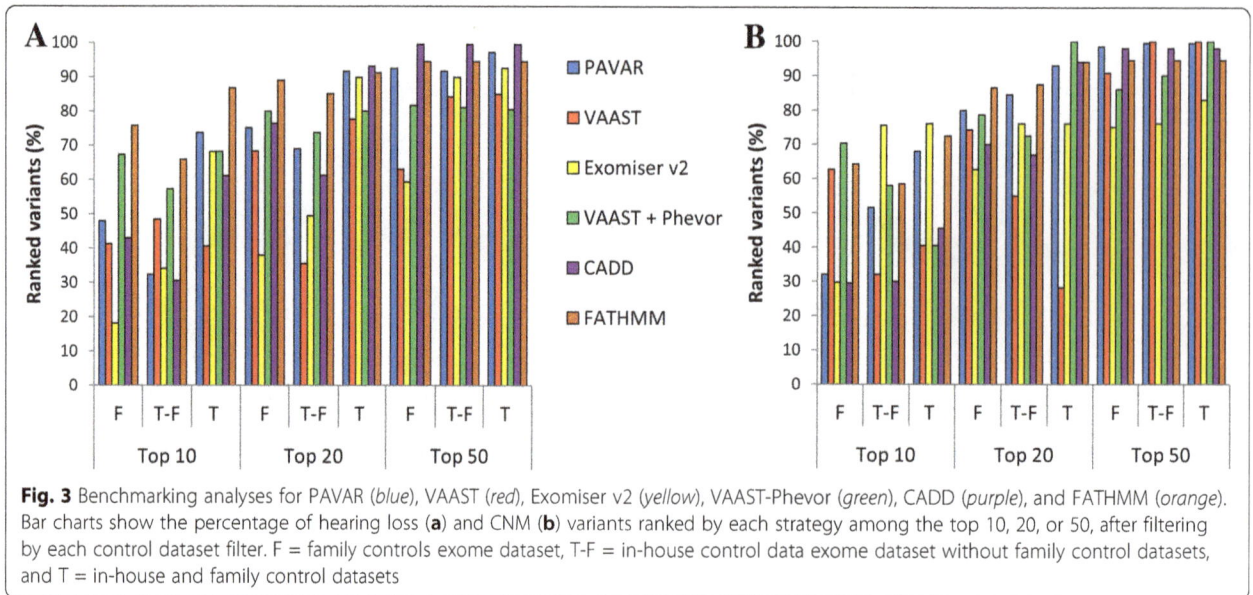

Fig. 3 Benchmarking analyses for PAVAR (*blue*), VAAST (*red*), Exomiser v2 (*yellow*), VAAST-Phevor (*green*), CADD (*purple*), and FATHMM (*orange*). Bar charts show the percentage of hearing loss (**a**) and CNM (**b**) variants ranked by each strategy among the top 10, 20, or 50, after filtering by each control dataset filter. F = family controls exome dataset, T-F = in-house control data exome dataset without family control datasets, and T = in-house and family control datasets

(AUC) for each system was calculated to assess the precision and accuracy to identify candidate variants for both diseases in several families (Additional file 1: Table S3). On average, the combination of PAVAR, Exomiser v2, VAAST-Phevor, CADD, and FATHMM predicts potentially pathogenic variants associated with the phenotype between 68 and 71% of times in top 10, for both diseases (Fig. 5a, b). These results were statistically significantly better than any single method (*p* values shown in Additional file 1: Table S3).

Discussion

The combination of linkage analysis and WES in large multicase pedigrees has shown a high effectiveness to

identify disease-causing variants in rare Mendelian disorders [4, 5]. However, small pedigrees with a few available cases are the most common clinical scenario and a challenge for the genetic diagnosis of dominant disorders, mainly those with overlapping phenotypes or incomplete penetrance such as AD-SNHL [31, 32], CNM [33], and MD [20]. Despite the increasing number of bioinformatics tools to analyze WES data [34, 35], the list of genes that must be experimentally validated for these diseases is too large.

The first issue to resolve for variant identification is the alignment of reads and variant calling algorithms. Current approaches have developed pipelines that combine tools to obtain consistent identification of variant

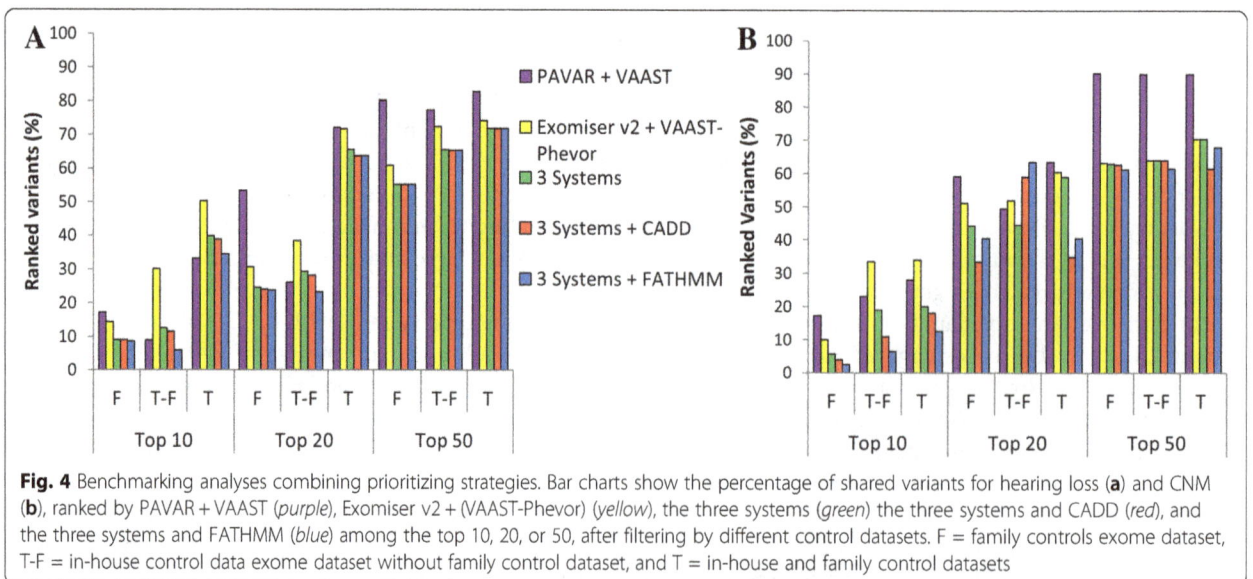

Fig. 4 Benchmarking analyses combining prioritizing strategies. Bar charts show the percentage of shared variants for hearing loss (**a**) and CNM (**b**), ranked by PAVAR + VAAST (*purple*), Exomiser v2 + (VAAST-Phevor) (*yellow*), the three systems (*green*) the three systems and CADD (*red*), and the three systems and FATHMM (*blue*) among the top 10, 20, or 50, after filtering by different control datasets. F = family controls exome dataset, T-F = in-house control data exome dataset without family control dataset, and T = in-house and family control datasets

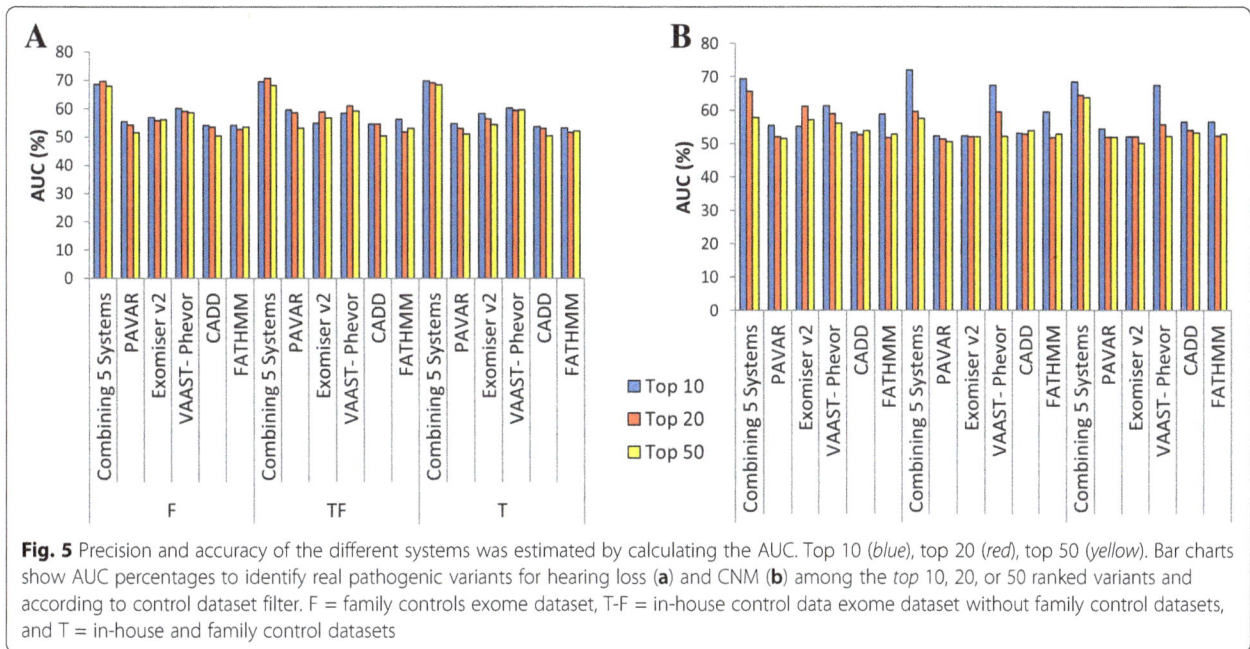

Fig. 5 Precision and accuracy of the different systems was estimated by calculating the AUC. Top 10 (*blue*), top 20 (*red*), top 50 (*yellow*). Bar charts show AUC percentages to identify real pathogenic variants for hearing loss (**a**) and CNM (**b**) among the *top* 10, 20, or 50 ranked variants and according to control dataset filter. F = family controls exome dataset, T-F = in-house control data exome dataset without family control datasets, and T = in-house and family control datasets

and facilitate the process [36, 37]. However, these pipelines do not provide functional annotation. Other pipelines go further, and they implement user-friendly graphic interface and include Annovar-based functional annotation [38]. However, our results show that the combination of multiple bioinformatics tools is a reliable strategy to reduce the list of candidate variants and to facilitate the identification of the disease-causing variants in small pedigrees. These results are consistent with previous studies designed to improve the yield of several prioritizing tools [39, 40].

The list of candidate variants generated by each system is usually too large to be validated experimentally (Table 1). So, the most common strategy is to filter by familiar controls to eliminate private familial variants and by controls' dataset from the same population to eliminate population-specific variants. However, the clinical evidence of incomplete penetrance or late age of onset of the disease should exclude the use of familial control datasets. Our results show that by combining five tools (PAVAR, Exomiser v2, VAAST-Phevor, CADD, FATHMM), the list of candidate variants is reduced and this facilitates the identification of potential disease-causing variants (Fig. 5).

Discrepancies between all the prioritization systems evaluated (PAVAR, VAAST, Exomiser v2, VAAST-Phevor, CADD, FATHMM) were found in the ranked results for all the diseases tested (Table 1 and Fig. 3). Consequently, systems based on the same criteria, protein structure, and sequence conservation or Phenotype Ontology information, were clustered to analyze the concordance between them in the top 10, 20, and 50

ranked variants. Although PAVAR and VAAST use a different methodology, both prioritize variants according to the intrinsic effect on the protein of the variants. Of note, MD, AD-SNHL, and CNM showed similar correlation scores between PAVAR and VAAST for top 10 and 20 ranked variants. Both systems were more concordant when in-house control datasets or the merge of in-house and family control datasets were used to filter. Although familial controls are important to filter private variants, a large control dataset of the same population is more effective to reduce the list of candidate variants list.

In contrast, the concordance between VAAST-Phevor and Exomiser v2 varies depending on the disease studied. Although both systems are based on phenotype, VAAST-Phevor has a balanced score between potential pathogenicity and the association with the phenotype whereas Exomiser v2 assigns more weight to the phenotype than the potential pathogenicity. Diseases with a well-characterized phenotype by several HPO terms or diseases with known involved genes show a high correlation between VAAST-Phevor and Exomiser v2, as our results confirm for AD-SNHL and CNM. However, since MD only has few HPO terms and no gene associated in public databases, our data show a reduced concordance. In particular, our results show that the correlation between both systems in well diseases with many HPO terms is twice than in disorders with limited phenotypic information such as diseases of the ear for all top 10, 20, and 50 ranked variants. Nevertheless, a high concordance between both systems does not indicate that those variants selected are really disease-causing variants. The degree of concordance between both systems only demonstrates that the candidate genes are associated with the phenotype, but not necessarily its pathogenicity.

Initially, our pipeline joins both approaches by the identification of variants ranked as potentially pathogenic by the PAVAR score and associated them with the phenotype by both Exomiser v2 and VAAST-Phevor. The combination of the three strategies gives few variants ranked in the top 10 or 20 and produces a short list of candidate variants to be validated experimentally [9, 10, 41]. In addition, other combined systems were added and the list was reduced. Logit regression models and benchmarking analyses show that the combination of PAVAR, Exomiser v2, VAAST-Phevor CADD, and FATHMM not only reduced the list of candidate variants to be validated; this combined approach is more efficient to predict potential diseases-causing variants than each system separately. This enhanced efficiency is observed independently of the type of control datasets used. Our results confirm previous studies showing that prioritizing tools have less ability to rank variants in disorders with no previously known candidate gene [42]. In addition, we demonstrate that the addition of more HPO terms improves the ranking of candidate genes. So, our pipeline allows to obtain a reduced list of variants when incomplete penetrance is found and familial control datasets cannot be used.

This combined strategy has a major limitation: a reduced phenotypic characterization of AD disorders (such as AD-SNHL or MD) will decrease the precision of the pipeline. So, a deep phenotyping and updating of HPO terms in major databases will improve the yield of the system. Although HPO project has been updated in 2017, ear diseases and, particularly, vestibular disorders still have a limited phenotype vocabulary and disease-phenotype annotations [43]. In addition, further improvements in the pipeline should be needed to include structural variants such as frameshift (insertions and deletions), synonymous variants, and copy number variants.

Conclusion

These results demonstrate that our pipeline combining multiple variant-prioritization algorithms is useful in small family-based analyses. We also showed that the model can reduce the number of variants in synthetic exome datasets with incomplete phenotypes without using familial controls. This approach will be useful when controls are not available or when incomplete penetrance is observed.

Material and methods

Patients

Four Spanish AD families with at least two patients with definite MD and a fifth family with monozygotic twins with MD, according to the diagnostic criteria of the Barany Society for familial MD [17], were selected for this study. The clinical phenotype and the pattern of inheritance in these families and their pedigrees were previously reported [10, 20, 41]. The number of asymptomatic relatives selected for WES in each family depended upon two criteria: (a) size and structure of the family, since some families showed patients with incomplete phenotype (i.e., SNHL without episodic vertigo), and (b) the availability to obtain samples from older asymptomatic relatives, which could be used as controls. All the procedures described were performed in accordance with the highest ethical standards on human experimentation, the Helsinki Declaration of 1975 and the EU regulations on biomedical research. In addition, this study was approved by the Review Board for Clinical Research of Instituto Biosanitario de Granada, and a written informed consent to donor biological samples was obtained from all subjects.

Whole exome sequencing (WES)

DNA was isolated from peripheral blood samples as previously described [9, 10] Exons and flanking intron regions were captured according to the methods previously described [9, 10]. Library products were sequenced with SOLiD 5500xl platform with Exact Call Chemistry and 200× of sequencing depth. A mean of 50–60 million of reads were obtained per sample. The quality of the reads was analyzed with SAMtools [44], MAQtools [45], and FastQC software (Babraham Bioinformatics), and shorter reads (<25) as well as all duplicate reads were deleted. The reads were aligned with the reference genome (GRCh37 hg19) with Bioscope™ (Applied Biosystems, Foster City, CA, USA) using the default settings. Results from Bioscope™ were filtered by depth >30 reads [46] and quality of the assigned genotype ≥100. This analysis identified SNVs, copy number variants, and frameshift variants (insertion and deletions). However, we only considered SNVs for this study.

Bioinformatics analysis

For each family, heterozygous SNVs found in all the affected cases with complete phenotype of the family were selected. The 1000 genome project [47], ExaAC database [48], and Exome Variant Server (EVS) were used to annotate the MAF and function for each variant (Additional file 1: Table S4). All SNVs were filtered by MAF. For MD and AD-SNHL, variants with MAF ≥0.001 were discarded, since MD has a prevalence of 10–225 cases/100,000 individuals [49, 50] and the low prevalence described for AD-SNHL [51]. For CNM, variants with MAF ≥0.0001 were also discarded, since CNM is considered as a rare disease with a very low prevalence (1/25,000 males).

The pipeline was designed using different strategies to filter and prioritize SNVs: (a) the calculation of a pathogenic variant (PAVAR) risk composite score; (b) Exomiser v2 software [26]; (c) VAAST annotation tool [25]; and (d) a combination of VAAST and Phevor tools [27]. However, Phevor returns the same results than VAAST, but ranked by phenotype. In addition, other composite algorithms were used CADD [28] and FATHMM [29]. So, the shared candidate variants were selected. All variants were considered as potentially pathogenic according to the ACMG Standards and Guidelines [52], and all digital resources used are listed in Additional file 1: Table S5.

In some AD diseases, incomplete penetrance was found; subsequently, familial controls could not be used to filter variants. Different control datasets collected for previous projects were used to evaluate the efficiency of our pipeline despite of the observed incomplete penetrance. F = family controls exome dataset, T-F = in-house control data exome dataset without familial control datasets, and T = in-house and family control datasets.

a) Pathogenic variant risk composite score (PAVAR score)

Functional annotation was used to prioritize SNVs, according to the effect on protein structure and phylogenetic conservation. Sequence conservation across species is a major criterium to assess the variant, and the number of compared species varies according to the tool. To estimate the risk of a SNV to become a pathogenic variant, we used a seven-point scoring system based upon open-access prediction bioinformatics tools. ANNOVAR and SeattleSeq Annotation tools were used to achieve the score of SIFT (Sort Intolerant from Tolerant) [53], PolyPhen2 (Polymorphism Phenotyping v2) [54], Grantham's Matrix [55], GERP++ (Genomic Evolutionary Rate Profiling) [56], Mutation taster [57], PhastCons, and PhyloP [58]. The threshold to consider each variant as pathogenic is described in Additional file 1: Table S6, according to the default settings suggested for each software developer. PAVAR score is calculated as the sum of the score obtained by seven systems. Each system adds one point if the variant is considered as potentially damaging and zero if it is benign. So, the higher the score is, the high the risk of pathogenicity for a given variant. PAVAR score cannot be calculated for nonsense variants, since protein structure tools cannot assign any value. Since nonsense variants can modify dramatically the sequence of the protein, they were considered directly as the maximum PAVAR score = 7. All the variants with a score ≥5 were not filtered, and they were considered as candidate variants.

b) Exomiser v2 software

Exomiser v2 prioritizes SNVs by comparing the phenotype across species, according to the inheritance pattern, using the mouse and fish as a model organism phenotype [26]. Variant Call Format (VCF) files were analyzed with the following parameters: (a) HPO terms, Vertigo (HP:0002321), Tinnitus (HP:0000360), and Hearing Impairment (HP:0000365), were selected for Clinical Phenotype and (b) AD inheritance model. Since there are only three HPO terms associated with MD according to the public Human Phenotype Ontology database, but no gene is still included on it, the "Exomiser Gene Combined Score" generated very low values. So, variants with a threshold ≥1.46×10^{-5} were considered as candidate variants. Exomiser v2 allows the use of several HPO terms, but Phevor only allows five HPO terms. To compare both systems, only five HPO terms were selected for the benchmarking analyses. The five HPO terms most commonly associated with each disease were selected (Additional file 1: Table S7 and S8).

c) VAAST annotation tool

The third approach was to annotate and filter SNVs, according to the dominant inheritance pattern by VAAST software [25]. All case and control VCF files were processed according to the manual provided in the official website. Case files from the same pedigree were combined by the VAAST selection tool (VST) into a single condenser file; SNVs found in all the affected cases were selected. The quality of the resulting files was measured using the background provided: 1KGv3_CG_Div_NHLBI_dbSNP_RefSeq. cdr. A p value >0.05 indicates that there is no significant difference between the files (Additional file 1: Table S9). The next step was to search for candidate genes and their potential disease-causing variants. Each family dataset was filtered with the following parameters: (a) dominant inheritance, (b) incomplete penetrance, (c) maximum combined population frequency for the disease-causing alleles >0.0005 [51], and (d) 1×10^{6} permutations per analysis to achieve a significant p value after Bonferroni correction. Variants with an alpha error ≤1 were considered as possibly pathogenic.

d) Phevor tool

In the fourth approach, the list of the resulting genes generated by VAAST tool was uploaded to the Phevor Webtool (phenotype driven variant ontological re-ranking tool) to prioritize candidate genes, according to phenotype and HPO terms [30]. To run the analyses for MD, AD-SNHL, and CNM, the phenotypes were generated in Phevor using HPO term described in Additional file 1: Tables S7 and S8. Exomiser v2 only admits HPO term so to compare with Phevor; Disease Ontology Terms and Gene Ontology Terms were not used. No threshold value was applied in these analyses since the list of variants is generated from pre-filtered variants from VAAST.

e) Combined Annotation-Dependent Depletion (CADD)

CADD v1.3 [28, 59] is pre-computed score database that is based on classifier algorithms. The major goal of CADD is to predict the deleterious, functionally significant and pathogenic variants from diversified class of variants by integrative annotations. For each variant, CADD generates the combined annotation score (c-score) as an output and all scores were referenced against the pre-computed c-scores of 8.6 billion possible human SNPs. In CADD scoring criteria, functional variants should possess c-score greater than or equal to 10, whereas damaging variants show the c-score greater than or equal to 20 and the most lethal human variants show the c-score of greater than or equal to 30. To identify causal variants, a score ≥15 was considered as potentially pathogenic.

f) Functional Analysis through Hidden Markov Models (FATHMM)

FATHMM [29] predict the functional effects of protein missense mutations by combining sequence conservation within hidden Markov models (HMMs), representing the alignment of homologous sequences and conserved protein domains, with "pathogenicity weights", representing the overall tolerance of the protein/domain to mutations. The prediction outputs are scored, and the majority of disease-associated AASs fell below −3 and −1.5 threshold. To identify potential causal variants, a score ≤−1.5 was considered as potentially pathogenic.

Benchmarking procedures

The efficiency of the workflow was tested by benchmarking procedures in different synthetic family datasets with MD. In addition, a group of no familial healthy controls was tested to identify any bias caused for MD that could influence in the analysis. Moreover, two AD disorders were selected: (a) autosomal dominant sensorineural hearing loss (AD-SNHL) and (b) Central nuclear myopathy (CNM). AD-SNHL has 33 genes diseases, but the phenotype could overlap with MD. To avoid the bias of analyzing AD-SNHL and MD, we selected another disease (CNM) with no overlap in the phenotype with MD. CNM was selected because it has five different genes to perform the benchmarking analysis. The best characterized genes available for AD-SNHL included in the Hereditary Hearing Loss Homepage and CNM genes described in Orphanet were selected (Additional file 1: Table S5). For these genes, exome sequencing data of all

exonic variants, in VCF format, were obtained from the public ESP database. Next, 200 variants for each disease were randomly selected to perform benchmarking analyses, but we also checked that at least part of them were described as pathogenic or associated with the disease in human mutation database (HGMD) (Additional file 1: Table S1 and S2). To perform the analyses, the synthetic files were built inserting two random variants into real cases VCF files of each family. These synthetic family files for both diseases were analyzed with the six systems. The top 10, 20, and 50 ranked variants for AD-SNHL and CNM were analyzed by each separate system and by all combined strategies.

Statistical analysis

Logit regression model was built to assess the accuracy to predict correctly pathogenic variants associated with the phenotype. Firstly, variants selected for benchmarking analysis were classified as pathogenic or benign according to HGMD. The ranks conferred by each system were converted into ranks predictor-wise and normalized in [0, 1], according to top 10, 20, or 50. ROC curves were generated to determine the ability to predict real causal variants based on models consisting of the combination of the five systems (PAVAR, Exomiser v2, VAAST-Phevor, CADD, and FATHMM) and each individual system. In all the cases, the analyses were performed for the top 10, 20, and 50 ranked variants and using different control datasets to filter for private variants. AUCs were calculated for each ROC curves (Additional file 1: Table S3). The statistical differences between AUCs were calculated by analysis of variance. The logit regression models obtained, according to the different combinations and ROC curves, were analyzed with R version 3.0.3 and RStudio version 0.98.1102.

Additional file

Additional file 1: Additional Tables for Requena et al., "A pipeline combining multiple strategies for prioritizing heterozygous variants for the identification of candidate genes in exome datasets". **Table S1** Two hundred randomly selected SNV located in genes causing autosomal dominant sensorineural hearing loss. **Table S2** Two hundred randomly selected SNV located in genes causing Centro Nuclear Myopathy. **Table S3** Logit regression model to predict pathogenic variants is based on models consisting of single or multiple prediction tools for the top 10, 20, and 50 ranked variants for each tool, respectively. **Table S4** Number of SNV obtained in 21 exome datasets according to its effect on protein sequence and position on the reference genome (GRCh37 hg19). **Table S5** Web Resources, the URLs for software presented. **Table S6** Pathogenic Variants Scoring System (PAVAR). **Table S7** HPO terms used to describe the AD-SNHLs. **Table S8** HPO terms used to describe the CNMs. **Table S9** VAAST files. p value of quality, no significant differences were found between WES data and the background.

Abbreviations

AD: Autosomical dominant; AD-SNH: Non-syndromic autosomal dominant sensorineural hearing loss; CADD: Combined annotation-dependent depletion; CNM: Centro Nuclear Myopathy; FATHMM: Functional analysis through hidden markov models; FMD: Familial MD; HPO term: Human Phenotype Ontology term; MAF: Minor allelic frequency; MD: Meniere's disease; PAVAR: Pathogenic variant risk composite score; SNV: Single-nucleotide variants; UTR: Untranslated regions; VAAST: Variant annotation analysis and search tool; WES: Whole-exome sequencing

Acknowledgements

The authors would like to thank all patients and their relatives that participated in this study. We also thank the assistance of the Genomic and Bioinformatics staff at Genyo. Teresa Requena was a PhD student, and this work is part of her Doctoral Thesis.

Funding

This study was funded by EU-FEDER Funds for R+D+i by the Grants 2013-1242 from Instituto de Salud Carlos III and 2016-WES from The Meniere Society, UK.

Authors' contributions

This study was conceived and designed by TR and JALE. Selection of samples was performed by TR. NGS libraries were prepared by TR. The bioinformatics pipeline and the NGS analysis were performed by TR and AGM. The manuscript was written by TR, AGM, and JALE. All aspects of the study were supervised by JALE. All authors read and approved the final manuscript.

Competing interests

The authors declare that they have no competing interests.

Consent for publication

Not applicable.

Author details

[1]Otology & Neurotology Group CTS495, Department of Genomic Medicine, GENYO - Centre for Genomics and Oncological Research – Pfizer/University of Granada/Junta de Andalucía, PTS, 18016 Granada, Spain. [2]Department of Otolaryngology, Complejo Hospitalario Universidad de Granada (CHUGRA), ibs.granada, 18014 Granada, Spain.

References

1. Ng SB, Turner EH, Robertson PD, Flygare SD, Bigham AW, Lee C, Shaffer T, Wong M, Bhattacharjee A, Eichler EE, et al. Targeted capture and massively parallel sequencing of 12 human exomes. Nature. 2009;461(7261):272–6.
2. Hodges E, Xuan Z, Balija V, Kramer M, Molla MN, Smith SW, Middle CM, Rodesch MJ, Albert TJ, Hannon GJ, et al. Genome-wide in situ exon capture for selective resequencing. Nat Genet. 2007;39(12):1522–7.
3. Robinson PN, Krawitz P, Mundlos S. Strategies for exome and genome sequence data analysis in disease-gene discovery projects. Clin Genet. 2011;80(2):127–32.
4. Ng SB, Buckingham KJ, Lee C, Bigham AW, Tabor HK, Dent KM, Huff CD, Shannon PT, Jabs EW, Nickerson DA, et al. Exome sequencing identifies the cause of a mendelian disorder. Nat Genet. 2010;42(1):30–5.
5. Rabbani B, Tekin M, Mahdieh N. The promise of whole-exome sequencing in medical genetics. J Hum Genet. 2014;59(1):5–15.
6. Do R, Kathiresan S, Abecasis GR. Exome sequencing and complex disease: practical aspects of rare variant association studies. Hum Mol Genet. 2012; 21(R1):R1–9.
7. Girard SL, Gauthier J, Noreau A, Xiong L, Zhou S, Jouan L, Dionne-Laporte A, Spiegelman D, Henrion E, Diallo O, et al. Increased exonic de novo mutation rate in individuals with schizophrenia. Nat Genet. 2011;43(9):860–3.
8. O'Roak BJ, Deriziotis P, Lee C, Vives L, Schwartz JJ, Girirajan S, Karakoc E, Mackenzie AP, Ng SB, Baker C, et al. Exome sequencing in sporadic autism spectrum disorders identifies severe de novo mutations. Nat Genet. 2011; 43(6):585–9.
9. Requena T, Cabrera S, Martin-Sierra C, Price SD, Lysakowski A, Lopez-Escamez JA. Identification of two novel mutations in FAM136A and DTNA genes in autosomal-dominant familial Meniere's disease. Hum Mol Genet. 2014. 15;24(4):1119–26.
10. Martin-Sierra C, Requena T, Frejo L, Price SD, Gallego-Martinez A, Batuecas-Caletrio A, Santos-Perez S, Soto-Varela A, Lysakowski A, Lopez-Escamez JA. A novel missense variant in PRKCB segregates low-frequency hearing loss in an autosomal dominant family with Meniere's disease. Hum Mol Genet. 2016;25(16):3407–15.
11. Haghighi A, Tiwari A, Piri N, Nurnberg G, Saleh-Gohari N, Haghighi A, Neidhardt J, Nurnberg P, Berger W. Homozygosity mapping and whole exome sequencing reveal a novel homozygous COL18A1 mutation causing Knobloch syndrome. PloS one. 2014;9(11):e112747.
12. Zhao S, Choi M, Heuck C, Mane S, Barlogie B, Lifton RP, Dhodapkar MV. Serial exome analysis of disease progression in premalignant gammopathies. Leukemia. 2014;28(7):1548–52.
13. Vermeer S, Hoischen A, Meijer RP, Gilissen C, Neveling K, Wieskamp N, de Brouwer A, Koenig M, Anheim M, Assoum M, et al. Targeted next-generation sequencing of a 12.5 Mb homozygous region reveals ANO10 mutations in patients with autosomal-recessive cerebellar ataxia. Am J Hum Genet. 2010; 87(6):813–9.
14. Koboldt DC, Larson DE, Sullivan LS, Bowne SJ, Steinberg KM, Churchill JD, Buhr AC, Nutter N, Pierce EA, Blanton SH, et al. Exome-based mapping and variant prioritization for inherited Mendelian disorders. Am J Hum Genet. 2014;94(3):373–84.
15. Smith RJH, Shearer AE, Hildebrand MS, et al. Deafness and Hereditary Hearing Loss Overview. 1999 Feb 14 [Updated 2014 Jan 9]. In: Pagon RA, Adam MP, Ardinger HH, et al., editors. GeneReviews® [Internet]. Seattle (WA): University of Washington, Seattle; 1993-2017. Available from: https://www.ncbi.nlm.nih.gov/books/NBK1434/.
16. Vona B, Nanda I, Hofrichter MA, Shehata-Dieler W, Haaf T. Non-syndromic hearing loss gene identification: a brief history and glimpse into the future. Mol Cell Probes. 2015;29(5):260–70.
17. Lopez-Escamez JA, Carey J, Chung WH, Goebel JA, Magnusson M, Mandala M, Newman-Toker DE, Strupp M, Suzuki M, Trabalzini F, et al. Diagnostic criteria for Meniere's disease. J Vestib Res. 2015;25(1):1–7.
18. Vrabec JT. Genetic investigations of Meniere's disease. Otolaryngol Clin N Am. 2010;43(5):1121–32.
19. Morrison AW, Bailey ME, Morrison GA. Familial Meniere's disease: clinical and genetic aspects. J Laryngol Otol. 2009;123(1):29–37.
20. Requena T, Espinosa-Sanchez JM, Cabrera S, Trinidad G, Soto-Varela A, Santos-Perez S, Teggi R, Perez P, Batuecas-Caletrio A, Fraile J, et al. Familial clustering and genetic heterogeneity in Meniere's disease. Clin Genet. 2014;85(3):245–52.
21. Klar J, Frykholm C, Friberg U, Dahl N. A Meniere's disease gene linked to chromosome 12p12.3. Am J Med Genet B Neuropsychiatr Genet. 2006; 141B(5):463–7.
22. Arweiler-Harbeck D, Horsthemke B, Jahnke K, Hennies HC. Genetic aspects of familial Meniere's disease. Otol Neurotol. 2011;32(4):695–700.
23. Nadeau JH. Modifier genes and protective alleles in humans and mice. Curr Opin Genet Dev. 2003;13(3):290–5.
24. Nadeau JH. Modifier genes in mice and humans. Nat Rev Genet. 2001;2(3): 165–74.
25. Kennedy B, Kronenberg Z, Hu H, Moore B, Flygare S, Reese MG, Jorde LB, Yandell M, Huff C: Using VAAST to identify disease-associated variants in next-generation sequencing data. Current protocols in human genetics/editorial board, Jonathan L Haines [et al]. 2014;81:6 14 11-16 14 25.
26. Robinson PN, Kohler S, Oellrich A, Sanger Mouse Genetics P, Wang K, Mungall CJ, Lewis SE, Washington N, Bauer S, Seelow D et al. Improved exome prioritization of disease genes through cross-species phenotype comparison. Genome Res. 2014. doi:10.1101/gr.160325.113.
27. Singleton MV, Guthery SL, Voelkerding KV, Chen K, Kennedy B, Margraf RL, Durtschi J, Eilbeck K, Reese MG, Jorde LB, et al. Phevor combines multiple biomedical ontologies for accurate identification of disease-causing alleles in single individuals and small nuclear families. Am J Hum Genet. 2014;94(4): 599–610.
28. Kircher M, Witten DM, Jain P, O'Roak BJ, Cooper GM, Shendure J. A general framework for estimating the relative pathogenicity of human genetic variants. Nat Genet. 2014;46(3):310–5.
29. Shihab HA, Gough J, Cooper DN, Stenson PD, Barker GL, Edwards KJ, Day IN,

Gaunt TR. Predicting the functional, molecular, and phenotypic consequences of amino acid substitutions using hidden Markov models. Hum Mutat. 2013; 34(1):57–65.

30. Kohler S, Doelken SC, Mungall CJ, Bauer S, Firth HV, Bailleul-Forestier I, Black GC, Brown DL, Brudno M, Campbell J, et al. The Human Phenotype Ontology project: linking molecular biology and disease through phenotype data. Nucleic Acids Res. 2014;42(Database issue):D966–74.

31. Schrijver I. Hereditary non-syndromic sensorineural hearing loss: transforming silence to sound. J Mol Diagn. 2004;6(4):275–84.

32. Chan DK, Schrijver I, Chang KW. Connexin-26-associated deafness: phenotypic variability and progression of hearing loss. Genet Med. 2010;12(3):174–81.

33. Bitoun M, Romero NB, Guicheney P. Mutations in dynamin 2 cause dominant centronuclear myopathy. Med Sci. 2006;22(2):101–2.

34. Bao R, Huang L, Andrade J, Tan W, Kibbe WA, Jiang H, Feng G. Review of current methods, applications, and data management for the bioinformatics analysis of whole exome sequencing. Cancer Informat. 2014;13 Suppl 2:67–82.

35. Precone V, Del Monaco V, Esposito MV, De Palma FD, Ruocco A, Salvatore F, D'Argenio V. Cracking the code of human diseases using next-generation sequencing: applications, challenges, and perspectives. Biomed Res Int. 2015;2015:161648.

36. Guo Y, Ding X, Shen Y, Lyon GJ, Wang K. SeqMule: automated pipeline for analysis of human exome/genome sequencing data. Sci Rep. 2015;5:14283.

37. Hwang S, Kim E, Lee I, Marcotte EM. Systematic comparison of variant calling pipelines using gold standard personal exome variants. Sci Rep. 2015;5:17875.

38. D'Antonio M, D'Onorio De Meo P, Paoletti D, Elmi B, Pallocca M, Sanna N, Picardi E, Pesole G, Castrignano T. WEP: a high-performance analysis pipeline for whole-exome data. BMC Bioinf. 2013;14 Suppl 7:S11.

39. Dong C, Wei P, Jian X, Gibbs R, Boerwinkle E, Wang K, Liu X. Comparison and integration of deleteriousness prediction methods for nonsynonymous SNVs in whole exome sequencing studies. Hum Mol Genet. 2015;24(8):2125–37.

40. Smedley D, Kohler S, Czeschik JC, Amberger J, Bocchini C, Hamosh A, Veldboer J, Zemojtel T, Robinson PN. Walking the interactome for candidate prioritization in exome sequencing studies of Mendelian diseases. Bioinformatics. 2014;30(22):3215–22.

41. Martin-Sierra C, Gallego-Martinez A, Requena T, Frejo L, Batuecas-Caletrio A, Lopez-Escamez JA. Variable expressivity and genetic heterogeneity involving DPT and SEMA3D genes in autosomal dominant familial Meniere's disease. Eur J hum genet. 2016;25(2):200–7.

42. Javed A, Agrawal S, Ng PC. Phen-Gen: combining phenotype and genotype to analyze rare disorders. Nat Methods. 2014;11(9):935–7.

43. Kohler S, Vasilevsky NA, Engelstad M, Foster E, McMurry J, Ayme S, Baynam G, Bello SM, Boerkoel CF, Boycott KM, et al. The Human Phenotype Ontology in 2017. Nucleic Acids Res. 2017;45(D1):D865–76.

44. Li H, Handsaker B, Wysoker A, Fennell T, Ruan J, Homer N, Marth G, Abecasis G, Durbin R. The sequence alignment/map format and SAMtools. Bioinformatics. 2009;25(16):2078–9.

45. Li H, Ruan J, Durbin R. Mapping short DNA sequencing reads and calling variants using mapping quality scores. Genome Res. 2008;18(11):1851–8.

46. Sims D, Sudbery I, Ilott NE, Heger A, Ponting CP. Sequencing depth and coverage: key considerations in genomic analyses. Nat Rev Genet. 2014; 15(2):121–32.

47. Genomes Project C, Auton A, Brooks LD, Durbin RM, Garrison EP, Kang HM, Korbel JO, Marchini JL, McCarthy S, McVean GA, et al. A global reference for human genetic variation. Nature. 2015;526(7571):68–74.

48. Lek M, Karczewski K, Minikel E, Samocha K, Banks E, Fennell T, O'Donnell-Luria A, Ware J, Hill A, Cummings B et al. Analysis of protein-coding genetic variation in 60,706 humans. bioRxiv. 2015. 18;536(7616):285–91.

49. Merchant SN, Adams JC, Nadol Jr JB. Pathophysiology of Meniere's syndrome: are symptoms caused by endolymphatic hydrops? Otol Neurotol. 2005;26(1):74–81.

50. Alexander TH, Harris JP. Current epidemiology of Meniere's syndrome. Otolaryngol Clin N Am. 2010;43(5):965–70.

51. Shearer AE, Eppsteiner RW, Booth KT, Ephraim SS, Gurrola 2nd J, Simpson A, Black-Ziegelbein EA, Joshi S, Ravi H, Giuffre AC, et al. Utilizing ethnic-specific differences in minor allele frequency to recategorize reported pathogenic deafness variants. Am J Hum Genet. 2014;95(4):445–53.

52. Richards S, Aziz N, Bale S, Bick D, Das S, Gastier-Foster J, Grody WW, Hegde M, Lyon E, Spector E, et al. Standards and guidelines for the interpretation of sequence variants: a joint consensus recommendation of the American College of Medical Genetics and Genomics and the Association for Molecular Pathology. Genet Med. 2015;17(5):405–23.

53. Choi Y, Sims GE, Murphy S, Miller JR, Chan AP. Predicting the functional effect of amino acid substitutions and indels. PloS one. 2012;7(10):e46688.

54. Adzhubei IA, Schmidt S, Peshkin L, Ramensky VE, Gerasimova A, Bork P, Kondrashov AS, Sunyaev SR. A method and server for predicting damaging missense mutations. Nat Methods. 2010;7(4):248–9.

55. Grantham R. Amino acid difference formula to help explain protein evolution. Science. 1974;185(4154):862–4.

56. Davydov EV, Goode DL, Sirota M, Cooper GM, Sidow A, Batzoglou S. Identifying a high fraction of the human genome to be under selective constraint using GERP++. PLoS Comput Biol. 2010;6(12):e1001025.

57. Schwarz JM, Rodelsperger C, Schuelke M, Seelow D. MutationTaster evaluates disease-causing potential of sequence alterations. Nat Methods. 2010;7(8):575–6.

58. Pollard KS, Hubisz MJ, Rosenbloom KR, Siepel A. Detection of nonneutral substitution rates on mammalian phylogenies. Genome Res. 2010;20(1):110–21.

59. Mather CA, Mooney SD, Salipante SJ, Scroggins S, Wu D, Pritchard CC, Shirts BH. CADD score has limited clinical validity for the identification of pathogenic variants in noncoding regions in a hereditary cancer panel. Genet Med. 2016; 18(12):1269–75.

Single nucleotide polymorphisms in the angiogenic and lymphangiogenic pathways are associated with lymphedema caused by Wuchereria bancrofti

Linda Batsa Debrah[1,2†], Anna Albers[3†], Alexander Yaw Debrah[4], Felix F. Brockschmidt[5,6], Tim Becker[7], Christine Herold[7], Andrea Hofmann[3], Jubin Osei-Mensah[1], Yusif Mubarik[1], Holger Fröhlich[8], Achim Hoerauf[3*] and Kenneth Pfarr[3*]

Abstract

Background: Lymphedema (LE) is a chronic clinical manifestation of filarial nematode infections characterized by lymphatic dysfunction and subsequent accumulation of protein-rich fluid in the interstitial space—lymphatic filariasis. A number of studies have identified single nucleotide polymorphisms (SNPs) associated with primary and secondary LE. To assess SNPs associated with LE caused by lymphatic filariasis, a cross-sectional study of unrelated Ghanaian volunteers was designed to genotype SNPs in 285 LE patients as cases and 682 infected patients without pathology as controls. One hundred thirty-one SNPs in 64 genes were genotyped. The genes were selected based on their roles in inflammatory processes, angiogenesis/lymphangiogenesis, and cell differentiation during tumorigenesis.

Results: Genetic associations with nominal significance were identified for five SNPs in three genes: vascular endothelial growth factor receptor-3 (VEGFR-3) rs75614493, two SNPs in matrix metalloprotease-2 (MMP-2) rs1030868 and rs2241145, and two SNPs in carcinoembryonic antigen-related cell adhesion molecule-1 (CEACAM-1) rs8110904 and rs8111171. Pathway analysis revealed an interplay of genes in the angiogenic/lymphangiogenic pathways. Plasma levels of both MMP-2 and CEACAM-1 were significantly higher in LE cases compared to controls. Functional characterization of the associated SNPs identified genotype GG of CEACAM-1 as the variant influencing the expression of plasma concentration, a novel finding observed in this study.

Conclusion: The SNP associations found in the MMP-2, CEACAM-1, and VEGFR-3 genes indicate that angiogenic/lymphangiogenic pathways are important in LE clinical development.

Keywords: Lymphatic filariasis, Angiogenesis, Lymphangiogenesis, Single nucleotide polymorphisms, Genotypes

Background

Worldwide, more than 850,000 people live in areas endemic for *Wuchereria bancrofti*, *Brugia malayi*, and *Brugia timori* filial nematodes that cause lymphatic filariasis, a disease of severe morbidity [1]. Lymphatic disease symptoms are characterized by a cascade of events that leads to lymphatic dysfunction with associated fibrosis

[2]. Lymphedema (LE) and hydrocele are pathologies that can develop in *Wuchereria bancrofti* infected individuals. These clinical symptoms are usually preceded by dilated and tortious lymphatic vessels and scrotal lymphangiectasia [3, 4]. Of these two pathologies, LE is the most debilitating, affecting about 7% of the population in a lymphatic filariasis (LF) endemic community even though all individuals in the endemic area may be inoculated with the parasite and the majority (80%) may be infected [5, 6].

LE is a condition caused by the leakage of plasma from the arterial blood capillaries that is then trapped in the

* Correspondence: achim.hoeraruf@ukbonn.de; kenneth.pfarr@ukbonn.de
†Equal contributors
3Institute for Medical Microbiology, Immunology and Parasitology, University Hospital Bonn, Sigmund-Freud-Str. 25, 53127 Bonn, Germany

soft tissues as a result of the dysfunction of the lymphatic vessel that originates from the infection with the filarial parasites *Wuchereria bancrofti* or *Brugia* spp. [7]. The global burden of LE in 2000 was 14.84 million [8]. After 13 years of treatment with ivermectin and albendazole or diethylcarbamazine, to eliminate the infection [1], and morbidity management procedures, there still remained 14.41 million LE cases [8], although an estimated 116–250 million DALYS have been averted within that period. This highlights the need for alternative strategies to current morbidity management procedures to help prevent or even ameliorate LE in the affected persons.

Individuals infected with lymphatic filariasis parasites do not show recognizable clinical symptoms. However, a third of those infected developed a clinical disease. What causes the expression of clinical disease is not well understood. Several reasons have been given to explain the differences in the cause(s) of heterogeneity in infection and disease of filarial infection. These include the immune interaction between the human host and the parasite [9–12], transmission potential of the mosquito vector [13], in utero exposure to parasite antigens [14, 15], and secondary bacterial/fungal infections superimposed on the lymphatic dysfunction [16].

The contribution of host immunogenetics to this heterogeneity has also been investigated, leading to the finding that susceptibility to infection, parasite load and pathology cluster in families [17–21], indicating an underlying genetic component is involved in the disease. Gene polymorphisms such as the variant Leu10Pro of transforming growth factor-β-1 (TGFβ-1) was found to be associated with both lack of microfilariae and differential microfilarial loads [22]. In that study, it was shown that the differential microfilaria loads and the lack of circulating microfilariae (Mf) in the blood exhibited by people in endemic areas have genetic propensity. Hence, some people in endemic areas may be infected with the adult worm but would have no Mf in the peripheral blood. Also, polymorphisms in TLR-2 (+ 597 > C, 1450T > C and −96 to −173 deletion) were found to be associated with higher asymptomatic bancroftian filariasis [23]. Association has also been found in the HH variant of Chitinase-1 (CHIT-1) that correlated with decreased activity as well as levels of chitotriosidase and susceptibility of filarial infection. The XX genotype in the mannose-binding lectin-2 (MBL-2) genes has been associated with susceptibility to bancroftian infection [24]. Positive association was reported for all variants of rs733618 of cytotoxic T-lymphocyte-associated protein 4 (CTLA-4) gene among asymptomatic amicrofilaremic cases [25]. IL-10 promoter haplotypes and *IL-10 RA* S138G polymorphisms have also been identified as possible genetic determinants of susceptibility to lymphatic filariasis [26]. All the above SNPs that

have been found to be associated with filarial infections were the basis for our study.

We were among the first to show that angiogenic/lymphangiogenic molecules such as vascular endothelial growth factors (VEGFs) may be involved in the development of LE and hydrocele in humans [27, 28]. In these studies, we showed that VEGF-C and its receptor VEGFR-3 are elevated in the plasma of LE patients and treating them with antiangiogenic drugs such as doxycycline reduced the factors prior to ameliorating early stages of pathology [27]. We went further to show that another angiogenic molecule, VEGF-A, is genetically associated with hydrocele caused by bancroftian infections. Treatment with doxycycline again reversed the pathology in men with early stages of hydrocele [28]. Other authors have also shown the involvement of angiogenic/lymphangiogenic molecules in the clinical manifestations of LF [29, 30].

SNPs in FOXC-2 and FLT-4 genes have been identified to be involved in lymphedema progression [31].

While LE is clinically well described, there have been few investigations of host genetic contributions to filarial LE. In this study, we have further shown an association of SNPs in genes of the angiogenic/lymphangiogenic pathways with LE. Identified SNPs could contribute to the search of biomarkers for diagnosis of LE and potential methods to ameliorate LE symptoms.

Results
Demographic and pathology information of study participants
The mean age of study participants was not statistically different between cases and controls (Table 1). Predominantly, 71% of them were females and 29% were males. In the control group, the majority were males (57%). The volunteers had stayed in the study community from a year to over 50 years. In the cases group, 171 people (60%) had been a resident for more than 40 years. A greater number of cases had stages 2 and 3 (32 and 37%, respectively) pathology according to Dreyer et al. [32, 33], while stages 4 and 7 (2% each) were the least frequent stage of pathology among the cases (Table 1).

Single marker analysis
One hundred and forty-seven (147) single nucleotide polymorphisms (SNPs) were initially selected for genotyping (Additional file 1). Sixteen (16) were rejected during the assay design because the primer sequences produced were prone to primer dimerization or the masses of the sequences were too similar to be distinguished by mass spectrometry. Eight out of the 16 rejected SNPs are in the coding region resulting in

Table 1 Demographic and pathology profile of study participants

Variable	Cases N = 285	Controls N = 682
Mean age/years (range)	44.4 (16–73)	40.8 (16–93)
Gender		
Male % (N)	29.5 (84)	57 (389)[a]
Female % (N)	70.5 (201)	43 (293)
Duration in community/years		
1–10	2	55
11–20	10	90
21–30	42	178
31–40	57	132
41–50	79	109
> 50	92	118
Stages of lymphedema		
Stage 1	11	–
Stage 2	90	–
Stage 3	106	–
Stage 4	5	–
Stage 5	20	–
Stage 6	49	–
Stage 7	4	–

[a]Fisher's exact test, $P \leq 0.05$ controls compared to cases

amino acid changes, three were in the promoter region with no amino acid change, and six were in non-coding regions (Additional file 1). With the exception of tumor necrosis factor-α (TNF-α), CTLA-4, and interleukin-4 (IL-4), all the rejected genes were represented by at least one other SNP in the Sequenom data. Thus, 131 SNPs in 64 genes were genotyped.

The single marker analysis compared 285 LE patients (cases) and 682 infected patients without LE pathology (controls). Of the 131 SNPs genotyped, 5 SNPs in three genes were associated with LE with nominal significance (Table 2): 2 SNPs in matrix metalloprotease-2 (MMP-2 rs1030868, $P = 0.0094$; rs2241145, $P = 0.0116$), 2 SNPs in carcinoembryonic antigen-related cell adhesion molecule-1 (CEACAM-1 rs8110904, $P = 0.024$; rs8111171, $P = 0.026$), and 1 SNP in vascular endothelial growth factor receptor-3 (VEGFR-3 rs75614493, $P = 0.034$). None of the nominally associated SNPs withstood correction for multiple testing (Benjamini-Hochberg). All the associated SNPs were in Hardy-Weinberg equilibrium (HWE, $P > 0.05$) with the exception of CEACAM-1 rs8110904 (controls $P = 2.93E-10$).

The risk alleles for MMP-2 SNPs rs1030868 and rs2241145 were A and C, respectively, each conferring a 1.3-fold risk to LE development (Table 2). Both alleles fit in a recessive model of association (Additional file 2: Table S1).

No individual in the cohort was homozygous for the T allele in VEGFR-3 SNP rs75614493, and only three people were heterozygous (Table 2). The participants with the C allele of the VEGFR-3 SNP had a 3.4-fold risk of LE development. Due to the lack of homozygosity for the T allele in the cohort, no model of association, whether dominant or recessive, could be assigned (Additional file 2: Table S1). The risk alleles for CEACAM-1 SNPs rs8110904 and rs8111171 were A and T and confer a 1.2- and 1.3-fold risk, respectively, of LE development (Table 2). Both alleles fit a dominant model of association (Additional file 2: Table S1).

Haplotype analysis

Two or more SNPs in a gene or on the same chromosome can form haplotypes that are inherited together [34]. Analysis of haplotype association with LE was done using the FamHap software package [35]. A likelihood ratio test with one degree of freedom was used to assess the significance of haplotype frequencies among SNPs on the same gene. Only haplotypes that were significantly associated at one degree of freedom were reported (Table 3).

Two CEACAM-1 SNPs, rs8110904 and rs8111171, were associated with LE in a single marker analysis with nominal significance. From these SNPs, three haplotypes were generated. The frequency of haplotype GG was significantly higher in the controls than the cases ($P = 0.026$); there was a trend in haplotype AT ($P = 0.055$) but there was no difference in haplotype GT between cases and controls (Table 3).

Six different haplotypes comprising SNPs rs11643630, rs1030868, rs2241145, and rs1992116 in the MMP-2 gene (GACG, GGCA, GGGG, TACG, TGCG, and TGGG) were predicted by the FamHap analysis. Haplotype TACG was significantly higher in cases than controls ($P = 0.046$), and this significance was even strengthened after multiple testing with 200,000 simulations, ($P = 0.03$, Table 3). The remaining haplotypes were not significant in either cases or controls.

The VEGFR-3 SNPs rs75614493 and rs3587489 formed three haplotypes (CC, CT, and TT). The TT haplotype was rare in this population and was significantly associated with controls ($P = 0.023$), but was lost after correcting for multiple testing ($P = 0.08$, Table 3).

Plasma levels of angiogenic/lymphangiogenic molecules

Plasma concentrations of the proteins encoded by the genes associated with LE development were measured to evaluate the functional phenotypes. CEACAM-1 and MMP-2 were measured using commercially available kits to compare the plasma levels between LE patients and infected controls. The plasma levels of CEACAM-1 were significantly elevated in LE patients ($P < 0.02$, Fig. 1a).

Table 2 Genotype frequencies and odds ratio of SNPs associated with lymphedema patients and infected controls

Gene (dbSNP rs#)	Functional category of SNP	Genotypes	Cases (%)	Controls (%)	P_{ATT}[a]	Adjusted[b]	OR[c] (95% CI)
CEACAM-1 (rs8110904)	Missense	AA	26 (9)	57 (9)	0.024	0.370	(A) 1.2 (0.99–1.49)
		AG	182 (67)	385 (59)			
		GG	65 (24)	213 (32)			
CEACAM-1 (rs8111171)	Missense	GG	61 (22)	205 (31)	0.026	0.370	(T) 1.3 (1.03–1.56)
		GT	145 (52)	311 (46)			
		TT	73 (26)	157 (23)			
FLT-4/VEGFR-3 (rs75614493)	Missense	CC	282 (99)	658 (96)	0.034	0.232	(C) 3.4 (1.02–11.29)
		CT	3 (1)	24 (4)			
		TT	0 (0)	0 (0)			
MMP-2 (rs1030868)	Intron	AA	69 (24)	113 (17)	0.0094	0.232	(A) 1.3 (1.07–1.58)
		AG	134 (47)	337 (49)			
		GG	82 (29)	232 (34)			
MMP-2 (rs2241145)	Intron	CC	86 (30)	158 (23)	0.0116	0.232	(C) 1.3 (1.06–1.57)
		CG	136 (48)	334 (49)			
		GG	63 (22)	190 (28)			

[a]Cochrane-Armitage test for trend
[b]Adjusted P values according to Benjamini-Hochberg
[c]Odds ratio with 95% confidence intervals for the risk allele

MMP-2 protein concentration was also significantly higher in the LE patients ($P = 0.025$, Fig. 1b).

To functionally characterize the genotypes, plasma levels of CEACAM-1 and MMP-2 were correlated with the respective genotypes. CEACAM-1 plasma level was higher in people with the GG genotype in both rs8110904 and rs8111171 SNPs (Fig. 2). Plasma levels of MMP-2 did not correlate with any of the SNP genotypes in this population.

Pathway interaction of lymphedema associated genes
The MetaCore™ software package was used to analyze the genotyped genes for pathways of protein interaction.

Table 3 SNP haplotypes in genes associated with lymphedema

Gene	dbSNPrs#	Allele	Distribution cases (%)	Distribution controls (%)	P value (1df)	Global P value[a]
CEACAM-1	rs8110904	G	48.3	54.0	*0.026*	0.092
	rs8111171	G				
	rs8110904	A	48.0	39	*0.046*	0.092
	rs8111171	T				
VEGFR-3	rs75614493	T	0.5	1.8	*0.023*	0.080
	rs3587489	T				
MMP-2	rs2241145	T	33.1	28.3	*0.046*	*0.030*
	rs1030868	A				
	rs11643630	C				
	rs1992116	G				

Italic text indicates a significant association
[a]Calculated by an omnibus statistic with 200,000 simulations in FamHap version 19

Because of the candidate gene approach of this case-control study, it is not surprising that the associated genes are in the angiogenesis pathway. Nevertheless, candidates for further study for their role in LE development are identified by this analysis. The three genes (gray circles) with SNPs associated with filarial LE interact with each other (CEACAM-1 and VEGFR-3) and 11 proteins in the angiogenesis pathway (Fig. 3). CEACAM-1 is predicted to up-regulate VEGFR-3 expression directly and up-regulates PROX-1, VEGF-C, and VEGF-D, which also up-regulate VEGFR-3. CEACAM-1 can also up-regulate MMP-2 via the up-regulation of TALIN. It down-regulates beta-catenin, a protein that up-regulates MMP-2 directly and also indirectly via up-regulation of VEFG-A, and up-regulates VEGFR-3 indirectly via PROX-1. MMP-2 is predicted to up-regulate TGF-β1, MMP-9, and VEGFR-1.

Discussion
Pathological effects of lymphatic filariasis such as LE and hydrocele are observed in a fraction of the individuals in endemic areas even though up to 80% may be infected with *W. bancrofti* [5, 36]. LE, the most debilitating pathology, occurs in ~ 7% of the endemic population [6, 8].

Different studies have been undertaken to unravel the genetic basis of this heterogeneity, but they concentrated on infection and hydrocele development [24, 28]. Filarial LE is the single largest cause of secondary lymphedema

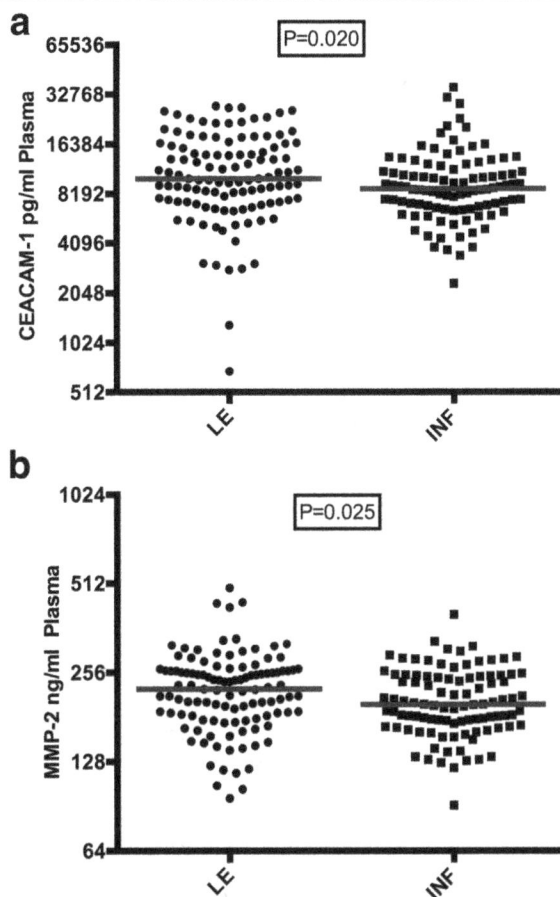

Fig. 1 Plasma concentrations of CEACAM-1 and MMP-2 are higher in lymphedema patients. **a** Plasma concentration of CEACAM-1 gene. **b** Plasma concentration of MMP-2 gene. EDTA Plasma was collected from 101 LE patients and 99 infected patients without disease symptoms for measurement of protein levels of CEACAM-1 and MMP-2 using kits from R&D Systems (Wiesbaden, Germany). ELISA and quantitative analyses were performed according to the manufacturer's protocol. The Mann-Whitney test (Statview software version 5.0) was performed to check for differences in plasma concentrations between the genotypes of the indicated SNPs, $P < 0.05$ considered significant. The red lines indicate the median of the plasma concentrations

[37], an inflammatory disease resulting from the destruction of the lymphatic vessel with associated fibrosis as a result of the presence and death of the adult filarial worms, larval death, and the release of *Wolbachia* endosymbionts [38]. Since the immune response of the host plays an integral role in disease etiology by inducing the expression of particular genes [38], host immunogenetics was exploited in this study to answer the question as to which SNPs could be causative variants in filarial LE development.

Five SNPs in three genes were identified to be associated with LE. The associations did not withstand correction for multiple testing, which is probably attributable to the low sample size and/or the small contributing effects of the SNPs on the disease.

Carcinoembryonic antigen-related cell adhesion molecule 1 (CEACAM-1) is a type 1 transmembrane protein involved in cell-to-cell adhesion [39]. It has been shown to be a potent stimulator of vascular endothelial growth factor (VEGF) mediated angiogenesis [40, 41]. It also stimulates microvascular endothelial cell growth in the presence of VEGF [40]. However, the overexpression of CEACAM-1 is associated with cancers such as thyroid cancer, gastric cancer, and metastasizing malignant melanomas [42].

Two SNPs in the CEACAM-1 gene (rs8110904 and rs8111171) were associated with LE development. The frequencies for the minor alleles A and T are consistent with the values reported for rs8111171 and rs8110904 from the Yorubian population (rs8110904 G = 57% A = 43%, rs8111171 G = 56% T = 44%) using 120 and 48 participants, respectively [43]. The minor allele A in rs8110904 was higher in the cases than in the controls, and patients with the A allele had an odds ratio of 1.2 (CI 0.99–1.49). The minor T allele frequency in rs8111171 was also higher in the cases than in the controls with a 1.3-fold risk (CI 1.03–1.56) of developing LE. The significant haplotype association of the

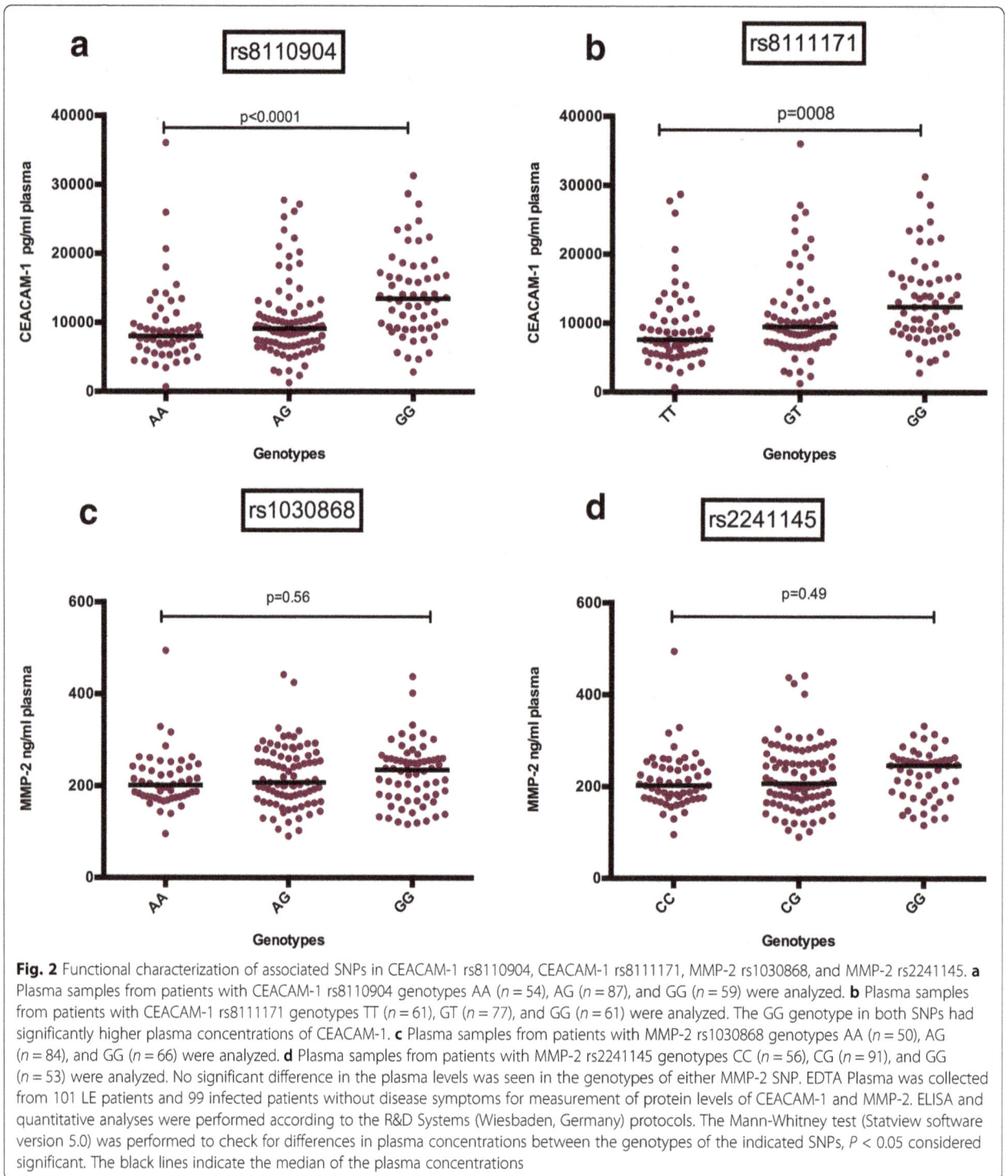

Fig. 2 Functional characterization of associated SNPs in CEACAM-1 rs8110904, CEACAM-1 rs8111171, MMP-2 rs1030868, and MMP-2 rs2241145. **a** Plasma samples from patients with CEACAM-1 rs8110904 genotypes AA ($n = 54$), AG ($n = 87$), and GG ($n = 59$) were analyzed. **b** Plasma samples from patients with CEACAM-1 rs8111171 genotypes TT ($n = 61$), GT ($n = 77$), and GG ($n = 61$) were analyzed. The GG genotype in both SNPs had significantly higher plasma concentrations of CEACAM-1. **c** Plasma samples from patients with MMP-2 rs1030868 genotypes AA ($n = 50$), AG ($n = 84$), and GG ($n = 66$) were analyzed. **d** Plasma samples from patients with MMP-2 rs2241145 genotypes CC ($n = 56$), CG ($n = 91$), and GG ($n = 53$) were analyzed. No significant difference in the plasma levels was seen in the genotypes of either MMP-2 SNP. EDTA Plasma was collected from 101 LE patients and 99 infected patients without disease symptoms for measurement of protein levels of CEACAM-1 and MMP-2. ELISA and quantitative analyses were performed according to the R&D Systems (Wiesbaden, Germany) protocols. The Mann-Whitney test (Statview software version 5.0) was performed to check for differences in plasma concentrations between the genotypes of the indicated SNPs, $P < 0.05$ considered significant. The black lines indicate the median of the plasma concentrations

"protective" haplotype GG ($P = 0.0256$) and haplotype of the case-associated alleles AT ($P = 0.046$) supports a role for this gene in disease development.

CEACAM-1 is up-regulated in some cancers such as thyroid and gastric cancers. The initial metastasis of these cancers is through the lymphatic vessels to the regional lymph nodes similar to the pathogenesis of LE

which mainly occurs after dilation of the lymphatic vessel with associated fibrosis [44]. Multicellular activities such as angiogenesis have been attributed to encoded proteins of CEACAM-1. The serum of CEACAM-1 served as a useful indicator for the presence of pancreatic cancer [45]. CEACAM-1 is a potent inducer of VEGFs. The receptor 3 of VEGF-C and D has been

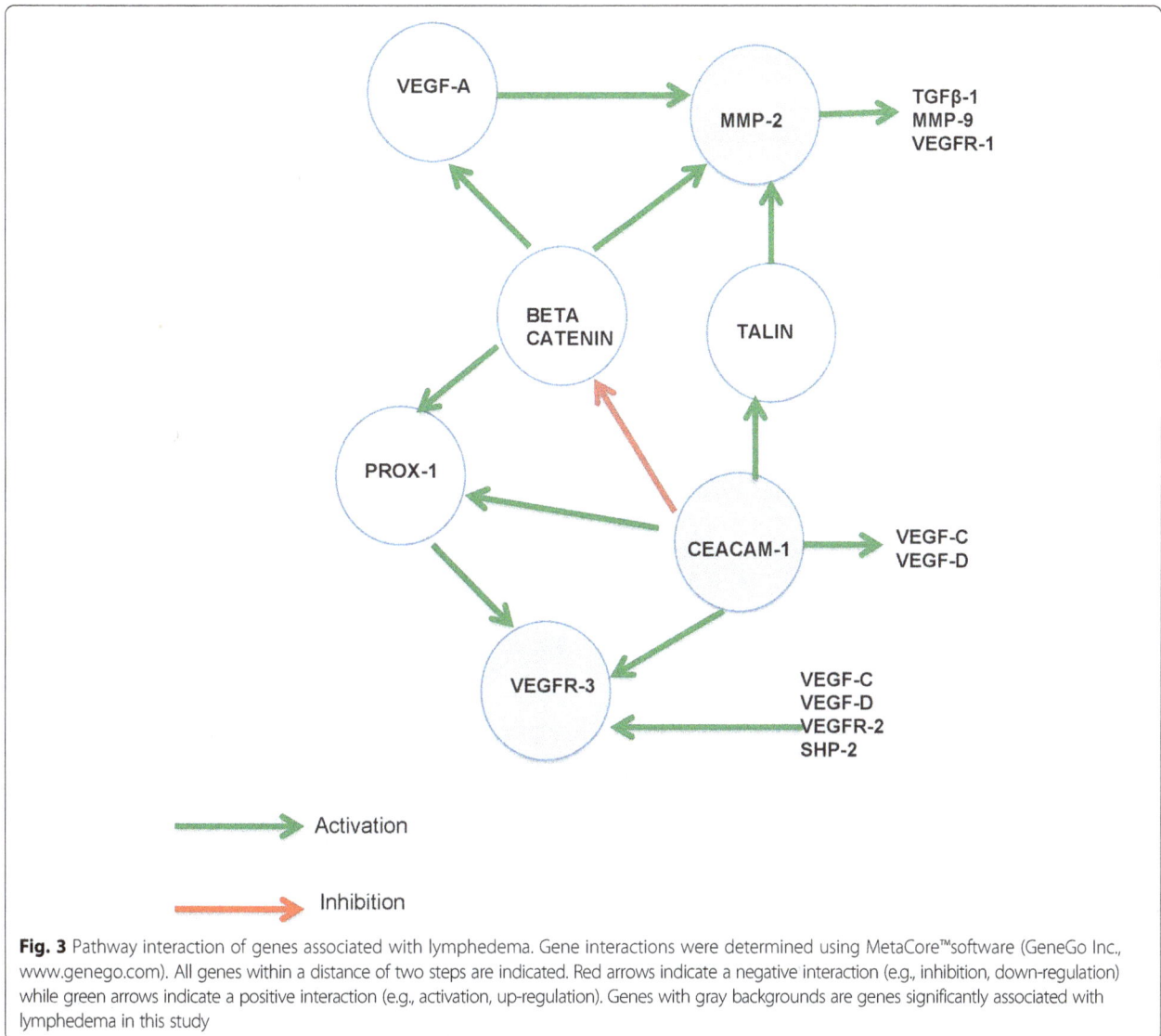

Fig. 3 Pathway interaction of genes associated with lymphedema. Gene interactions were determined using MetaCore™software (GeneGo Inc., www.genego.com). All genes within a distance of two steps are indicated. Red arrows indicate a negative interaction (e.g., inhibition, down-regulation) while green arrows indicate a positive interaction (e.g., activation, up-regulation). Genes with gray backgrounds are genes significantly associated with lymphedema in this study

associated with LE, and its plasma was found to be elevated in LE [27]. If CEACAM-1 stimulates VEGFs and the plasma levels of the receptors influence LE then higher plasma levels of CEACAM-1 in LE could play a role in the development of the disease. The significant increase in plasma protein concentration of CEACAM-1 gene in LE patients (Fig. 1a) correlates with the genotype and is an indication that CEACAM-1 might play a role in LE development. The plasma protein level was also observed to be higher with the GG genotype compared with the other genotypes indicating that the genotype GG directly or indirectly impacts the expression of CEACAM-1 plasma proteins and could be a variant for LE development.

Matrix metalloprotease-2 (MMP2) is a known angiogenic factor whose activity involves the breakdown of extracellular matrix in physiological processes such as embryonic development, reproduction, and tissue remodeling [46]. Mutations in the MMP-2 gene lead to a number of disease processes, such as arthritis and metastasis, tumor growth vascular aneurysmal disease development, Winchester syndrome, and nodulosis-arthropathy-osteolysis (NAO) syndrome [47–50].

The significant association of MMP-2 SNPs rs1030868 ($P = 0.0094$) and rs2241145 ($P = 0.0116$) is an indication that this gene might be involved in LE development. Patients with rs1030868 (minor allele A) and rs2241145 (minor allele C) SNPs have a 1.3-fold (CI 1.07–1.58 and CI 1.06–1.57, respectively) risk of developing LE than those who do not have these alleles (Table 2). In National Center for Biotechnology Information database of single nucleotide polymorphism (NCBI dbSNP), these minor alleles have a frequency of C = 52% (rs2241145) and A = 42% (rs1030868) in the Yorubian population ($n = 120$) and are similar to the values calculated from our larger sample size [43]. These SNPs have also been found to be associated with development of lacunar stroke [51], and higher levels of MMP-2 protein and activity have been described. The authors hypothesize that more MMP-2 alters and remodels the extracellular matrix around the vessels that contribute to the development of edema [52, 53], a hypothesis supported by the finding that MMP-2 also disrupts tight junctions [54]. Thus, extravasation of fluid occurs and contributes to stroke. A similar phenomenon is seen in LE development in which the lymph vessels dilate, reducing lymph flow. With destruction/remodeling of the vessel architecture, here hypothesized to be in part caused by MMP-2, lymph fluid enters the surrounding tissue causing lymphedema. The affected limb then enlarges progressively due to fibroadipose deposition [7].

MMP-2 mRNA levels are known to be higher in lymphedematous specimens compared to non-lymphedematous specimens of progenitor cells [7], and a blockage or down-regulation of this gene leads to reduced lymphangiogenesis [55]. Therefore, the significant increase in the plasma concentration of MMP-2 in LE patients (Fig. 1b) is an indication that this gene might have a role in the LE development. Even though the plasma concentration was evenly distributed among the genotypes (Fig. 2c, d), the identified associations of these intronic SNPs seem to account for another genetic effect which is independent of the plasma level. Thus, these intronic SNPs may act as proxy markers for another, yet to be identified, functional SNP in this chromosomal region.

Tetracycline and its derivatives have been shown to profoundly inhibit mammalian MMPs by a mechanism that is independent of their antimicrobial activity, thereby reducing excessive degradation or remodeling at the healing enthesis after rotator cuff repair [56, 57]. It has been shown by Debrah et al. [27, 58] that doxycycline improves the condition of disease symptoms of LF patients with early stages of LE. However, the mechanism of action was not clear. This study supports that the effect might be a direct effect on MMP-2 and explains why doxycycline is able to ameliorate LE pathology even though most LE patients do not have active infections. Additionally, pathway analysis with MetaCore shows that MMP-2 positively influences the expression of VEGFs, and therefore, inhibition of MMP-2 may result in additive or synergistic effects with other factors in this pathway.

VEGFR-3, a tyrosine-protein kinase, emerged as one of the genes associated with LE development in this study. A single marker association was found between the VEGFR-3 SNP (rs75614493) in the exon region of chromosome 5 and LE development (Table 2, P = 0.034). There was no patient homozygous TT in the study population even though 1 and 2% of patients carried the T allele in the cases and control groups, respectively. The allelic distribution from this study is consistent with earlier work done in sub-Saharan Africa

(Yorubian population) involving 118 patients (NCBI dbSNP). The allelic frequency of the study was 97.5% for the C allele and 2.5% for the T allele [59].

From this present study, the C allele was significantly more frequent in the cases compared to the controls with an odds ratio of 3.4 (CI 1.02–11.29). Even though the genotype frequencies are 99 and 96% in cases vs controls, a statistical difference was observed and a significant difference in haplotype frequencies between rs75614493 and rs3587489 was also observed in the TT haplotype with one degree of freedom.

VEGFR-3 is restricted largely to the lymphatic endothelium and acts as a cell surface receptor for VEGF-C and VEGF-D [60]. The other two receptors of VEGFs that have been identified, VEGFR-1 and VEGFR-2, are expressed mainly in the blood vascular endothelium [61]. Studies on the molecular mechanisms controlling the lymphatic vessels have shown that vascular endothelial growth factors C and D specifically control lymphangiogenesis in humans by activating the VEGF receptor-3 (VEGFR-3) [62, 63] [61, 64]. VEGFR-3 has also been linked to human hereditary LE [65].

VEGFs and VEGFR3 are needed for the development of lymphatic vessels. However, their overproduction leads to lymphatic dilation and LE development [66]. In animal models, overexpression of VEGF-C in the skin of transgenic mice resulted in lymphatic endothelial proliferation and dilation of lymph vessels [64] with a resemblance to lymphatics infected with filarial parasites. These transgenic mice then developed a lymphedema-like phenotype characterized by swelling of feet, edema, and dermal fibrosis [67], similar to what is observed in humans.

Several studies by Debrah et al. [27, 28, 58], involving our study participants, showed that plasma levels of VEGFs and a VEGF soluble receptor, sVEGFR-3, are significantly elevated in patients infected with filarial worms, and a correlation was found between sVEGFR-3, lymphatic dilation, and pathology development. Targeting the filarial worms by doxycycline reduced the levels of VEGFs/sVEGFR-3, with amelioration of dilated supratesticular lymphatic vessels and reduction in LE and hydrocele stages. A mechanism that could be due to the non-antimicrobial activity of tetracyclines. The fact that the VEGF and sVEGFR-3 reduction preceded the improvement of pathology indicates a possible causal interaction between lymphangiogenic factors and lymphatic pathology, rather than only a coincidence or an epiphenomenon.

Pathway interaction of associated genes was done using the MetaCore software package [68]. Genes in the angiogenic pathway were shown to be involved in a complex relationship (Fig. 3). Genes in gray background had SNPs that were directly associated with LE in this study (MMP-2, CEACAM-1, and VEGFR-3). However, during pathway analysis, other genes were found to either activate or inhibit those genes that were found to be directly associated with LE.

CEACAM-1, MMP-2, and VEGFR-3 genes are directly found to be involved in LE development. The pathway interaction of LE associated genes provides information on the involvement of other genes and probably other SNPs in the development of LE. CEACAM-1 is known to be involved in angiogenesis, and its overexpression in human dermal microvascular endothelial cells (HDMEC) leads to an up-regulation of VEGF-C, VEGF-D, and VEGFR-3 [39, 69]. The down-regulation of CEACAM-1 results in deregulation of beta-catenin, which is known to be associated with malignant transformation [70, 71]. TALIN interacts with CEACAM-1 and increases its activity [72].

Prospero homeobox protein-1 (PROX-1) activates VEGFR-3 with a subsequent increase in receptor expression [73]. Activation of PROX-1 is positively regulated by beta-catenin signaling [74]. VEGF-A is essential for cancer neovascularization and cancer invasion by promoting endothelial mitogenesis and permeability. However, the overexpression is known to increase MMP-2 levels in glioblastoma [75]. At the same time, MMP-2 up-regulates TGF beta-1, MMP-9, and VEGFR-1 (Fig. 3). The interaction of these genes can, therefore, be said to contribute to the disease.

This study is the first to determine the genotype frequencies of CEACAM-1 and MMP-2 in our study population. We have gone a further step to confirm the involvement of lymphangiogenic/angiogenic factors in the development of pathology of LF. Our pathway analysis also supports the assertion that LE due to *W. bancrofti* infection is a complex disease and not caused by a single genetic factor.

Conclusion

SNPs in the angiogenic and lymphangiogenic pathway contribute to the development of filarial LE. The genes whose SNPs were found to be associated with LE (CEACAM-1, MMP-2, and VEGFR-3) have an influence in the vascular endothelial growth factors either directly or indirectly, supporting the fact that VEGFs are major functional proteins in filarial LE development and that the identified angiogenic/lymphangiogenic factors function through influencing the VEGFs. The direct activity of MMP-2 on the extracellular matrix which results in the progressive damage of vascular walls in vascular diseases when elevated could also have direct influence in the development of LE. The outcomes of this study are important in the diagnosis of the disease as well as the development of future vaccine. Although the associations were not as strong as anticipated, they underscore

the fact that LE is a complex disease caused by multiple genetic markers. This complex interaction of genes therefore calls for a first-stage genome-wide association study (GWAS) to identify all genes associated with LE development that could serve as markers for diagnosis of the disease, and identify pathways that could be targeted by chemotherapeutics to prevent/reduce lymphedema, providing amelioration of disease to the thousands of people with LE.

The 1000 Genome haplotypes is a valuable resource to infer information on further SNPs via genotype imputation. Unfortunately, the number of SNPs per linkage disequilibrium/gene region available in our study was not sufficient to enable the application of the IMPUTE2 software. In our ongoing GWAS study with a larger sample size, we hope to be able to achieve denser genotyping to be able to make use of public resources like 1000 Genomes or the haplotype reference panel (http://www.haplotype-reference-consortium.org/).

Methods

Study design

The study participants were selected from the Nzema East and Ahanta West districts in the Western Region of Ghana, which are the LF endemic districts in Ghana.

Participants numbering 967—comprising 285 lymphedema patients (Cases) and 682 infected patients without pathology (controls) were enrolled into the study. All volunteers included in the study underwent finger prick blood collection at night for assessment and quantification of microfilariae (Mf) in the peripheral blood. Circulating filarial antigen (CFA) test to identify infected patients who did not have Mf was also done. The procedures for microfilaria and CFA determination were done as described [27, 76]. LE patients were examined separately by a clinician conversant with the symptoms of LF, and the staging was done according to Dreyer et al. [32].

Genotyping

DNA extraction

In the field lab, volunteer's blood was mixed with equal volumes of 8 M urea for preservation at ambient temperature. The Chemagen platform (Chemagen Biopolymer-Technologies AG, Baesweiler, Germany) in Bonn was used for the DNA extraction as per kit instructions. After the genomic DNA was isolated, a DNA stock concentration of 100 ng/ml was diluted to 15 ng/μl working concentration with Tris-EDTA buffer. Quality-checked DNA samples with an A260/A280 ratio between 1.7 and 2.0 were pipetted into aliquots of 2 μl (30 ng DNA) per well in a 384-well plate for genotyping.

SNP genotyping

To investigate several candidate genes conferring susceptibility or protection to LE, the SNP databases National Center for Biotechnology Information (NCBI) and Online Mendelian Inheritance in Man (OMIM) were used [77, 78]. These databases provide information on genetic variations. SNP Annotation and Proxy Search (SNAP) (http://www.broadinstitute.org) was also used to find proxy SNPs based on linkage disequilibrium, physical distance, and/or membership in selected commercial genotyping arrays. Pair-wise linkage disequilibrium was pre-calculated based on phased genotype data from the International HapMap Project (www.hapmap.org), which has since been replaced by the 1000 Genomes Project (www.1000genomes.org).

In all, 64 genes of interest known to have a role in inflammation, angiogenesis/lymphangiogenesis, extravasation of fluid, and also in other mechanisms, such as cell differentiation during tumorigenesis, were selected. A total of 131 functional variants from the 64 genes were successfully genotyped and analyzed (Additional file 2: Table S1). Genotyping was done in multiplex reactions using the MassARRAY (Sequenom Inc., San Diego, USA) platform. Identified monoallelic SNPs and SNPs with genotyping call rates of < 95% were excluded from the analysis.

Determination of plasma levels of angiogenic/ lymphangiogenic molecules

Plasma concentrations of the angiogenic/lymphangiogenic molecules associated SNPs were assessed as a measure of the functional phenotypes in the genotyped samples. Blood from cases and controls was taken with ethylenediaminetetraacetic acid (EDTA) tubes, and plasma was collected for measurement of protein levels of CEACAM-1 and MMP-2 using commercially available kits (R&D Systems, Wiesbaden, Germany) according to the manufacturer's protocol. ELISA plates were read with a Wallac VICTOR^2 1420 (PerkinElmer Inc., Waltham MA, USA) at 450 nm and corrected with a second read at 540 nm. A standard curve was created for each plate using a four-parameter logistic (4-PL) curve fit. Only plates with a standard curve r^2 > 0.99 were evaluated.

Statistical analyses

FamHap version 19 software was used for single marker analysis as well as haplotype analysis of association of the SNPs with cases or controls [79]. Genotype and haplotype frequencies were summarized as percentages. Statistical significance for the single marker SNP analysis was calculated using the Cochrane-Armitage test for trend with $P_{ATT} \leq 0.05$ considered significant. The Armitage test is less influenced by deviation from Hardy-Weinberg equi-

librium (HWE), and the result obtained is valid and acceptable even when a group is not in HWE [35].

Genotype-specific risks were estimated as odds ratios (ORs) with 95% confidence intervals (CIs). Analysis of dominant or recessive association was done using the DeFinetti formula at ihg.gsf.de/cgi-bin/hw/hwa1.pl.

Haplotype analyses calculated an omnibus statistic using 200,000 simulations of the case-control data correcting for multiple testing (global P value). MetaCore™ software (GeneGo Inc., St. Joseph, MI, USA) was used to test for SNP interaction and network analysis [68]. Unpaired t test with GraphPad Prism version 6 (La Jolla, California, USA, www.graphpad.com) software was used for comparing the differences in the plasma concentration of the samples and for plotting the graphs from data generated. $P \leq 0.05$ were considered statistically significant.

Abbreviations

CEACAM-1: Carcinoembryonic antigen-related cell adhesion molecule-1; CHIT-1: Chitinase-1; CTLA-4: Cytotoxic T-lymphocyte-associated protein 4; EDTA: Ethylenediaminetetraacetic acid; ELISA: Enzyme-linked immunosorbent assay; IL-4: Interleukin-10; LE: Lymphedema; LF: Lymphatic filariasis; MBL-2: Mannose-binding lectin-2; MHC: Major histocompatibility complex; MMP-2: Matrix metalloprotease-2; PROX-1: Prospero homeobox protein-1; SNP: Single nucleotide polymorphism; TGFβ-1: Transforming growth factor-β-1; TNF-α: tumor necrosis factor-alpha; VEGFR-3: Vascular endothelial growth factor receptor-3

Acknowledgements

We thank all the volunteers in the study, as well as the Ahanta West and Nzema East District Health Directorates in the Western Region of Ghana for their cooperation. We are grateful to the staff of filariasis team at Kumasi Centre for Collaborative Research into Tropical Medicine for their support and cooperation. We are again grateful to the DFG, the Bill and Melinda Gates Foundation, the European Commission (No. 242121, EPIAF), and the European Foundation Initiative for African Research into Neglected Tropical Diseases for funding this work.
The technical assistance of Lydia Lust, Institute for Medical Microbiology, Immunology and Parasitology is gratefully appreciated.
We are also thankful for the helpful discussions of the results with Dr. Johannes Schumacher and Dr. Kerstin Ludwig of the Institute of Human Genetics, University of Bonn.

Funding

This work was funded through the Deutsche Forschungsgemeinschaft (DFG, German Research Foundation) within the German-African Cooperation Projects in Infectology (PF 673/2-1 and PF 673/4-1). Volunteers were recruited, and samples for genotyping were collected with support through grants to Achim Hoerauf from the Liverpool School of Tropical Medicine as part of the A-WOL (Anti-Wolbachia) Consortium funded by the Bill and Melinda Gates Foundation, the European Commission (No. 242121, EPIAF), and a grant to Alexander Yaw Debrah from the European Foundation Initiative for African Research into Neglected Tropical Diseases Grant (No. 1/81995 and 8652).

Authors' contributions

AYD, AH, and KP conceptualized the study and obtained funding. AYD, LBD, JOM, and YM performed the field work and obtained patient samples. LBD and AA organized patient samples. LBD, AA, and FFB performed the experiments. LBD, TB, CH, and JHF analyzed the data. LBD, AA, and KP wrote the manuscript. AH contributed critical suggestions for the manuscript. All authors read and approved the final manuscript.

Ethics approval and consent to participate

The study was approved by the Committee on Human Research, Publications and Ethics of the School of Medical Sciences of the Kwame Nkrumah University of Science and Technology (KNUST), and Komfo Anokye Teaching Hospital, Kumasi, Ghana, as well as by the Ethics Committee of the University Hospital of Bonn, Germany. Permission was also obtained from the Nzema East and Ahanta West District Health Directorates, Ghana.
Before commencement of recruitment and sample collection, meetings were held in the study communities to explain in detail the purpose and procedures of the study. The inconveniences involved, such as blood sampling, were also explained to the participants. Verbal consent to perform the study in the villages was obtained from community leaders, i.e., chiefs and elders of the selected communities, and written informed consent was obtained from all participants. The study was undertaken according to the principles of the Helsinki Declaration of 1975 (as revised 2008).

Consent for publication

Consent to publish was obtained from the participants and the guardians of participants below 18 years of age who are considered in Ghana as minors at the time of recruitment into the study.

Competing interests

The authors declare that they have no competing interests.

Author details

[1]Kumasi Centre for Collaborative Research in Tropical Medicine, Kumasi, Ghana. [2]Department of Clinical Microbiology, Kwame Nkrumah University of Science and Technology, Kumasi, Ghana. [3]Institute for Medical Microbiology, Immunology and Parasitology, University Hospital Bonn, Sigmund-Freud-Str. 25, 53127 Bonn, Germany. [4]Faculty of Allied Health Sciences of Kwame Nkrumah University of Science and Technology, Kumasi, Ghana. [5]Institute of Human Genetics, University of Bonn, Bonn, Germany. [6]Department of Genomics, Life and Brain Center, University of Bonn, Bonn, Germany. [7]Institute for Medical Biometry, Informatics and Epidemiology, University of Bonn, Bonn, Germany. [8]Bonn-Aachen International Center for Information Technology (B-IT), University of Bonn, Bonn, Germany.

References

1. WHO. Global programme to eliminate lymphatic filariasis: progress report, 2016. Wkly Epidemiol Rec. 2017;92(40):594–607.
2. Hoerauf A, Pfarr K, Mand S, Debrah AY, Specht S. Filariasis in Africa—treatment challenges and prospects. Clin Microbiol Infect. 2011; 17(7):977–85.
3. Noroes J, Addiss D, Santos A, Medeiros Z, Coutinho A, Dreyer G. Ultrasonographic evidence of abnormal lymphatic vessels in young men with adult Wuchereria bancrofti infection in the scrotal area. J Urol. 1996; 156(2 Pt 1):409–12.
4. Noroes J, Addiss D, Amaral F, Coutinho A, Medeiros Z, Dreyer G. Occurrence of living adult Wuchereria bancrofti in the scrotal area of men with microfilaraemia. Trans R Soc Trop Med Hyg. 1996;90(1):55–6.
5. Nutman TB, Kumaraswami V. Regulation of the immune response in lymphatic filariasis: perspectives on acute and chronic infection with Wuchereria bancrofti in South India. Parasite Immunol. 2001;23(7): 389–99.
6. Tisch DJ, Hazlett FE, Kastens W, Alpers MP, Bockarie MJ, Kazura JW. Ecologic and biologic determinants of filarial antigenemia in bancroftian filariasis in Papua New Guinea. J Infect Dis. 2001;184(7):898–904.
7. Couto RA, Kulungowski AM, Chawla AS, Fishman SJ, Greene AK. Expression of angiogenic and vasculogenic factors in human lymphedematous tissue. Lymphat Res Biol. 2011;9(3):143–9.
8. Ramaiah KD, Ottesen EA. Progress and impact of 13 years of the global programme to eliminate lymphatic filariasis on reducing the burden of filarial disease. PLoS Negl Trop Dis. 2014;8(11):e3319.
9. Steel C, Nutman TB. CTLA-4 in filarial infections: implications for a role in diminished T cell reactivity. J Immunol. 2003;170(4):1930–8.
10. Babu S, Blauvelt CP, Kumaraswami V, Nutman TB. Regulatory networks induced by live parasites impair both Th1 and Th2 pathways in patent lymphatic filariasis: implications for parasite persistence. J Immunol. 2006; 176(5):3248–56.
11. Taylor MJ, Cross HF, Bilo K. Inflammatory responses induced by the filarial nematode Brugia malayi are mediated by lipopolysaccharide-like activity from endosymbiotic Wolbachia bacteria. J Exp Med. 2000;191(8):1429–36.
12. Harnett W, Harnett MM, Leung BP, Gracie JA, McInnes IB. The anti-inflammatory potential of the filarial nematode secreted product, ES-62. Curr Top Med Chem. 2004;4(5):553–9.
13. King CL, Connelly M, Alpers MP, Bockarie M, Kazura JW. Transmission intensity determines lymphocyte responsiveness and cytokine bias in human lymphatic filariasis. J Immunol. 2001;166(12):7427–36.
14. Steel C, Guinea A, McCarthy JS, Ottesen EA. Long-term effect of prenatal exposure to maternal microfilaraemia on immune responsiveness to filarial parasite antigens. Lancet. 1994;343(8902):890–3.
15. Malhotra I, Ouma JH, Wamachi A, Kioko J, Mungai P, Njzovu M, Kazura JW, King CL. Influence of maternal filariasis on childhood infection and immunity to Wuchereria bancrofti in Kenya. Infect Immun. 2003;71(9):5231–7.
16. Mahanty S, Ravichandran M, Raman U, Jayaraman K, Kumaraswami V, Nutman TB. Regulation of parasite antigen-driven immune responses by interleukin-10 (IL-10) and IL-12 in lymphatic filariasis. Infect Immun. 1997; 65(5):1742–7.
17. Cuenco KT, Halloran ME, Louis-Charles J, Lammie PJ. A family study of lymphedema of the leg in a lymphatic filariasis-endemic area. Am J Trop Med Hyg. 2004;70(2):180–4.
18. Cuenco KT, Halloran ME, Lammie PJ. Assessment of families for excess risk of lymphedema of the leg in a lymphatic filariasis-endemic area. Am J Trop Med Hyg. 2004;70(2):185–90.
19. Terhell AJ, Price R, Koot JW, Abadi K, Yazdanbakhsh M. The development of specific IgG4 and IgE in a paediatric population is influenced by filarial endemicity and gender. Parasitology. 2000;121(5):535–43.
20. Wahyuni S, Houwing-Duistermaat JJ, Syafruddin ST, Yazdanbakhsh M, Sartono E. Clustering of filarial infection in an age-graded study: genetic, household and environmental influences. Parasitology. 2004;128(3):315–21.
21. Chesnais CB, Sabbagh A, Pion SD, Missamou F, Garcia A, Boussinesq M. Familial aggregation and heritability of Wuchereria bancrofti infection. J Infect Dis. 2016;
22. Debrah AY, Batsa L, Albers A, Mand S, Toliat MR, Nurnberg P, Adjei O, Hoerauf A, Pfarr K. Transforming growth factor-beta1 variant Leu10Pro is associated with both lack of microfilariae and differential microfilarial loads in the blood of persons infected with lymphatic filariasis. Hum Immunol. 2011;72(11):1143–8.
23. Junpee A, Tencomnao T, Sanprasert V, Nuchprayoon S. Association between toll-like receptor 2 (TLR2) polymorphisms and asymptomatic bancroftian filariasis. Parasitol Res. 2010;107(4):807–16.
24. Choi EH, Zimmerman PA, Foster CB, Zhu S, Kumaraswami V, Nutman TB, Chanock SJ. Genetic polymorphisms in molecules of innate immunity and susceptibility to infection with Wuchereria bancrofti in South India. Genes Immun. 2001;2(5):248–53.
25. Idris ZM, Miswan N, Muhi J, Mohd TA, Kun JF, Noordin R. Association of CTLA4 gene polymorphisms with lymphatic filariasis in an east Malaysian population. Hum Immunol. 2011;72(7):607–12.
26. Yasmeen Sheik1 SFQ, Ananthapur Venkateshwari SN, BMAP N. Association of IL-10 & IL-10RA polymorphisms with lymphatic Filariasis in south Indian population. Int J Trop Dis Health. 2012:2.
27. Debrah AY, Mand S, Specht S, Marfo-Debrekyei Y, Batsa L, Pfarr K, Larbi J, Lawson B, Taylor M, Adjei O, et al. Doxycycline reduces plasma VEGF-C/sVEGFR-3 and improves pathology in lymphatic filariasis. PLoS Pathog. 2006; 2(9):e92.

28. Debrah AY, Mand S, Toliat MR, Marfo-Debrekyei Y, Batsa L, Nurnberg P, Lawson B, Adjei O, Hoerauf A, Pfarr K. Plasma vascular endothelial growth factor-a (VEGF-A) and VEGF-A gene polymorphism are associated with hydrocele development in lymphatic filariasis. Am J Trop Med Hyg. 2007;77(4):601–8.

29. Panda AK, Sahoo PK, Kerketta AS, Kar SK, Ravindran B, Satapathy AK. Human lymphatic filariasis: genetic polymorphism of endothelin-1 and tumor necrosis factor receptor II correlates with development of chronic disease. J Infect Dis. 2011;204(2):315–22.

30. Weinkopff T, Mackenzie C, Eversole R, Lammie PJ. Filarial excretory-secretory products induce human monocytes to produce lymphangiogenic mediators. PLoS Negl Trop Dis. 2014;8(7):e2893.

31. Sheik Y, Qureshi SF, Mohhammed B, Nallari P. FOXC2 and FLT4 gene variants in lymphatic Filariasis. Lymphat Res Biol. 2015;13(2):112–9.

32. Dreyer GDA, Dreyer P, Noroes J. Basic lymphoedema management: treatment and prevention of problems associated with lymphatic filariasis. London: Royal Free and University College Medical School; 2002.

33. Mand S, Debrah AY, Klarmann U, Batsa L, Marfo-Debrekyei Y, Kwarteng A, Specht S, Belda-Domene A, Fimmers R, Taylor M, et al. Doxycycline improves filarial lymphedema independent of active filarial infection: a randomized controlled trial. Clin Infect Dis. 2012;55(5):621–30.

34. Clark AG. The role of haplotypes in candidate gene studies. Genet Epidemiol. 2004;27(4):321–33.

35. Armittage P. Tests for linear trends in proportions and frequencies. Biometrics. 1995;11:375–86.

36. Kumaraswami V. The clinical manifestation of lymphatic filariasis. In: Lymphatic Filariasis. Volume 1, edn. Edited by Nutman T. London: Imperial College Press; 2000.

37. Karpanen T, Alitalo K. Molecular biology and pathology of lymphangiogenesis. Annu Rev Pathol. 2008;3:367–97.

38. Pfarr KM, Debrah AY, Specht S, Hoerauf A. Filariasis and lymphoedema. Parasite Immunol. 2009;31(11):664–72.

39. Gu A, Tsark W, Holmes KV, Shively JE. Role of Ceacam1 in VEGF induced vasculogenesis of murine embryonic stem cell-derived embryoid bodies in 3D culture. Exp Cell Res. 2009;315(10):1668–82.

40. Ergun S, Kilik N, Ziegeler G, Hansen A, Nollau P, Gotze J, Wurmbach JH, Horst A, Weil J, Fernando M, et al. CEA-related cell adhesion molecule 1: a potent angiogenic factor and a major effector of vascular endothelial growth factor. Mol Cell. 2000;5(2):311–20.

41. Oliveira-Ferrer L, Tilki D, Ziegeler G, Hauschild J, Loges S, Irmak S, Kilic E, Huland H, Friedrich M, Ergun S. Dual role of carcinoembryonic antigen-related cell adhesion molecule 1 in angiogenesis and invasion of human urinary bladder cancer. Cancer Res. 2004;64(24):8932–8.

42. Matsuda Y. CEACAM1 (carcinoembryonic antigen-related cell adhesion molecule 1 (biliary glycoprotein). Atlas Database. 2009;14(4):4.

43. http://www.ncbi.nlm.nih.gov/snp. AA.

44. Bennuru S, Maldarelli G, Kumaraswami V, Klion AD, Nutman TB. Elevated levels of plasma angiogenic factors are associated with human lymphatic filarial infections. Am J Trop Med Hyg. 83(4):884–90.

45. Simeone DM, Ji B, Banerjee M, Arumugam T, Li D, Anderson MA, Bamberger AM, Greenson J, Brand RE, Ramachandran V, et al. CEACAM1, a novel serum biomarker for pancreatic cancer. Pancreas. 2007;34(4):436–43.

46. Bennuru S, Nutman TB. Lymphangiogenesis and lymphatic remodeling induced by filarial parasites: implications for pathogenesis. PLoS Pathog. 2009;5(12):e1000688.

47. Zankl A, Pachman L, Poznanski A, Bonafe L, Wang F, Shusterman Y, Fishman DA, Superti-Furga A. Torg syndrome is caused by inactivating mutations in MMP2 and is allelic to NAO and Winchester syndrome. J Bone Miner Res. 2007;22(2):329–33.

48. Bedi A, Fox AJ, Kovacevic D, Deng XH, Warren RF, Rodeo SA. Doxycycline-mediated inhibition of matrix metalloproteinases improves healing after rotator cuff repair. Am J Sports Med. 38(2):308–17.

49. Candelario-Jalil E, Thompson J, Taheri S, Grossetete M, Adair JC, Edmonds E, Prestopnik J, Wills J, Rosenberg GA. Matrix metalloproteinases are associated with increased blood-brain barrier opening in vascular cognitive impairment. Stroke. 2011;42(5):1345–50.

50. Peng ZH, Wan DS, Li LR, Chen G, ZH L, XJ W, Kong LH, Pan ZZ. Expression of COX-2, MMP-2 and VEGF in stage II and III colorectal cancer and the clinical significance. Hepato-Gastroenterology. 2011;58(106):369–76.

51. Fatar M, Stroick M, Steffens M, Senn E, Reuter B, Bukow S, Griebe M, Alonso A, Lichtner P, Bugert P, et al. Single-nucleotide polymorphisms of MMP-2 gene in stroke subtypes. Cerebrovasc Dis. 2008;26(2):113–9.

52. Shigemori Y, Katayama Y, Mori T, Maeda T, Kawamata T. Matrix metalloproteinase-9 is associated with blood-brain barrier opening and brain edema formation after cortical contusion in rats. Acta Neurochir Suppl. 2006;96:130–3.

53. Kelly MA, Shuaib A, Todd KG. Matrix metalloproteinase activation and blood-brain barrier breakdown following thrombolysis. Exp Neurol. 2006; 200(1):38–49.

54. Yang Y, Estrada EY, Thompson JF, Liu W, Rosenberg GA. Matrix metalloproteinase-mediated disruption of tight junction proteins in cerebral vessels is reversed by synthetic matrix metalloproteinase inhibitor in focal ischemia in rat. J Cereb Blood Flow Metab. 2007;27(4):697–709.

55. Detry B, Erpicum C, Paupert J, Blacher S, Maillard C, Bruyere F, Pendeville H, Remacle T, Lambert V, Balsat C, et al. Matrix metalloproteinase-2 governs lymphatic vessel formation as an interstitial collagenase. Blood. 2012;119(21):5048–56.

56. Pasternak B, Fellenius M, Aspenberg P. Doxycycline impairs tendon repair in rats. Acta Orthop Belg. 2006;72(6):756–60.

57. Lo IK, Marchuk LL, Hollinshead R, Hart DA, Frank CB. Matrix metalloproteinase and tissue inhibitor of matrix metalloproteinase mRNA levels are specifically altered in torn rotator cuff tendons. Am J Sports Med. 2004;32(5):1223–9.

58. Debrah AY, Mand S, Marfo-Debrekyei Y, Batsa L, Pfarr K, Lawson B, Taylor M, Adjei O, Hoerauf A. Reduction in levels of plasma vascular endothelial growth factor-a and improvement in hydrocele patients by targeting endosymbiotic Wolbachia sp. in Wuchereria bancrofti with doxycycline. Am J Trop Med Hyg. 2009;80(6):956–63.

59. http://www.ncbi.nlm.nih.gov/projects/SNP/snp_ref.cgi?rs=75614493#locus. Accessed 14 May 2010.

60. Spiegel R, Ghalamkarpour A, Daniel-Spiegel E, Vikkula M, Shalev SA. Wide clinical spectrum in a family with hereditary lymphedema type I due to a novel missense mutation in VEGFR3. J Hum Genet. 2006;51(10):846–50.

61. Veikkola T, Jussila L, Makinen T, Karpanen T, Jeltsch M, Petrova TV, Kubo H, Thurston G, McDonald DM, Achen MG, et al. Signalling via vascular endothelial growth factor receptor-3 is sufficient for lymphangiogenesis in transgenic mice. EMBO J. 2001;20(6):1223–31.

62. Korpelainen El, Alitalo K. Signaling angiogenesis and lymphangiogenesis. Curr Opin Cell Biol. 1998;10(2):159–64.

63. Achen MG, Jeltsch M, Kukk E, Makinen T, Vitali A, Wilks AF, Alitalo K, Stacker SA. Vascular endothelial growth factor D (VEGF-D) is a ligand for the tyrosine kinases VEGF receptor 2 (Flk1) and VEGF receptor 3 (Flt4). Proc Natl Acad Sci USA. 1998;95(2):548–53.

64. Jeltsch M, Kaipainen A, Joukov V, Meng X, Lakso M, Rauvala H, Swartz M, Fukumura D, Jain RK, Alitalo K. Hyperplasia of lymphatic vessels in VEGF-C transgenic mice. Science. 1997;276(5317):1423–5.

65. Yu Z, Wang J, Peng S, Dong B, Li Y. Identification of a novel VEGFR-3 missense mutation in a Chinese family with hereditary lymphedema type I. J Genet Genomics. 2007;34(10):861–7.

66. Makinen T, Jussila L, Veikkola T, Karpanen T, Kettunen MI, Pulkkanen KJ, Kauppinen R, Jackson DG, Kubo H, Nishikawa S, et al. Inhibition of lymphangiogenesis with resulting lymphedema in transgenic mice expressing soluble VEGF receptor-3. Nat Med. 2001;7(2):199–205.

67. Kaipainen A, Korhonen J, Mustonen T, van Hinsbergh VW, Fang GH, Dumont D, Breitman M, Alitalo K. Expression of the fms-like tyrosine kinase 4 gene becomes restricted to lymphatic endothelium during development. Proc Natl Acad Sci USA. 1995;92(8):3566–70.

68. Froehlich H, Fellmann M, Sueltmann H, Poustka A, Beissbarth T. Large scale statistical inference of signaling pathways from RNAi and microarray data. BMC Bioinformatics. 2007;8:386.

69. Horst AK, Ito WD, Dabelstein J, Schumacher U, Sander H, Turbide C, Brummer J, Meinertz T, Beauchemin N, Wagener C. Carcinoembryonic antigen-related cell adhesion molecule 1 modulates vascular remodeling in vitro and in vivo. J Clin Invest. 2006;116(6):1596–605.

70. Jin L, Li Y, Chen CJ, Sherman MA, Le K, Shively JE. Direct interaction of tumor suppressor CEACAM1 with beta catenin: identification of key residues in the long cytoplasmic domain. Exp Biol Med. 2008;233(7):849–59.

71. Reyes M, Rojas-Alcayaga G, Maturana A, Aitken JP, Rojas C, Ortega AV. Increased nuclear beta-catenin expression in oral potentially malignant lesions: a marker of epithelial dysplasia. Med Oral Patol Oral Cir Bucal. 2015; 20(5):e540–6.

72. Muller MM, Singer BB, Klaile E, Obrink B, Lucka L. Transmembrane CEACAM1 affects integrin-dependent signaling and regulates extracellular matrix

protein-specific morphology and migration of endothelial cells. Blood. 2005;
105(10):3925–34.

73. Kilic N, Oliveira-Ferrer L, Neshat-Vahid S, Irmak S, Obst-Pernberg K,
Wurmbach JH, Loges S, Kilic E, Weil J, Lauke H, et al. Lymphatic
reprogramming of microvascular endothelial cells by CEA-related cell
adhesion molecule-1 via interaction with VEGFR-3 and Prox1. Blood. 2007;
110(13):4223–33.

74. Karalay O, Doberauer K, Vadodaria KC, Knobloch M, Berti L, Miquelajauregui
A, Schwark M, Jagasia R, Taketo MM, Tarabykin V, et al. Prospero-related
homeobox 1 gene (Prox1) is regulated by canonical Wnt signaling and has
a stage-specific role in adult hippocampal neurogenesis. Proc Natl Acad Sci
USA. 2011;108(14):5807–12.

75. Gong J, Zhu S, Zhang Y, Wang J. Interplay of VEGFa and MMP2 regulates
invasion of glioblastoma. Tumour Biol. 2014;35(12):11879–85.

76. Debrah AY, Mand S, Marfo-Debrekyei Y, Batsa L, Albers A, Specht S,
Klarmann U, Pfarr K, Adjei O, Hoerauf A. Macrofilaricidal activity in
Wuchereria bancrofti after 2 weeks treatment with a combination of
rifampicin plus doxycycline. J Parasitol Res. 2011;2011:201617.

77. http://www.ncbi.nlm.nih.gov/snp. Accessed 14 May 2010.

78. http://omim.org/entry/109770. Accessed 14 May 2010.

79. Herold C, Becker T. Genetic association analysis with FAMHAP: a major
program update. Bioinformatics. 2009;25(1):134–6.

12

Identification of functional single nucleotide polymorphisms in the branchpoint site

Hung-Lun Chiang[1,2], Jer-Yuarn Wu[2,3] and Yuan-Tsong Chen[1,2,4*]

Abstract

Background: The human genome contains millions of single nucleotide polymorphisms (SNPs); many of these SNPs are intronic and have unknown functional significance. SNPs occurring within intron branchpoint sites, especially at the adenine (A), would presumably affect splicing; however, this has not been systematically studied. We employed a splicing prediction tool to identify human intron branchpoint sites and screened dbSNP for identifying SNPs located in the predicted sites to generate a genome-wide branchpoint site SNP database.

Results: We identified 600 SNPs located within branchpoint sites; among which, 216 showed a change in A. After scoring the SNPs by counting the As in the ± 10 nucleotide region, only four SNPs were identified without additional As (rs13296170, rs12769205, rs75434223, and rs67785924). Using minigene constructs, we examined the effects of these SNPs on splicing. The three SNPs (rs13296170, rs12769205, and rs75434223) with nucleotide substitution at the A position resulted in abnormal splicing (exon skipping and/or intron inclusion). However, rs67785924, a 5-bp deletion that abolished the branchpoint A nucleotide, exhibited normal RNA splicing pattern, presumably using two of the downstream As as alternative branchpoints. The influence of additional As on splicing was further confirmed by studying rs2733532, which contains three additional As in the ± 10 nucleotide region.

Conclusions: We generated a high-confidence genome-wide branchpoint site SNP database, experimentally verified the importance of A in the branchpoint, and suggested that other nearby As can protect branchpoint A substitution from abnormal splicing.

Keywords: RNA splicing, Single nucleotide polymorphism, Branchpoint site, Minigene

Background

Precursor messenger RNA (pre-mRNA) splicing is essential for gene expression in eukaryotes [1–3]. Splicing comprises a two-step trans-esterification reaction of intron removal and exon ligation. Splicing depends on the spliceosome, which is a large complex of small nuclear ribonucleoproteins (snRNPs; U1, U2, U4/U6, and U5) and non-snRNPs; these components recognize the target sequence and assemble on the pre-mRNA [4]. The intronic target sequences include a 5′ donor site, a 3′ acceptor site, a polypyrimidine tract (PPT) upstream of the 3′ acceptor, and a branchpoint site upstream of the PPT. The branchpoint contains a conserved splicing signal important for spliceosome assembly and lariat intron formation, with a consensus sequence (YNCTRAY, which differs slightly between species; Y is pyrimidine, N is any nucleotide, and R is purine) [5]. Tools to predict branchpoint sites based on the consensus sequence have been developed [6–10]; more recently, an NGS-based genome-wide study of splicing branchpoints was published [11–13].

Within the consensus branchpoint site sequence YNCTRAY, the well conserved A appears to be the most important one. A previous report showed that IVS4,-22A>G in the *LCAT* gene, which is an A to G change at the splicing branchpoint, resulted in intron inclusion and exon skipping of the mRNA and caused the Fish-eye disease [14]. There is also a report suggesting that mutations in the branchpoint sequence, especially the adenine (A) may result in aberrant pre-mRNA splicing and give rise to human genetic disorders [15].

---

* Correspondence: chen0010@ibms.sinica.edu.tw
[1]Institute of Clinical Medicine, National Yang-Ming University, Taipei, Taiwan
[2]Institute of Biomedical Sciences, Academia Sinica, Taipei, Taiwan
Full list of author information is available at the end of the article

There are millions of SNPs in the human genome; many are intronic, and have unknown functional significance. SNPs at the intron branchpoint sites, especially the adenine (A) nucleotide, would presumably affect splicing; however, this has not been systematically studied. It is therefore desirable to create a genome-wide branchpoint site SNP database, and perform functional analysis.

In the present study, we used an in silico splicing prediction program for branchpoint site prediction and combined its predictions with dbSNP data, to create a genome-wide branchpoint site SNP dataset. We experimentally verified the importance of A in the branchpoint, and further suggested that other nearby As may also influence RNA splicing.

Methods

Creating a dataset of SNPs located within branchpoint sites

All exon (n = 404,454) and intron (n = 363,190) sequences of the human genes were collected (human 1000genome v37), and the SROOGLE tool, which is based on two different algorithms, was used to predict branchpoint sites [8]. We were able to predict 338,787 (93.3%) branchpoint sites as output. Next, we screened NCBI's dbSNP for candidate SNPs located within the set of predicted branchpoint sites. Because adenine is the most important nucleotide at the branchpoint site, and 90% of branchpoint sites are upstream 19–37 bp from the 3′ acceptor [12, 13], we scored each SNP by the number of adenines found in the ± 10 nucleotide region (20 nucleotides total) surrounding the SNP. The SNPs identified in the predicted branchpoint sites and reported lariat sequences associated with these SNPs [12] are tabulated in Additional file 1.

Cell lines and genotyping

293T cells were obtained from The Bioresource Collection and Research Center (Hsinchu, Taiwan). Randomly selected EBV-transformed normal control B cell lines (n = 96) were obtained from the Taiwan Han Chinese Cell and Genome Bank [16]. Genomic DNA was extracted from the cell lines using the Gentra Puregene® Blood Kit (Gentra Systems, MN, USA) and genotyped for the SNPs of interest (*XPC* rs2733532, *PIP5KL1* rs13296170, *CYP2C19* rs12769205, *MYH11* rs75434223, and *KLC3* rs67785924), to identify cell lines carrying different branchpoint site SNP alleles (Table 1). The primer sequences are provided in Additional file 2.

Minigene constructs

Minigene constructs (Fig. 1) encompassing exons/introns of interest were prepared by amplifying introns and exons from genomic DNA; the amplified regions comprised *PIP5KL1* (chr9:130688147-130689612; 1466 bp), *CYP2C19* (chr10:96534815-96535296; 482 bp), *MYH11* (chr16:15833924-15835748; 1825 bp), and *KLC3* (chr19:45853898-45854704; 807 bp). The amplified minigenes were cloned into pJET1.2/blunt cloning vector (Thermo Scientific, Waltham, MA, USA), and subsequently sub-cloned into pEGFP-C1 vector (Additional file 2 indicates each restriction enzyme site). *PIP5KL1* and *MYH11*'s SNPs and *KLC3*'s seventh and eighth adenine substitution were used the GeneArt™ Site-Directed Mutagenesis System (Thermo Scientific, Waltham, MA, USA) with mutagenesis primers (Additional file 2). The complete sequences of the minigene constructs were confirmed by Sanger sequencing. Transient transfections of minigene constructs in 293T cells were performed using TransIT®-2020 transfection reagent (Mirus Bio, Inc., Madison, WI, USA). To isolate total RNA, the cells were harvested in TRIzol® reagent 24 h later, following the manufacturer's instructions to isolate total RNA.

Reverse transcription-PCR (RT-PCR)

Each cDNA was prepared from 2 μg total RNA, which was extracted from different minigenes of transfected cells and EVB-transformed B cells, using SuperScript®III reverse transcriptase, with oligo(dT)12–18 as primer, following the manufacturer's protocol (Invitrogen, CA, USA). All RT-PCR products were gel extracted and sequenced to confirm normal splicing, intron inclusion, and exon skipping forms.

Results

We identified 600 SNPs at the branchpoint sites; among these SNPs, 216 showed a change in adenine. After scoring the SNPs by counting the As in the ± 10 nucleotide region, only four SNPs were identified without any additional As; 17 SNPs had one additional A, and 29 SNPs had two additional As (Additional file 1).

The four SNPs identified without any additional As in the ± 10 nucleotide region were rs13296170, rs12769205, rs75434223, and rs67785924; these SNPs were the candidates most likely to affect RNA splicing (Table 1).

Table 1 Selected splice-site SNPs for functional studies

Chromosome	Position	Position	Gene name	SNP ID	Alleles	SNP ± 10 nucleotides sequence	Allele frequency
3	14187698	14187699	*XPC*	rs2733532	A/G	TCTGATTACT*A*ACCCTCGCCT	A = 0.363 G = 0.637
9	130689507	130689508	*PIP5KL1*	rs13296170	A/C	GGCCTCCCTC*A*CTCCCTGTCC	A = 1 C = 0
10	96535124	96535125	*CYP2C19*	rs12769205	A/G	TCTCCCTCCT*A*GTTTCGTTTC	A = 0.670 G = 0.330
16	15835555	15835556	*MYH11*	rs75434223	A/C	CGTGGGGCTC*A*CCCGCCTCCT	A = 1 C = 0
19	45854507	45854508	*KLC3*	rs67785924	–/ACCTC	CTTGCCCCTC*A*CCTCCCCTCC	– = 0.079 ACCTC = 0.921

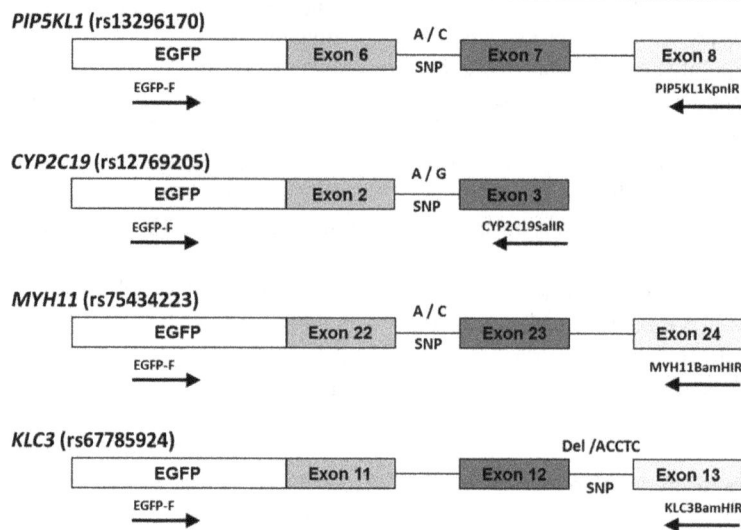

Fig. 1 Schematic representation of minigene constructs. SNPs in the branching point sites are indicated. Locations and orientations of RT-PCR primers (arrows) are shown. See Additional file 2: Table S2 for primer sequences

rs13296170, rs12769205, rs75434223, and rs67785924 are located on *PIP5KL1* intron6, *CYP2C19* intron2, *MYH11* intron22, and *KLC3* intron12, respectively (Fig. 1). These SNPs were further investigated for their functional significance. Minigenes containing the SNPs of interest were built using 3 exons and 2 introns, except for the *CYP2C19* SNP, for which two exons and one intron were used, because intron 3 of *CYP2C19* is large in size (4.9 kb).

RT-PCR of cDNAs prepared from 293T cells transfected with different minigene constructs showed that the rs12769205 A allele produced three bands (normal spliced, intron inclusion, and hybrid forms) when A was substituted with guanine (G) in *CYP2C19*, which spliced majorly in the intron inclusion form with lesser normal form (Fig. 2a). Since this construct comprised two exons and one intron, to make sure there was no exon skipping, we examined mRNA from EBV-transformed B cells carrying different genotypes for the spliced forms. The results showed that B cells had genotype AA spliced in the normal form, AG spliced equally in the normal and intron inclusion forms, and GG spliced mostly in the intron inclusion form (Fig. 2b). The results were further confirmed by using another set of primers such that the forward primer was located on intron 2, and it was noted that AG and GG genotypes spliced in the intron inclusion form (Fig. 2c).

We also studied the minigene constructs for the other three SNPs. When A was substituted with cytosine (C) in rs13296170, *PIP5KL1* spliced mostly into the exon skipping form and somewhat into the intron inclusion form, but not into the normal-spliced RNA form (Fig. 3a, lanes 5 and 6). While rs75434223 substituted A with C, *MYH11* spliced into the intron inclusion form, and not into the normal spliced form (Fig. 3b, lanes 8 and 9).

The SNP rs67785924 in *KLC3* has a normal (wild type) allele containing A and a deletion allele with five missing nucleotides, ACCTC. Both alleles produced normal spliced form, and some intron inclusion form. The level of intron inclusion form in the deletion allele was actually less than that in the normal A allele (Fig. 3b, lanes 2 and 3).

To understand why the deletion allele that did not contain branchpoint A still produces the normally spliced form, we checked the nearby intron sequence and found two other As located at the seventh and eighth nucleotides from the branchpoint A (Fig. 4a). We performed the branchpoint site prediction analysis using SROOGLE and Human Splicing Finder [9]; both tools predicted that these two nearby As also lie within the potential consensus branchpoint site sequence and can be used as alternative branchpoints in the deletion allele. We then tested the influence of these two nearby As on splicing using minigene constructs (Fig.4). In the wild-type allele, when the two nearby AA were changed to AG or GA, RNA spliced majorly in the normal form; when changed to GG, there was a decrease in the normal form and an increase in the intron inclusion form. In the deletion allele, when both AA were changed to GG, there was a further decrease in the normal form accompanied with a further increase in the intron inclusion form (Fig.4b). These results suggested other As nearby may serve as alternative branchpoints.

The influence of additional As on splicing was further examined for the branchpoint site SNP rs2733532 A/G, which contains three additional As in the ±10 nucleotide (Fig. 5a). This SNP is located in *XPC* and is reportedly associated with susceptibility to air pollution and childhood bronchitis [17]. In this case, EBV-transformed B cell lines from subjects carrying different genotypes at

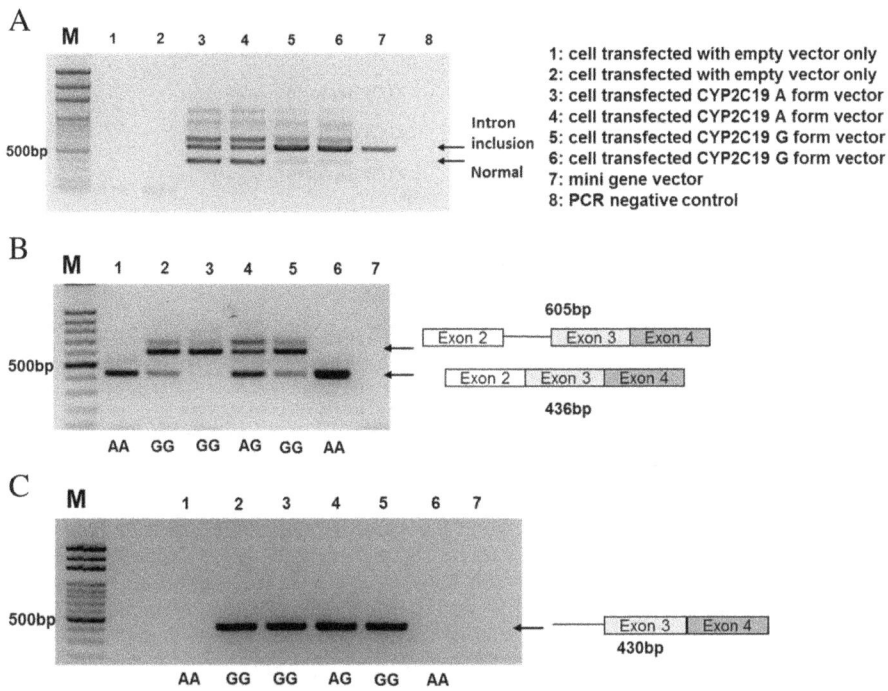

Fig. 2 *CYP2C19* alternative splicing forms in minigene-transfected 293T and in EBV-transformed B cells carrying different genotypes at SNP rs12769205. *CYP2C19* RT-PCR was performed with **a** 293T cells transfected with minigene of rs12769205, genotype A or G, using EGFP-F and *CYP2C19* SalIR primers, and **b** cDNA from B cells carrying different genotypes AA, AG, and GG at rs12769205 position using *CYP2C19* ex2F and *CYP2C19* ex4R primers or **c** cDNA from B cells using *CYP2C19* in2F and *CYP2C19* ex4R as primer set. Marker represents the 100-bp DNA ladder and indicated 500-bp site. See Additional file 2: Table S2 for primer sequences

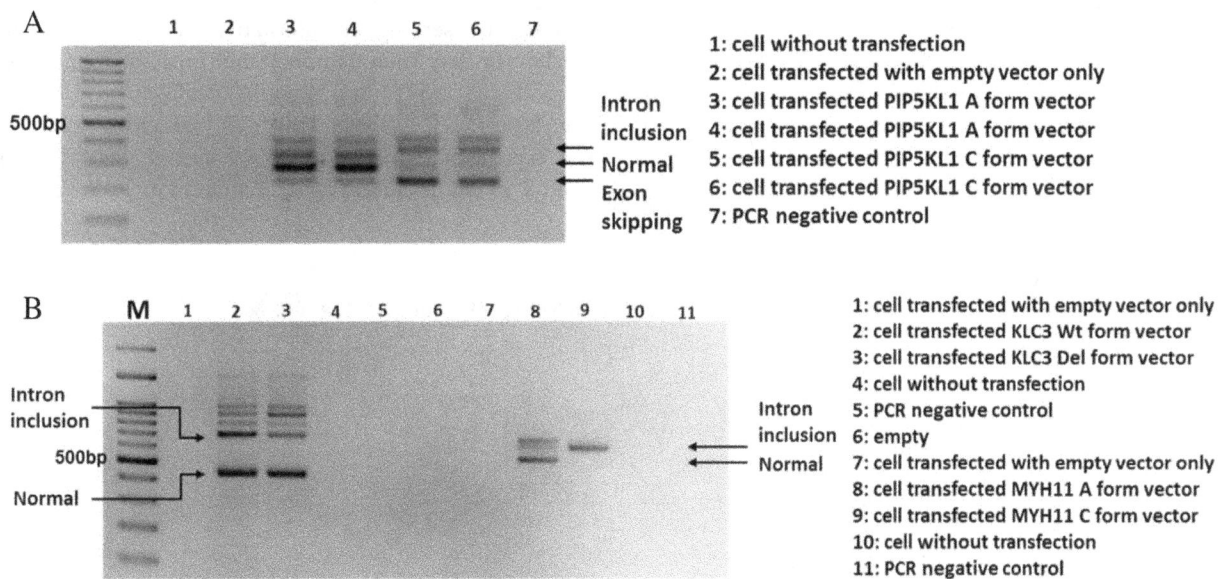

Fig. 3 *PIP5KL1*, *KLC3*, and *MYH11* alternative splicing in minigene-transected 293T cell. RT-PCR was performed in minigene-transfected 293T cells. **a** *PIP5KL1*; rs13296170, genotype A or C, was using EGFP-F and *PIP5KL1* KpnIR primers. **b** *KLC3* (land 1~5); rs67785924, genotype ACCTC (Wt) or deletion (Del), with EGFP-F and *KLC3* BamHIR primers, and *MYH11* (land 7~11); rs75434223, genotype A or C, was using EGFP-F and *MYH11* BamHIR primers. Marker represents the 100-bp DNA ladder and indicated 500-bp site. See Additional file 2: Table S2 for primer sequences

A

-gagctggagggtggatgtaacacttgccctcAcctcccctccAAccatcccctgtgcctgtctccag   [Exon 13]

↓

cttgccctc*ccctccAAcc   **KLC3 Del form**

cttgccctcAcctcccctccAGcc   **KLC3 AG form**

cttgccctcAcctcccctccGAcc   **KLC3 GA form**

cttgccctcAcctcccctccGGcc   **KLC3 GG form**

cttgccctc*ccctccGGcc   **KLC3 Del with GG form**

B

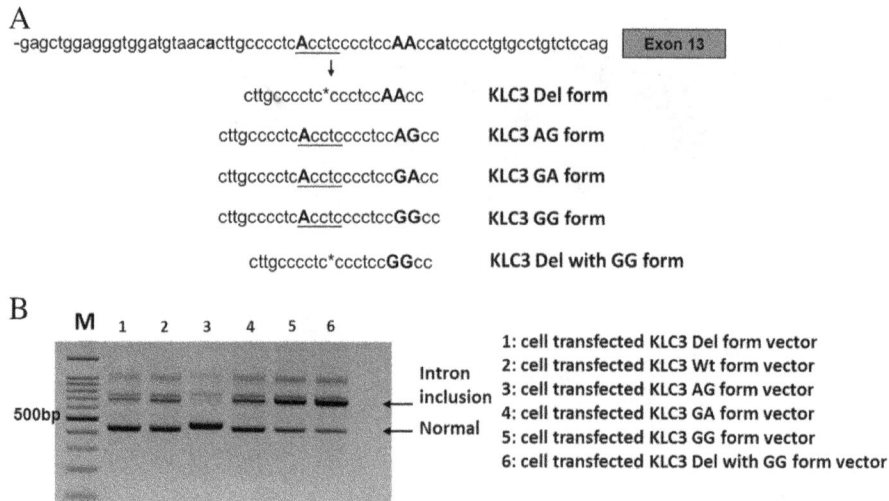

1: cell transfected KLC3 Del form vector
2: cell transfected KLC3 Wt form vector
3: cell transfected KLC3 AG form vector
4: cell transfected KLC3 GA form vector
5: cell transfected KLC3 GG form vector
6: cell transfected KLC3 Del with GG form vector

Fig. 4 Nucleotide sequences of *KLC3* intron 12 and alternative splicing forms in minigene-transected 293T cell. **a** Nucleotide sequences at 3′part of the *KLC3* intron 12. Possible splicing branchpoint adenine is indicated by bold and uppercase, SNP rs67785924 is underlined, and asterisk shown after deletion sequence. The following shown each A substituted minigene. **b** RT-PCR was performed in 293T cells transfected with different *KLC3* adenine substitution minigene constructs: ACCTC (Wt); ACCTC deletion (Del); Wt with AG (AG form); Wt with GA (GA form); Wt with GG (GG form) and Del with GG form. Marker represents the 100-bp DNA ladder and indicated 500-bp site. See Additional file 2: Table S2 for primer sequences

the branchpoint site, regardless of genotype (AA, AG, or GG), showed only the normally spliced form (Fig. 5b), suggesting that other As can serve as a branchpoint site.

Discussion

In the present study, we used SROOGLE to predict splice branchpoints and screened dbSNP for SNPs located within the branchpoint sites. Using minigene constructs and, when available, EBV-transformed cell lines carrying different SNP alleles, we experimentally verified that SNPs comprising a change to branchpoint A resulted in abnormal splicing, suggesting that the predicted sites are indeed involved in pre-mRNA splicing, and further confirming the functional importance of A. However, only 20% of the branchpoint sites that we identified had a reported corresponding lariat sequence

[12](see Additional file 1). This observation may be understandable given that the number of reported lariat sequences based on next generation sequencing represents only 28% of all introns in the genome [12].

We found only three branchpoint site SNPs that have a single A at the branchpoint site, without additional As nearby. It is possible that organisms evolved to have additional As in the branchpoint site to ensure proper splicing. Additional As in the ± 10 nucleotide region may protect SNPs at the branchpoint A from abnormal splicing, by serving as alternative branchpoints. This mechanism has been demonstrated in the present study for SNP rs67785924 and SNP rs2733532 (Figs. 4 and 5). The latter SNP is located on *XPC* on chromosome 3, and has been reported to be associated with diseases related to air pollution and childhood bronchitis [17].

A

-ggggcttcctggtatctgAttActAAccctcgcctgtgtcctcccaccactgccacctgtccag   [Exon 16]

B

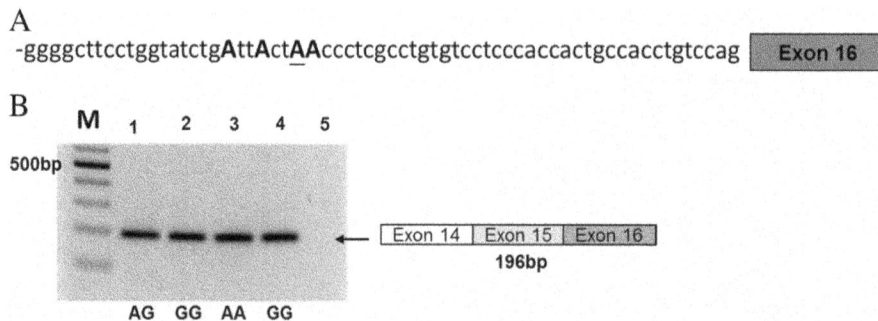

Exon 14 | Exon 15 | Exon 16
196bp

AG   GG   AA   GG

Fig. 5 *XPC* alternative splicing in different genotypes of EBV-transformed B cells. **a** Graphic representation 3′part of the *XPC* intron 15 sequences, possible splicing branchpoint adenine is indicated by bold and uppercase, underlining is rs2733532. **b** cDNA from normal control B cells of different genotypes AA, AG, and GG at rs2733532 using *XPC* ex14F and *XPC* ex16R primers. Marker represents the 100-bp DNA ladder and indicated 500-bp site. See Additional file 2: Table S2 for primer sequences

The risk allele G of the branchpoint SNP A/G resides in the population at a frequency of ~ 0.637. Both the lariat sequence database and our prediction algorithms classified it as a branchpoint site. However, our experiments demonstrated that this branchpoint A to G SNP did not influence splicing; this observation presumably results from the presence of additional nearby As, which serve as alternative splicing sites; this explanation implies that other mechanisms may be involved in this disease association.

Several algorithms and tools have been used to predict branchpoints [6–10], and surprisingly, no branchpoint site SNP database has been reported. Because minigene constructs are time-consuming and not all SNPs in the branchpoint sites have cell lines available for study, in the present study we have tested only five SNPs, and verified their significance. More functional studies are needed to examine the functional significance of other SNPs, especially those SNPs that do not involve A changes at branchpoints.

In conclusion, we have generated a high-confidence genome-wide branchpoint site SNP database, experimentally verified the importance of A in the branchpoint, and suggested that other nearby As may serve as alternative branchpoints and ensure proper pre-mRNA splicing. These results improve upon the prediction of functional SNPs at branchpoint sites, and inform the study of the SNPs at intron branchpoint sites.

Abbreviations
A: Adenine; C: Cytosine; G: Guanine; PPT: Polypyrimidine tract; pre-mRNA: Precursor messenger RNA; SNP: Single nucleotide polymorphism; snRNPs: Small nuclear ribonucleoproteins

Acknowledgements
Not applicable.

Funding
This work was supported by Academia Sinica Genomic Medicine Multicenter Study, Taiwan [40-05-GMM] and National Resource Center for Genomic Medicine MOST 105-2319-B-001-001]. The funding organization had no role in the design or conduct of this research. The authors alone are responsible for the content and writing of the paper.

Authors' contributions
HLC, JYW, and YTC conceived and designed the experiments. HLC performed the experiments and wrote the paper. All authors analyzed the data and approved the final manuscript.

Consent for publication
Not applicable.

Competing interests
The authors are no competing of interest to declare.

Author details
[1]Institute of Clinical Medicine, National Yang-Ming University, Taipei, Taiwan. [2]Institute of Biomedical Sciences, Academia Sinica, Taipei, Taiwan. [3]Graduate Institute of Chinese Medical Science, China Medical University, Taichung, Taiwan. [4]Department of Pediatrics, Duke University Medical Center, Durham, USA.

References
1. Berget SM, Moore C, Sharp PA. Spliced segments at the 5' terminus of adenovirus 2 late mRNA. Proc Natl Acad Sci U S A. 1977;74(8):3171–5.
2. Chow LT, Gelinas RE, Broker TR, Roberts RJ. An amazing sequence arrangement at the 5' ends of adenovirus 2 messenger RNA. Cell. 1977;12(1):1–8.
3. Gilbert W. Why genes in pieces? Nature. 1978;271(5645):501.
4. Will CL, Luhrmann R. Spliceosome structure and function. Cold Spring Harb Perspect Biol. 2011;3(7). doi:10.1101/cshperspect.a003707.
5. Kramer A. The structure and function of proteins involved in mammalian pre-mRNA splicing. Annu Rev Biochem. 1996;65:367–409.
6. Kol G, Lev-Maor G, Ast G. Human-mouse comparative analysis reveals that branch-site plasticity contributes to splicing regulation. Hum Mol Genet. 2005;14(11):1559–68.
7. Schwartz SH, Silva J, Burstein D, Pupko T, Eyras E, Ast G. Large-scale comparative analysis of splicing signals and their corresponding splicing factors in eukaryotes. Genome Res. 2008;18(1):88–103.
8. Schwartz S, Hall E, Ast G. SROOGLE: webserver for integrative, user-friendly visualization of splicing signals. Nucleic Acids Res. 2009;37(Web Server issue): W189–92.
9. Desmet FO, Hamroun D, Lalande M, Collod-Beroud G, Claustres M, Beroud C. Human Splicing Finder: an online bioinformatics tool to predict splicing signals. Nucleic Acids Res. 2009;37(9):e67.
10. Faber K, Glatting KH, Mueller PJ, Risch A, Hotz-Wagenblatt A. Genome-wide prediction of splice-modifying SNPs in human genes using a new analysis pipeline called AAS sites. BMC bioinformatics. 2011;12(Suppl 4):S2.
11. Taggart AJ, DeSimone AM, Shih JS, Filloux ME, Fairbrother WG. Large-scale mapping of branchpoints in human pre-mRNA transcripts in vivo. Nat Struct Mol Biol. 2012;19(7):719–21.
12. Mercer TR, Clark MB, Andersen SB, Brunck ME, Haerty W, Crawford J, Taft RJ, Nielsen LK, Dinger ME, Mattick JS. Genome-wide discovery of human splicing branchpoints. Genome Res. 2015;25(2):290–303.
13. Taggart AJ, Lin CL, Shrestha B, Heintzelman C, Kim S, Fairbrother WG. Large-scale analysis of branchpoint usage across species and cell lines. Genome Res. 2017;27(4):639–49.
14. Kuivenhoven JA, Weibusch H, Pritchard PH, Funke H, Benne R, Assmann G, Kastelein JJ. An intronic mutation in a lariat branchpoint sequence is a direct cause of an inherited human disorder (fish-eye disease). J Clin Invest. 1996;98(2):358–64.
15. Kralovicova J, Lei H, Vorechovsky I. Phenotypic consequences of branch point substitutions. Hum Mutat. 2006;27(8):803–13.
16. Pan WH, Fann CS, Wu JY, Hung YT, Ho MS, Tai TH, Chen YJ, Liao CJ, Yang ML, Cheng AT, et al. Han Chinese cell and genome bank in Taiwan: purpose, design and ethical considerations. Hum Hered. 2006;61(1):27–30.
17. Ghosh R, Rossner P, Honkova K, Dostal M, Sram RJ, Hertz-Picciotto I. Air pollution and childhood bronchitis: interaction with xenobiotic, immune regulatory and DNA repair genes. Environ Int. 2016;87:94–100.

13

The peptidylglycine-α-amidating monooxygenase (PAM) gene rs13175330 A>G polymorphism is associated with hypertension in a Korean population

Hye Jin Yoo[1,2†], Minjoo Kim[3†], Minkyung Kim[3], Jey Sook Chae[3], Sang-Hyun Lee[4] and Jong Ho Lee[1,2,3*]

Abstract

Background: Peptidylglycine-α-amidating monooxygenase (PAM) may play a role in the secretion of atrial natriuretic peptide (ANP), which is a hormone involved in the maintenance of blood pressure (BP). The objective of the present study was to determine whether *PAM* is a novel candidate gene for hypertension (HTN).

Results: A total of 2153 Korean participants with normotension and HTN were included. Genotype data were obtained using the Korean Chip. The rs13175330 polymorphism of the *PAM* gene was selected from the ten single nucleotide polymorphisms (SNPs) most strongly associated with BP. The presence of the G allele of the *PAM* rs13175330 A>G SNP was associated with a higher risk of HTN after adjustments for age, sex, BMI, smoking, and drinking [OR 1.607 (95% CI 1.220–2.116), $p = 0.001$]. The rs13175330 G allele carriers in the HTN group treated without antihypertensive therapy (HTN w/o therapy) had significantly higher systolic and diastolic BP than the AA carriers, whereas the G allele carriers in the HTN group treated with antihypertensive therapy (HTN w/ therapy) showed significantly higher diastolic BP. Furthermore, rs13175330 G allele carriers in the HTN w/o therapy group had significantly increased levels of insulin, insulin resistance, and oxidized low-density lipoprotein (LDL) and significantly decreased LDL-cholesterol levels and LDL particle sizes compared to the AA carriers.

Conclusion: These results suggest that the *PAM* rs13175330 A>G SNP is a novel candidate gene for HTN in the Korean population. Additionally, the *PAM* rs13175330 G allele might be associated with insulin resistance and LDL atherogenicity in patients with HTN.

Keywords: Hypertension, Genetic polymorphisms, Genetic association, Peptidylglycine-α-amidating monooxygenase, Atrial natriuretic peptide, LDL atherogenicity

Background

Hypertension (HTN) is a significant contributor to the global burden of heart disease, stroke, kidney failure, and premature mortality and disability [1, 2]. HTN is a complex trait that is caused by both genetic and environmental factors [3]. Evidence from family studies indicates that more than 30% of blood pressure (BP) variation can be attributed to genetics [4, 5]. Recently, genome-wide association studies (GWASs) have identified more than 50 single nucleotide polymorphisms (SNPs) associated with an increased risk of HTN [6–8].

The neuroendocrine processing enzyme peptidylglycine-α-amidating monooxygenase (PAM) is highly concentrated in the atrium and may play a role in the secretion of atrial natriuretic peptide (ANP), which is a hormone involved in BP maintenance and fluid homeostasis [9–11]. Indeed, PAM and pro-ANP (the bioactive form of ANP) are the predominant membrane-associated proteins in atrial secretory granules [12]. Because close relationships exist

* Correspondence: jhleeb@yonsei.ac.kr
†Equal contributors
[1]National Leading Research Laboratory of Clinical Nutrigenetics/ Nutrigenomics, Department of Food and Nutrition, College of Human Ecology, Yonsei University, 50 Yonsei-ro, Seodaemun-gu, Seoul 03722, South Korea
[2]Department of Food and Nutrition, Brain Korea 21 PLUS Project, College of Human Ecology, Yonsei University, Seoul 03722, South Korea
Full list of author information is available at the end of the article

between ANP and BP and between ANP and PAM [9–11], specific *PAM* SNP genotypes in humans may be associated with BP alterations.

Since 2014, the Korean Chip (K-CHIP), which includes 833,535 SNPs and uses an oligomer as a probe, has been developed by the Korea Biobank Array project as a low-cost customized chip that is optimized for genetic studies of diseases and complex traits in Koreans (Additional file 1: Table S1). Since whole-genome sequencing requires very high calculation capacity and cost and commercial chips are designed for Western populations, whose genomic variants differ from Asian populations, the K-CHIP is more suitable for the discovery and identification of Korean population-specific SNPs related to disease occurrence [13, 14]. Although the K-CHIP has only been released recently, several published studies have used the K-CHIP [15, 16]. These studies have gained international recognition; thus, the K-CHIP has been shown to be an appropriate tool for analyzing SNPs associated with diseases in a Korean population.

To the best of our knowledge, this study was the first to investigate HTN-related SNPs using the K-CHIP in a Korean population. Therefore, the objective of the present study was to explore HTN-related SNPs using the K-CHIP, to identify the SNP most strongly associated with BP, and to determine whether *PAM* is a novel candidate gene for HTN among the Korean population.

Methods
Study population
A total of 2153 Korean male and female adult participants (male, $n = 866$; female, $n = 1287$; aged 20–86 years; median = 50 years) with nondiabetic normotension (systolic BP < 140 mmHg and diastolic BP < 90 mmHg) and HTN (systolic BP ≥ 140 mmHg or diastolic BP ≥ 90 mmHg) were recruited for this study from the Health Service Center (HSC) during routine checkups at the National Health Insurance Corporation Ilsan Hospital in Goyang, Korea (January 2010–March 2015). Based on the data screened from the HSC, potential subjects with HTN were referred to the Department of Family Medicine or Internal Medicine, where their health and BP were rechecked. Finally, subjects who did not meet the exclusion criteria were all included as study participants. The exclusion criteria were a current diagnosis and/or a history of diabetes, cardiovascular disease, liver disease, renal disease, pancreatitis, cancer, or any life- and health-threatening diseases; pregnancy or lactation; and regular use of any medication except HTN therapy. The aim of the study was carefully explained to all participants, whom provided written informed consent. The Institutional Review Board of Yonsei University and the National Health Insurance Corporation Ilsan Hospital approved the study protocol, which complied with the Declaration of Helsinki.

Anthropometric measurements
Body weight (UM0703581; Tanita, Tokyo, Japan) and height (GL-150; G-Tech International, Uijeongbu, Korea) were measured in lightly clothed subjects without shoes, and body mass index (BMI) values were calculated (kg/m^2). Waist circumference was measured directly on the skin at the umbilical level after normal expiration with the subject in an upright standing position. Hip circumference was measured at the protruding part of the hip in standing subjects using a plastic measuring tape with measurements to the nearest 0.1 cm. Waist to hip ratio values were obtained by dividing the waist circumference by the hip circumference. Systolic and diastolic BP were measured using a random-zero sphygmomanometer (HM-1101, Hico Medical Co., Ltd., Chiba, Japan) with appropriately sized cuffs after a rest period of at least 20 min in a seated position. BP was measured three times in both arms. The differences among the three systolic BP measurements were always less than 2 mmHg. Participants were instructed not to smoke or drink alcohol for at least 30 min before each BP measurement.

Sample collection
Fasting venous blood specimens were collected following an overnight fast of at least 12 h. The samples were collected in EDTA-treated tubes and serum tubes (BD Vacutainer; Becton, Dickinson and Company, Franklin Lakes, NJ, USA). The samples were placed in an ice box that was protected from light within approximately 30 min and then centrifuged (1200 rpm for 20 min at 4 °C) within 3 h to obtain plasma and serum. The plasma and serum aliquots were stored at − 80 °C prior to analysis.

Serum fasting lipid profiles
The serum fasting triglyceride (TG) and total-cholesterol (TC) levels were measured using enzymatic assays with the TG and CHOL Kits (Roche, Mannheim, Germany), respectively. Serum fasting high-density lipoprotein (HDL)-cholesterol was measured using a selective inhibition method with the HDL-C Plus Kit (Roche, Mannheim, Germany). The resulting color reactions of the assays were monitored using a Hitachi 7600 autoanalyzer (Hitachi, Tokyo, Japan). The Friedewald formula was used to indirectly calculate low-density lipoprotein (LDL)-cholesterol levels as follows: LDL-cholesterol = TC − [HDL-cholesterol + (TG/5)].

Serum fasting glucose, insulin, and insulin resistance (IR)
The serum fasting glucose level was measured using the hexokinase method with the GLU Kit (Roche, Mannheim, Germany), and the resulting color reaction was monitored with the Hitachi 7600 autoanalyzer (Hitachi, Tokyo, Japan). Serum fasting insulin was measured with an immunoradiometric assay using the Insulin IRMA

Kit (DIAsource, Louvain, Belgium), and the resulting color reaction was monitored with an SR-300 system (Stratec, Birkenfeld, Germany). The homeostatic model assessment (HOMA) equation was used to calculate the IR as follows: HOMA-IR = [fasting insulin (μIU/mL) × fasting glucose (mg/dL)]/405.

Plasma LDL particle size and oxidized (ox)-LDL level

Plasma LDL particles were isolated by sequential flotation ultracentrifugation, and the particle size distribution (1.019–1.063 g/mL) was assessed using a pore-gradient lipoprotein system (CBS Scientific Company, San Diego, CA, USA) on commercially available non-denaturing gels containing a linear 2–16% acrylamide gradient (CBS Scientific Company, San Diego, CA, USA). Latex bead (30 nm)-conjugated thyroglobulin (17 nm), ferritin (12.2 nm), and catalase (10.4 nm) standards were used to measure the relative band migration rates. The gels were scanned using a GS-800 Calibrated Imaging Densitometer (Bio-Rad Laboratories, Hercules, CA, USA). Plasma ox-LDL was estimated using an enzyme immunoassay (Mercodia AB, Uppsala, Sweden), and the resulting color reaction was determined at 450 nm on a Wallac Victor2 multilabel counter (Perkin-Elmer Life Sciences, Boston, MA, USA).

Affymetrix axiom™ KORV1.0–96 array hybridization and SNP selection

A total of 2167 samples were genotyped according to the manufacturer's protocol included in the Axiom® 2.0 Reagent Kit (Affymetrix Axiom® 2.0 Assay User Guide; Affymetrix, Santa Clara, CA, USA). Approximately 200 ng of genomic DNA (gDNA) was amplified and randomly fragmented into 25- to 125-base pair (bp) fragments. The initial gDNA amplification was performed in a 40-μL reaction volume containing 20 μL of genomic DNA at a 10 ng/μL concentration and 20 μL of the denaturation master mix. The initial amplification reaction included a 10-min incubation at room temperature; then, the incubated products were amplified with 130 μL of Axiom 2.0 Neutral Soln, 225 μL of Axiom 2.0 Amp Soln, and 5 μL of Axiom 2.0 Amp Enzyme. The amplification reactions were performed for 23 ± 1 h at 37 °C. The amplification products were analyzed in an optimized reaction to amplify fragments between 200 and 1100 bp in length. A fragmentation step reduced the amplified products to segments of approximately 25–50 bp in length, which were end-labeled using biotinylated nucleotides. Following hybridization, the bound target was washed under stringent conditions to remove non-specific background and to minimize the background noise caused by random ligation events. Each polymorphic nucleotide was investigated via a multicolor ligation event conducted on the array surface. After ligation, the arrays were stained and imaged using the GeneTitan MC Instrument (Affymetrix, Santa Clara, CA, USA). The images were analyzed using the Genotyping Console™ Software (Affymetrix, Santa Clara, CA, USA). Genotype data were produced using the K-CHIP, which is available through the K-CHIP consortium. The K-CHIP was designed by the Center for Genome Science at the Korea National Institute of Health (4845–301, 3000–3031).

Samples with the following thresholds were excluded: sex inconsistency, markers with a high missing rate (> 5%), individuals with a high missing rate (> 10%), a minor allele frequency < 0.01, and a significant deviation from the Hardy-Weinberg equilibrium (HWE) ($p <$ 0.001). Additionally, SNPs that were in linkage disequilibrium (LD, $r^2 \geq 0.5$) were excluded. The remaining 394,222 SNPs and 2159 samples were included in the subsequent association analysis.

Statistical analysis

HWE and the associations between SNPs and BP were analyzed with PLINK version 1.07 (http://zzz.bwh.harvard.edu/plink/); the associations were assessed using the linear regression analysis method. Descriptive statistical analyses were conducted using SPSS version 23.0 (IBM, Chicago, IL, USA). For the *PAM* rs13175330 polymorphism, because of the small number of rare allele homozygotes (GG), we pooled heterozygotes (AG) and rare allele homozygotes to increase the statistical power. Logarithmic transformation was used for skewed variables, and data are expressed as the mean ± standard error (SE). A two-tailed p value < 0.05 was considered statistically significant. An independent t test was performed on continuous variables to compare values between the normotensive group and each hypertensive subgroup and to compare the values between the genotypes within the normotensive group and each hypertensive subgroup. Frequencies were tested with a chi-square test. The association of HTN with a *PAM* rs13175330 genotype was calculated using the odds ratio (OR) [95% confidence intervals (CIs)] of a logistic regression model with adjustments for confounding factors.

Results

As mentioned above, 394,222 SNPs and 2159 samples were included in the analysis. The ten SNPs that were most strongly associated with BP were selected from the linear regression analysis to assess the association between SNPs and BP. Among them, we identified rs13175330 in the *PAM* gene. The first systolic and diastolic BP-related SNP did not have a reference SNP ID in the SNP database; therefore, we conducted an association analysis using rs13175330, which was the second diastolic BP-related SNP and the seventh systolic BP-related SNP (Additional file 1: Table S2). Among the 2159 genotyped subjects, 6 subjects did not possess a *PAM* variant; thus, only 2153 subjects were included in the final sample.

This was a large study with many samples, and the samples were run on multiple assay plates; thus, we checked inter-and intra-assay coefficient of variability (CV) to reduce multiple assay error although the experiments were not repeated. The mean inter- and intra-assay CV (%) of each variable was as follows (inter-assay CV; intra-assay CV): triglyceride (2.18; 0.95), total-cholesterol (1.19; 1.29), HDL-cholesterol (1.13; 1.16), LDL-cholesterol (0.98; 0.65), glucose (1.01; 0.71), insulin (4.25; 3.55), LDL particle size (2.29; 3.88), and ox-LDL (0.96; 6.78).

Clinical and biochemical characteristics according to the presence of hypertension

A total of 2153 subjects were divided into a normotensive control group ($n = 1610$) and an HTN group ($n = 543$). The HTN group was stratified according to their antihypertensive therapy [patients with HTN without antihypertensive therapy (HTN w/o therapy), $n = 377$; and patients with HTN with antihypertensive therapy (HTN w/ therapy), $n = 166$]. The clinical and biochemical characteristics of each group are shown in Table 1. Patients in the HTN group and all of the HTN subgroups were significantly older and heavier and had significantly higher systolic and diastolic BP, TG, glucose,

insulin, and HOMA-IR indices than the normotensive controls (Table 1). Conversely, the HTN group and all of the HTN subgroups had significantly decreased HDL-cholesterol levels compared with the normotensive controls (Table 1). Ox-LDL was significantly increased in the HTN and HTN w/o therapy groups but not in the HTN w/ therapy group compared to the normotensive controls (Table 1).

Distribution of the *PAM* rs13175330 A>G polymorphism

The genotype distributions of the *PAM* rs13175330 A>G polymorphism were in HWE in the entire population. Among the 1610 normotensive controls, 1377 subjects (85.5%) had the AA genotype, 228 subjects (14.2%) had the AG genotype, and 5 subjects (0.3%) had the GG genotype. The allele frequency of the G allele was 0.074 in the normotensive controls. Conversely, among the 543 patients with HTN, 434 subjects (79.9%) had the AA genotype, 102 (18.8%) subjects had the AG genotype, and 7 (1.29%) subjects had the GG genotype. The allele frequency of the G allele was 0.107 in the HTN group. The distribution of the *PAM* rs13175330 A>G genotype ($p = 0.001$) and the allele frequencies ($p = 0.001$) in the HTN group differed significantly from the values obtained in the normotensive controls (Additional file 1: Table S3).

Table 1 Clinical and biochemical characteristics in the normotensive controls and HTN patient subgroups according to the antihypertensive therapy

	Normotensive controls ($n = 1610$)		HTN group ($n = 543$)					
			Total ($n = 543$)		HTN w/o therapy ($n = 377$)		HTN w/ therapy ($n = 166$)	
Age (years)	48.0	± 0.27	54.4	± 0.50[***]	53.0	± 0.61[***]	57.7	± 0.83[***]
Weight (kg)	63.0	± 0.25	68.1	± 0.51[***]	68.9	± 0.64[***]	66.3	± 0.81[***]
BMI (kg/m^2)	23.7	± 0.07	25.4	± 0.14[***]	25.4	± 0.17[***]	25.2	± 0.22[***]
Waist (cm)	83.5	± 0.19	87.9	± 0.37[***]	87.9	± 0.46[***]	88.0	± 0.63[***]
Waist hip ratio	0.88	± 0.00	0.90	± 0.00[***]	0.90	± 0.00[***]	0.91	± 0.00[***]
Systolic BP (mmHg)	116.4	± 0.29	138.5	± 0.66[***]	145.2	± 0.62[***]	123.2	± 0.79[***]
Diastolic BP (mmHg)	72.7	± 0.22	87.4	± 0.46[***]	91.8	± 0.42[***]	77.4	± 0.66[***]
Triglyceride (mg/dL)[a]	119.6	± 1.84	148.6	± 3.76[***]	151.2	± 4.79[***]	142.8	± 5.72[***]
Total-cholesterol (mg/dL)[a]	198.1	± 0.90	198.3	± 1.54	200.2	± 1.87	193.9	± 2.68
HDL-cholesterol (mg/dL)[a]	53.9	± 0.34	50.4	± 0.55[***]	50.3	± 0.65[***]	50.5	± 1.03[**]
LDL-cholesterol (mg/dL)[a]	121.1	± 0.82	119.1	± 1.42	120.7	± 1.77	115.4	± 2.31
Glucose (mg/dL)[a]	95.6	± 0.51	103.9	± 1.11[***]	103.9	± 1.40[***]	103.9	± 1.73[***]
Insulin (μIU/dL)[a]	9.09	± 0.12	9.84	± 0.25[*]	10.1	± 0.33[*]	9.73	± 0.35[*]
HOMA-IR[a]	2.15	± 0.03	2.55	± 0.09[***]	2.65	± 0.12[***]	2.44	± 0.09[***]
LDL particle size (nm)[a]	23.9	± 0.03	24.0	± 0.05	23.9	± 0.06	24.0	± 0.08
Oxidized LDL (U/L)[a]	46.1	± 0.53	48.3	± 0.97[*]	50.8	± 1.15[***]	43.7	± 1.75

Mean ± SE

HTN hypertension, *HTN w/o therapy* HTN group treated without antihypertensive therapy, *HTN w/ therapy* HTN group treated with antihypertensive therapy, *BMI* body mass index, *BP* blood pressure, *HDL* high-density lipoprotein, *LDL* low-density lipoprotein, *HOMA-IR* homeostatic model assessment of insulin resistance

[*]$p < 0.05$, [**]$p < 0.01$, and [***]$p < 0.001$ derived from an independent t test between the normotensive controls and each HTN subgroup

[a]Tested following logarithmic transformation

Increased HTN risk associated with the *PAM* rs13175330 A>G polymorphism

Table 2 shows the unadjusted and adjusted odds ratios (OR) for all patients with HTN according to their *PAM* rs13175330 genotype. The presence of the GG genotype of the *PAM* rs13175330 A>G SNP was associated with a higher risk of HTN [OR 4.192 (95% CI 1.325–13.263), $p = 0.015$] (Table 2). The significance of the association remained after adjustments for confounding factors, including age, sex, BMI, smoking, and drinking [OR 7.826 (95% CI 2.228–27.484), $p = 0.001$]. Moreover, the rs13175330 G allele was associated with a higher risk of HTN before [OR 1.484 (95% CI 1.154–1.909), $p = 0.002$] and after adjustments for the confounding factors [OR 1.607 (95% CI 1.220–2.116), $p = 0.001$] (Table 2).

Association between BP and the *PAM* rs13175330 A>G genotype

No significant genotype-related differences were observed among the normotensive controls or HTN subjects treated with/without antihypertensive therapy according to the *PAM* rs13175330 A>G genotype with respect to age, sex, BMI, smoking, and drinking (data not shown). In the normotensive controls, rs13175330 G allele carriers tended to have higher systolic BP than the AA carriers ($p = 0.056$) (Table 3). In the HTN w/o therapy group, rs13175330 G allele carriers had significantly higher systolic BP ($p = 0.036$) and diastolic BP ($p = 0.048$) than the AA carriers. Additionally, the rs13175330 G allele carriers in the HTN w/ therapy group had significantly higher diastolic BP ($p < 0.001$); however, no significant difference in systolic BP between rs13175330 G allele and AA carriers was observed in the HTN w/ therapy group (Table 3).

Table 2 Unadjusted and adjusted OR for all patients with HTN according to the *PAM* rs13175330 genotypes

PAM rs13175330	HTN group ($n = 543$) OR (95% CI)	p values
Model 1		
A|| compared with G	1.498 (1.186, 1.892)	0.001
AA + AG|| compared with GG	4.192 (1.325, 13.263)	0.015
AA|| compared with AG + GG	1.484 (1.154, 1.909)	0.002
Model 2		
A|| compared with G	1.642 (1.272, 2.121)	< 0.001
AA + AG|| compared with GG	7.826 (2.228, 27.484)	0.001
AA|| compared with AG + GG	1.607 (1.220, 2.116)	0.001

||Reference

CI confidence interval, *Model 1* unadjusted, *Model 2* adjusted for age, sex, BMI, smoking, and drinking, *OR* odds ratio, *HTN* hypertension, *PAM* peptidylglycine-α-amidating monooxygenase

Lipid profiles, insulin levels, LDL particle sizes, and ox-LDL levels according to the *PAM* rs13175330 A>G genotype

In the HTN w/o therapy group, rs13175330 G allele carriers had significantly higher insulin levels ($p = 0.001$), HOMA-IR indices ($p = 0.002$), and ox-LDL levels ($p = 0.046$) than the AA carriers (Table 3) but significantly lower LDL-cholesterol levels ($p = 0.039$) and smaller LDL particle sizes ($p = 0.003$) than the AA carriers. These genotype effects on the lipid profile, insulin level, LDL particle size, and ox-LDL level were not observed in the normotensive controls or the HTN w/ therapy group (Table 3).

Discussion

To conduct statistical analysis, we pooled heterozygotes (AG) and rare allele homozygotes (GG) because the number of rare allele homozygotes was too small. Even though there are statistical procedures that can handle unbalanced sample size across compared groups, immoderate unbalanced samples can cause problems [17]. In many cases, a sample size of groups stratified by a genotype is unbalanced due to rare allele frequency. To solve the problem, researchers (1) should anticipate the number of study participants via estimation of genotype frequency according to allele frequency or (2) should balance the number of study participants across groups through a prescreening of genotypes [17]. However, in both cases, if rare allele frequency of a SNP in which researchers are interested is too low, the number of study participants increases too much and the loss of time and cost associated with genotype prescreening will be huge [17]. Therefore, as the other strategy for balancing sample size across the groups, we combined AG and GG genotypes; it is more balanced for statistical purposes [17] rather than analyzing the subjects according to the rs13175330 genotypes (AA vs. AG vs. GG).

The major finding of this study was that the minor G allele frequency of *PAM* rs13175330 A>G was significantly higher in the patients with HTN than in the normotensive controls, suggesting an association between *PAM* rs13175330 A>G and HTN. There are no previous publications regarding a relationship between polymorphisms of *PAM* and HTN development and the effects of PAM dysfunction caused by *PAM* polymorphisms on the risk of HTN; thus, this is the first study to suggest that the *PAM* rs13175330 G allele is related to an increase risk of HTN.

PAM rs13175330 A>G is an intronic SNP. Introns have several functions including regulation of alternative splicing and gene expression [18]; via regulating a rate of transcriptional elongation, RNA processing, or RNA turnover, intronic regions can influence RNA levels [19]. Studies support that intronic SNPs do affect RNA splicing [20] and mRNA expression [19, 21]. Finally, Zhou et al. [22] reported that an intronic SNP in *CD44* intron 1 is associated with breast cancer development. Many evidences prove that an intronic SNP can be a novel

Table 3 Clinical and biochemical characteristics in the normotensive controls and HTN patient subgroups according to the *PAM* rs13175330 genotype

PAM rs13175330	Normotensive controls (n = 1610)		HTN group (n = 543)					
			HTN w/o therapy (n = 377)		HTN w/ therapy (n = 166)			
	AA (n = 1377)	G allele (n = 233)	AA (n = 305)	G allele (n = 72)	AA (n = 129)	G allele (n = 37)		
Systolic BP (mmHg)	116.2 ± 0.31	117.7 ± 0.74^{t}	144.6 ± 0.68	147.9 ± 1.46*	123.2 ± 0.90	123.4 ± 1.70		
Diastolic BP (mmHg)	72.6 ± 0.23	73.3 ± 0.58	91.4 ± 0.45	93.8 ± 1.11*	75.7 ± 0.63	83.3 ± 1.67***		
Triglyceride (mg/dL)^{a}	119.4 ± 2.02	121.1 ± 4.29	148.2 ± 5.10	163.9 ± 12.7	141.3 ± 6.79	148.4 ± 9.93		
Total-cholesterol (mg/dL)^{a}	198.0 ± 0.96	198.6 ± 2.53	201.7 ± 2.06	194.1 ± 4.43	193.8 ± 3.01	194.5 ± 5.97		
HDL-cholesterol (mg/dL)^{a}	54.0 ± 0.37	53.5 ± 0.84	50.3 ± 0.73	50.4 ± 1.45	50.7 ± 1.20	49.7 ± 1.98		
LDL-cholesterol (mg/dL)^{a}	121.1 ± 0.88	121.1 ± 2.20	122.7 ± 1.99	112.0 ± 3.71*	115.5 ± 2.66	115.1 ± 4.63		
Glucose (mg/dL)^{a}	95.5 ± 0.55	96.5 ± 1.47	103.7 ± 1.49	104.8 ± 3.76	105.7 ± 2.08	97.5 ± 2.56*		
Insulin (μIU/dL)^{a}	9.07 ± 0.13	9.26 ± 0.28	9.60 ± 0.33	12.1 ± 0.99**	9.56 ± 0.38	10.4 ± 0.79		
HOMA-IR^{a}	2.14 ± 0.04	2.21 ± 0.08	2.50 ± 0.11	3.29 ± 0.41**	2.45 ± 0.11	2.40 ± 0.17		
LDL particle size (nm)^{a}	23.9 ± 0.04	24.0 ± 0.06	24.0 ± 0.07	23.5 ± 0.12**	24.0 ± 0.09	23.9 ± 0.17		
Oxidized LDL (U/L)^{a}	45.9 ± 0.56	47.4 ± 1.53	49.8 ± 1.27	55.0 ± 2.70*	43.7 ± 2.09	43.7 ± 2.85		

Mean ± SE

HTN hypertension, *HTN w/o therapy* HTN group treated without antihypertensive therapy, *HTN w/ therapy* HTN group treated with antihypertensive therapy, *BP* blood pressure, *HDL* high-density lipoprotein, *LDL* low-density lipoprotein, *HOMA-IR* homeostatic model assessment of insulin resistance, *PAM* peptidylglycine-α-amidating monooxygenase

$^{t}p < 0.1$, $^{*}p < 0.05$, $^{**}p < 0.01$, and $^{***}p < 0.001$ derived from an independent t test within the normotensive controls and each HTN subgroup

^{a}Tested following logarithmic transformation

candidate gene for a certain disease (in case of the present study, HTN) through various mechanisms. Likewise, our data in Tables 2 and 3 support that the *PAM* rs13175330 G allele is associated with a higher risk of HTN development compared with the AA genotype although it is an intronic SNP.

The primary function of the neuroendocrine processing enzyme PAM in the atrium may be to package ANP, a hormone involved in the control of BP and the regulation of sodium and water excretion [10, 11], into atrial secretory granules for storage. Additionally, PAM possibly functions in the presence of activated ANP, which results from the proteolytic processing of pro-ANP [9]. The exact mechanism underlying the association between *PAM* rs13175330 A>G and HTN is unknown; thus, it is difficult to ascertain which function of *PAM* listed in Additional file 1: Table S4 (NCBI gene database; http://www.ncbi.nlm.nih.gov/gene/) is related to BP alteration and HTN development. However, since PAM and pro-ANP are the predominant membrane-associated proteins in atrial secretory granules [12] and PAM plays a role in ANP secretion [9], the *PAM* rs13175330 polymorphism may be involved in the dysregulation of ANP secretion and thus cause HTN. Indeed, the significance of the present observations is underscored by the identification of a human polymorphism in the *PAM* locus that is associated with altered systolic and diastolic BP. The present study showed that normotensive control *PAM* rs13175330 G allele carriers showed a trend toward increased systolic BP, whereas *PAM* rs13175330 G allele carriers in the HTN w/o therapy group had significantly higher

systolic and diastolic BP than the AA carriers. Moreover, in the HTN w/ therapy group, rs13175330 G allele carriers also showed significantly increased diastolic BP even though antihypertensive medication significantly lowered systolic and diastolic BP in both AA and G allele carriers compared to those in the HTN w/o therapy group (p values of systolic and diastolic BP between AA carriers in the HTN w/o therapy and HTN w/ therapy groups: both $p < 0.001$; p values of systolic and diastolic BP between G allele carriers in the HTN w/o therapy and HTN w/ therapy groups: both $p < 0.001$). As shown in Table 2, G allele carriers had a significantly high risk of HTN development; therefore, alteration of BP can be partially explained by PAM dysfunction due to the *PAM* rs13175330 polymorphism.

PAM gene polymorphisms affect not only BP but also other HTN-related risk factors. Recently, two missense variants in *PAM* (p.Asp563Gly and p.Ser539Trp) were reported to be associated with a high risk of type 2 diabetes [23]. Additionally, Czyzyk et al. [12] showed that 10-month-old PAM-heterozygous mice had mild but significant glucose intolerance compared to wild-type mice. Although the *PAM* rs13175330 A>G SNP found in this study was different from those SNPs, rs13175330 G allele carriers showed significantly higher insulin and HOMA-IR indices than the AA carriers in the HTN w/o therapy group. Moreover, *PAM* rs13175330 G allele carriers showed significantly smaller LDL particle sizes and higher ox-LDL levels than the AA carriers in the HTN w/o therapy group even though the G allele carriers had significantly lower

LDL-cholesterol levels. Thus, the rs13175330 G allele may be associated with worse atherogenicity of LDL-cholesterol. To verify associations between *PAM* genotypes and each variable involved in HTN development (HOMA-IR, LDL particle size, and ox-LDL), we performed a logistic regression analysis (data not shown). The G allele carriers in the HTN group were significantly associated with high HOMA-IR before adjustment for confounding factors including age, sex, BMI, smoking, and drinking [OR 1.110 (95% CI 1.007–1.224), $p = 0.036$]; after adjustment for confounding factors, only tendency was remained [OR 1.094 (95% CI 0.993–1.205), $p = 0.069$]. LDL particle size was significantly small in the G allele carriers before [OR 0.579 (95% CI 0.390–0.861), $p = 0.007$] and after [OR 0.529 (95% CI 0.349–0.800), $p = 0.003$] adjustment. Ox-LDL did not show any association with *PAM* rs13175330 genotypes. In summary, HOMA-IR and LDL particle size are associated with HTN development along with the *PAM* rs13175330 polymorphism. Thus, the increased risk of HTN in the *PAM* rs13175330 G allele carriers can be partially explained by the association between the *PAM* rs13175330 mutation and alteration of glucose tolerance and atherogenicity of LDL-cholesterol.

Taken together, our results indicated a genotype effect from the *PAM* rs13175330 A>G SNP on systolic and diastolic BP, insulin level, the HOMA-IR index, LDL particle size, and ox-LDL level in the HTN w/o therapy group. Since type 2 diabetes [24], decreased LDL particle size [25], and increased ox-LDL level [26] are well-known atherogenic traits related to HTN, these results suggest that the *PAM* gene polymorphism may be involved in HTN development via complex mechanisms, including PAM dysfunction, alterations of glucose tolerance, and atherogenicity of LDL-cholesterol.

Our results share the limitations of cross-sectional observational studies, because we evaluated only associations rather than prospective predictions. Since we verified only the relationship between the *PAM* rs13175330 polymorphism and the risk of HTN, exact mechanisms regarding HTN development by the rs1317530 SNP cannot be fully explained; thus, further studies are needed to demonstrate an association between PAM dysfunction due to the *PAM* rs13175330 polymorphism and HTN. Additionally, we specifically focused on a representative group of Korean subjects in the present study. Therefore, our results cannot be generalized to other ethnic, age, or geographic groups. Moreover, the IR in the HTN group could exaggerate other cardiometabolic syndrome phenotypes and should be considered when interpreting the present findings. Despite these limitations, our results show an interesting association between the *PAM* rs13175330 G allele and an increased risk of HTN.

Conclusions

Although *PAM* has diverse functions (Additional file 1: Table S4), there are a lack of studies on the association between PAM dysfunction due to *PAM* polymorphisms and diseases. Therefore, the findings in the present study are valuable, as this study reported associations between *PAM* polymorphisms and HTN for the first time. Our study suggests that the *PAM* rs13175330 A>G SNP is a novel candidate gene for HTN among the Korean population. Additionally, the *PAM* rs13175330 G allele may be associated with IR and LDL atherogenicity in patients with HTN. To verify the exact *PAM* rs131753301 polymorphism-related mechanisms underlying HTN development, further studies are required.

Abbreviations

ANP: Atrial natriuretic peptide; BMI: Body mass index; bp: Base pair; BP: Blood pressure; CV: Coefficient of variability; gDNA: Genomic DNA; GWAS: Genome-wide association study; HDL: High-density lipoprotein; HOMA: Homeostatic model assessment; HTN: Hypertension; HWE: Hardy-Weinberg equilibrium; IR: Insulin resistance; K-CHIP: Korean Chip; LDL: Low-density lipoprotein; OR: Odds ratio; Ox-: Oxidized; PAM: Peptidylglycine-α-amidating monooxygenase; SNP: Single nucleotide polymorphism; TC: Total cholesterol; TG: Triglyceride

Acknowledgements

The genotype data were generated using the Korean Chip (K-CHIP), which is available through the K-CHIP consortium. The K-CHIP was designed by the Center for Genome Science at the Korea National Institute of Health, Korea (4845-301, 3000-3031).

Funding

This study was funded by the Bio-Synergy Research Project (NRF-2012M3A9C4048762) and the Mid-Career Researcher Program (NRF-2016R1A2B4011662) of the Ministry of Science, ICT and Future Planning through the National Research Foundation of Korea in the Republic of Korea.

Authors' contributions

All authors contributed to the conception and design of the study. HJY, MJK, and JHL contributed to the acquisition, analysis, and interpretation of the data and preparation of the manuscript. MKK, JSC, and S-HL contributed to the acquisition and analysis of the data. All authors contributed to the critical revisions of the paper and have approved the manuscript for publication.

Consent for publication

Not applicable.

Competing interests

The authors declare that they have no competing interests.

Author details

[1]National Leading Research Laboratory of Clinical Nutrigenetics/Nutrigenomics, Department of Food and Nutrition, College of Human Ecology, Yonsei University, 50 Yonsei-ro, Seodaemun-gu, Seoul 03722, South Korea. [2]Department of Food and Nutrition, Brain Korea 21 PLUS Project,

College of Human Ecology, Yonsei University, Seoul 03722, South Korea. [3]Research Center for Silver Science, Institute of Symbiotic Life-TECH, Yonsei University, Seoul 03722, South Korea. [4]Department of Family Practice, National Health Insurance Corporation, Ilsan Hospital, Goyang 10444, South Korea.

References

1. World Health Organization. A global brief on hypertension. Geneva: WHO Press; 2013.
2. Lim SS, Vos T, Flaxman AD, Danaei G, Shibuya K, Adair-Rohani H, et al. A comparative risk assessment of burden of disease and injury attributable to 67 risk factors and risk factor clusters in 21 regions, 1990-2010: a systematic analysis for the global burden of disease study 2010. Lancet. 2012;380:2224–60.
3. Xi B, Cheng H, Shen Y, Zhao X, Hou D, Wang X, et al. Physical activity modifies the associations between genetic variants and hypertension in the Chinese children. Atherosclerosis. 2012;225:376–80.
4. El Shamieh S, Visvikis-Siest S. Genetic biomarkers of hypertension and future challenges integrating epigenomics. Clin Chim Acta. 2012;414:259–65.
5. Van Rijn MJ, Schut AF, Aulchenko YS, Deinum J, Sayed-Tabatabaei FA, Yazdanpanah M, et al. Heritability of blood pressure traits and the genetic contribution to blood pressure variance explained by four blood-pressure-related genes. J Hypertens. 2007;25:565–70.
6. Levy D, Ehret GB, Rice K, Verwoert GC, Launer LJ, Dehghan A, et al. Genome-wide association study of blood pressure and hypertension. Nat Genet. 2009;41:677–87.
7. Padmanabhan S, Melander O, Johnson T, Di Blasio AM, Lee WK, Gentilini D, et al. Genome-wide association study of blood pressure extremes identifies variant near UMOD associated with hypertension. PLoS Genet. 2010;6:e1001177.
8. International consortium for blood pressure genome-wide association studies. Genetic variants in novel pathways influence blood pressure and cardiovascular disease risk. Nature. 2011;478:103–9.
9. O'Donnell PJ, Driscoll WJ, Bäck N, Muth E, Mueller GP. Peptidylglycine-alpha-amidating monooxygenase and pro-atrial natriuretic peptide constitute the major membrane-associated proteins of rat atrial secretory granules. J Mol Cell Cardiol. 2003;35:915–22.
10. Thibault G, Amiri F, Garcia R. Regulation of natriuretic peptide secretion by the heart. Annu Rev Physiol. 1999;61:193–217.
11. Sagnella GA. Atrial natriuretic peptide mimetics and vasopeptidase inhibitors. Cardiovasc Res. 2001;51:416–28.
12. Czyzyk TA, Ning Y, Hsu MS, Peng B, Mains RE, Eipper BA, et al. Deletion of peptide amidation enzymatic activity leads to edema and embryonic lethality in the mouse. Dev Biol. 2005;287:301–13.
13. Department of Infectious Disease Control, Korea Centers for Disease Control & Prevention (KCDC). Public health weekly report, PHWR 2015: Vol. 8 No. 29. In: The Korea Biobank Array Project. Korea Centers for Disease Control & Prevention. 2015. http://www.cdc.go.kr/CDC/info/CdcKrInfo0301.jsp?menuIds=HOME001-MNU1154-MNU0005-MNU0037&q_type=&year=2015&cid=64288&pageNum=. Accessed 6 Nov 2017.
14. Korea Centers for Disease Control & Prevention (KCDC). Korean Chip project. 2017. http://cdc.go.kr/CDC/eng/contents/CdcEngContentView.jsp?cid=74266&menuIds=HOME002-MNU0576-MNU0586. Accessed 6 Nov 2017.
15. Kim M, Kim M, Yoo HJ, Yun R, Lee SH, Lee JH. Estrogen-related receptor γ gene (ESRRG) rs1890552 A>G polymorphism in a Korean population: association with urinary prostaglandin $F_{2\alpha}$ concentration and impaired fasting glucose or newly diagnosed type 2 diabetes. Diabetes Metab. 2017;43:385–8.
16. Kim M, Yoo HJ, Kim M, Seo H, Chae JS, Lee SH, et al. Influence of estrogen-related receptor γ (ESRRG) rs1890552 A > G polymorphism on changes in fasting glucose and arterial stiffness. Sci Rep. 2017;7:9787.
17. Roth SM. Genetics primer for exercise science and health. In: Roth SM, editor. Issues in study design and analysis. Champaign: Human Kinetics Publishers; 2007. p. 89–90.
18. Jo BS, Choi SS. Introns: the functional benefits of introns in genomes. Genomics Inform. 2015;13:112–8.
19. Wang D, Guo Y, Wrighton SA, Cooke GE, Sadee W. Intronic polymorphism in CYP3A4 affects hepatic expression and response to statin drugs. Pharmacogenomics J. 2011;11:274–86.
20. Wang D, Sadee W. CYP3A4 intronic SNP rs35599367 (CYP3A4*22) alters RNA splicing. Pharmacogenet Genomics. 2016;26:40–3.
21. Xia Z, Yang T, Wang Z, Dong J, Liang C. GRK5 intronic (CA)n polymorphisms associated with type 2 diabetes in Chinese Hainan Island. PLoS One. 2014;9:e90597.
22. Zhou J, Nagarkatti PS, Zhong Y, Creek K, Zhang J, Nagarkatti M. Unique SNP in CD44 intron 1 and its role in breast cancer development. Anticancer Res. 2010;30:1263–72.
23. Steinthorsdottir V, Thorleifsson G, Sulem P, Helgason H, Grarup N, Sigurdsson A, et al. Identification of low-frequency and rare sequence variants associated with elevated or reduced risk of type 2 diabetes. Nat Genet. 2014;46:294–8.
24. Mehta JL, Rasouli N, Sinha AK, Molavi B. Oxidative stress in diabetes: a mechanistic overview of its effects on atherogenesis and myocardial dysfunction. Int J Biochem Cell Biol. 2006;38:794–803.
25. Maruyama C, Imamura K, Teramoto T. Assessment of LDL particle size by triglyceride/HDL-cholesterol ratio in non-diabetic, healthy subjects without prominent hyperlipidemia. J Atheroscler Thromb. 2003;10:186–91.
26. Witztum JL, Steinberg D. Role of oxidized low density lipoprotein in atherogenesis. J Clin Invest. 1991;88:1785–92.

Identification of compound heterozygous variants in the noncoding RNU4ATAC gene in a Chinese family with two successive foetuses with severe microcephaly

Ye Wang[1], Xueli Wu[2], Liu Du[3], Ju Zheng[3], Songqing Deng[1], Xin Bi[4], Qiuyan Chen[5], Hongning Xie[3], Claude Férec[6], David N. Cooper[7], Yanmin Luo[1*], Qun Fang[1*] and Jian-Min Chen[6,8*] ⓘ

Abstract

Background: Whole-exome sequencing (WES) over the last few years has been increasingly employed for clinical diagnosis. However, one *caveat* with its use is that it inevitably fails to detect disease-causative variants that occur within noncoding RNA genes. Our experience in identifying pathogenic variants in the noncoding *RNU4ATAC* gene, in a Chinese family where two successive foetuses had been affected by severe microcephaly, is a case in point. These foetuses exhibited remarkably similar phenotypes in terms of their microcephaly and brain abnormalities; however, the paucity of other characteristic phenotypic features had made a precise diagnosis impossible. Given that no external causative factors had been reported/identified during the pregnancies, we sought a genetic cause for the phenotype in the proband, the second affected foetus.

Results: A search for chromosomal abnormalities and pathogenic copy number variants proved negative. WES was also negative. These initial failures prompted us to consider the potential role of *RNU4ATAC*, a noncoding gene implicated in microcephalic osteodysplastic primordial dwarfism type-1 (MOPD1), a severe autosomal recessive disease characterised by dwarfism, severe microcephaly and neurological abnormalities. Subsequent targeted sequencing of *RNU4ATAC* resulted in the identification of compound heterozygous variants, one being the most frequently reported MOPD1-causative mutation (51G>A), whereas the other was a novel 29T>A variant. Four distinct lines of evidence (allele frequency in normal populations, evolutionary conservation of the affected nucleotide, occurrence within a known mutational hotspot for MOPD1-causative variants and predicted effect on RNA secondary structure) allowed us to conclude that 29T>A is a new causative variant for MOPD1.

Conclusions: Our findings highlight the limitations of WES in failing to detect variants within noncoding RNA genes and provide support for a role for whole-genome sequencing as a first-tier genetic test in paediatric medicine. Additionally, the identification of a novel *RNU4ATAC* variant within the mutational hotspot for MOPD1-causative variants further strengthens the critical role of the 5′ stem-loop structure of U4atac in health and disease. Finally, this analysis enabled us to provide prenatal diagnosis and genetic counselling for the mother's third pregnancy, the first report of its kind in the context of inherited *RNU4ATAC* variants.

(Continued on next page)

* Correspondence: luoyanm@mail.sysu.edu.cn; fang_qun@163.com; jian-min.chen@univ-brest.fr
[1]Fetal Medicine Centre, Department of Obstetrics and Gynaecology, The First Affiliated Hospital of Sun Yat-Sen University, Guangzhou, China
[6]UMR1078 "Génétique, Génomique Fonctionnelle et Biotechnologies", INSERM, EFS - Bretagne, Université de Brest, CHRU Brest, Brest, France
Full list of author information is available at the end of the article

(Continued from previous page)

Keywords: Genetic counselling, Microcephalic osteodysplastic primordial dwarfism type 1, MOPD1, Noncoding *RNU4ATAC* gene, Prenatal diagnosis, RNA secondary structure, Small nuclear RNA, Taybi-Linder syndrome, WES, Whole-exome sequencing

Background

Microcephaly is usually defined in terms of a head circumference more than two standard deviations below the mean for age and sex; it can occur in the womb or may develop during the first few years of life [1, 2]. Abnormal growth of the head may occur as a consequence of a number of factors, both genetic and environmental (e.g. exposure to certain viruses such as rubella, drugs and alcohol during pregnancy) [3]. The genetic causes are highly heterogeneous; thus, a search for microcephaly in the Human Phenotype Ontology database [4] yielded 652 genes. Depending on the precise nature of the condition involved, microcephaly may be associated with seizures, developmental delay, intellectual disability or other problems. It may even be associated with substantial physical disability and premature death; there is no treatment for microcephaly. Therefore, it is extremely important to identify the genetic causes of severe microcephaly in affected families with a view to providing prenatal diagnosis and genetic counselling in subsequent pregnancies.

With the decreased cost of next-generation sequencing, whole-exome sequencing (WES) has rapidly evolved from its original application as a tool for gene discovery in research settings to an important diagnostic tool in a clinical context [5–7], especially for diseases that are characterised by a significant level of genetic heterogeneity [8]. However, one *caveat* with WES is that disease-causative variants which occur within noncoding RNA genes will invariably be missed [9, 10]. Here, we highlight this issue by describing our experience of identifying novel compound heterozygous variants in the noncoding *RNU4ATAC* gene (OMIM #601428), in a Chinese family with two successive foetuses affected by severe microcephaly.

Results

Family description

A 30-year-old woman was referred to our centre at the First Affiliated Hospital of Sun Yat-Sen University after her second foetus (II:2) had been found to have severe microcephaly at 24 gestational weeks (GW), just as her first one (II:1; Fig. 1a) had previously. Clinical findings in the two affected foetuses, who were terminated at 36 GW (II:1) and 30 GW (II:2) respectively, are illustrated in Fig. 2 and summarised in Table 1. However, no precise diagnosis of the underlying abnormality could be made based upon these clinical findings owing to the

paucity of characteristic features beyond severe microcephaly. In the case of II:1, standard G-banding karyotyping using cord blood cells taken at 35 GW revealed a normal karyotype, whereas chromosomal microarray analysis failed to detect any pathogenic copy number variations; no further analyses were performed at the time.

The parents were of North Chinese origin, healthy and nonconsanguineous. Exposure to known causative environmental factors during pregnancy was neither reported nor identified. Taken together with the remarkably similar clinical phenotypes in the two affected foetuses (Fig. 2; Table 1), a genetic cause was considered to be likely. An extensive molecular genetic analysis was therefore performed on foetus II:2.

Extensive karyotyping and chromosomal microarray analysis failed to identify any chromosomal abnormality or pathogenic copy number variations in II:2

We first performed standard G-banding karyotyping using cord blood cells from II:2 (taken at 29 GW), but no chromosomal abnormalities were found. In the meantime, we also performed chromosomal microarray analysis using genomic DNA prepared from the cord blood cells taken from II:2. No pathogenic copy number variants were identified by reference to data available in OMIM [11], DGV [12] and DECIPHER [13].

WES also failed to reveal a genetic cause of the microcephaly in II:2

We further employed WES to search for putative causal variants in an unbiased and hypothesis-free manner. The resulting single-nucleotide variants (SNVs) and small insertions or deletions (indels) were subjected to the following prioritizations: (i) variants that cause nonsynonymous, frameshift and in-frame changes and variants that occurred at splice sites; (ii) variants with a minor allele frequency of less than 5% according to either the 1000 Genomes Project [14] or the ESP5400 data of the National Heart, Lung, and Blood Institute GO Exome Sequencing Project [15]; (iii) in case of missense variants, those predicted to be deleterious using the programs of PolyPhen-2 [16], SIFT [17] and Mutation Taster [18] and (iv) variants occurring in known microcephaly-causing or microcephaly-associated genes as well as in candidate genes selected on the basis of known biological, physiological or functional relevance

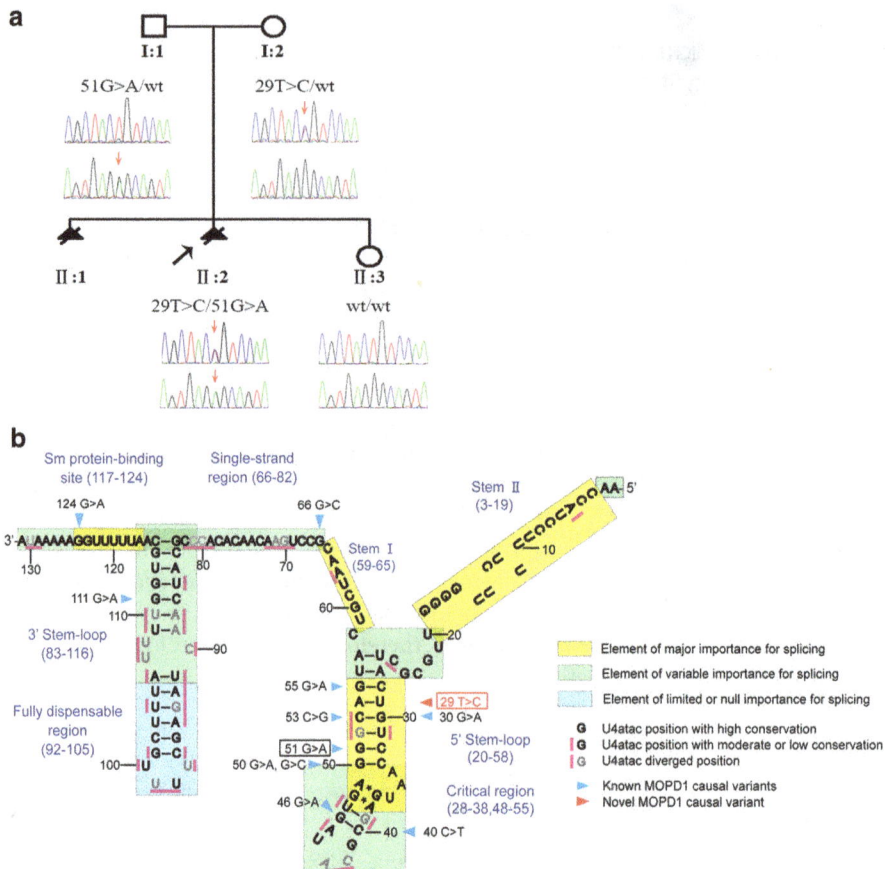

Fig. 1 Identification of the genetic cause of severe microcephaly in a Chinese family. **a** Family pedigree. Filled triangles with oblique lines indicate the two successive foetuses affected with severe microcephaly and terminated by therapeutic abortion. Arrow indicates the proband. Open symbols indicate clinically unaffected family members. Genotypes with respect to the *RNU4ATAC* gene are also provided where it was possible to determine them. wt, wild-type. **b** U4atac snRNA secondary structure elements, evolutionary conservation status of each nucleotide position and MOPD1-causative SNVs (adapted from [46]). The novel variant found in the present study, 29T>C, is highlighted in red and boxed. For a detailed description of the structure and function of U4atac, see Merico et al. [46] and references therein

to microcephaly. However, no variants survived this process of prioritisation.

Targeted sequencing of the noncoding *RNU4ATAC* gene identified causal variants in II:2

After failing to detect any pathogenic lesion by karyotyping, chromosomal microarray analysis and WES, we began to consider the potential involvement of noncoding RNA genes in the aetiology of microcephaly. An extensive literature research resulted in the recognition of two such genes. The first was the miR-17-92a-1 cluster host gene (*MIR17HG*; OMIM #609415). Large-scale copy number variants that serve either to delete or duplicate the entire *MIR17HG* locus cause Feingold syndrome 2 (OMIM #614326), a rare autosomal dominant disorder characterised by variable combinations of microcephaly, limb malformations, oesophageal and duodenal atresias and learning disability [19–26]. Although the disease entity under study here is most

consistent with a model of autosomal recessive inheritance, we nevertheless revisited our chromosomal microarray analysis data and confirmed the absence of large deletions or duplications involving the *MIR17HG* locus.

The second noncoding gene emerging from our literature search was *RNU4ATAC*, in which homozygous or compound heterozygous variants have been reported to cause microcephalic osteodysplastic primordial dwarfism type-1 (MOPD1; OMIM #210710) in two simultaneous papers in 2011 [27, 28]. MOPD1, also known as Taybi-Linder syndrome [29], is a severe autosomal recessive disease characterised by dwarfism, microcephaly and neurologic abnormalities; patients usually die within the first year of life [30]. *RNU4ATAC* is located on chromosome 2q14.2 and encodes the highly conserved, 130-bp small nuclear RNA (snRNA) U4atac (RefSeq NR_023343.1). U4atac is a component of the minor spliceosome that is responsible for the correct splicing of the U12-dependent class of introns [31–35]. To date, a

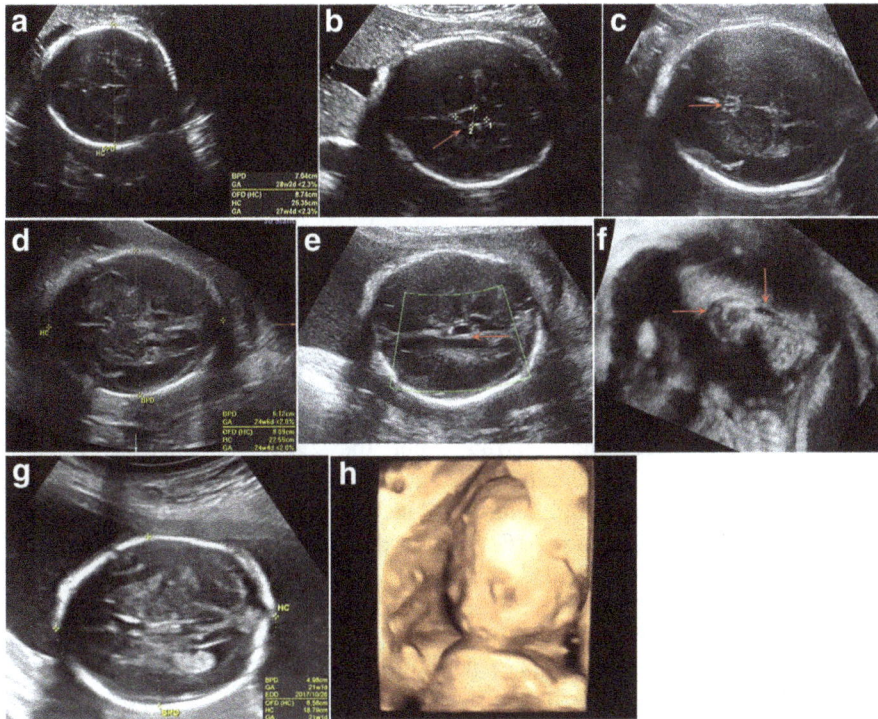

Fig. 2 Ultrasound images of the three foetuses. **a** Biparietal diameter (BPD) and head circumference (HC) of foetus II:1 measured at 35 GW indicating severe microcephaly. **b** Cross-section plane of the skull displaying arachnoid cysts (arrow) in foetus II:1. **c–f** Ultrasound images of foetus II:2: BPD and HC measured at 29 GW indicating severe microcephaly (**d**); cross-section and three-dimensional sagittal plane of the skull showing the absence of the septum pellucidum cavity (arrow; **c**), presence of intracranial cyst (arrows; **e**, **f**) and agenesis of corpus callosum (arrow; **f**). **g**, **h** Ultrasound images of the healthy foetus II:3: BPD and HC (measured at 21 GW) shown in cross-section plane (**g**) and three-dimensional rebuilt imaging for foetus face (**h**)

Table 1 Clinical data of the two affected foetuses

Case	Foetus 1 (II:1)			Foetus 2 (II:2)		
Growth (GW)	24	33	36	24	26	29
BPD (SD)	− 6.0	− 6.0	− 6.1	− 4.4	− 5.6	− 6.0
HC (SD)	− 4.0	− 5.6	− 7.5	− 3.7	− 4.2	− 5.5
FL (SD)	− 2.0	− 1.6	− 2.0	0	− 1.0	0
HL (SD)	− 1.5	/	− 1.0	0	0	0
AC (SD)	/	− 2.4	− 3.8	− 0.7	− 1.0	− 1.4
Weight	/			1.047 ± 0.153 kg (29 GW, < 50 centile)		
Brain anomalies						
Hypogenesis or agenesis of corpus callosum	−			−		
Absence of septum pellucidum cavity	−			+		
Intracranial cysts	+			+		
Small vermis	−			−		
Skeletal anomalies						
Short limb	+ (slightly short)			−		
Flexion contractures	−			−		
Cardiac abnormalities				−		
Skin and skin appendage	/			/		

BPD biparietal diameter, *HC* head circumference, *FL* femur length, *HL* humerus length, *AC* abdomen circumference, *GW* gestational week, *SD* standard deviation, / not evaluable, − negative, + positive

total of 12 *RNU4ATAC* variants including 11 SNVs (i.e. 30G>A, 40C>T, 46G>A, 50G>A, 50G>C, 51G>A, 53C>G, 55G>A, 66G>C, 111G>A and 124G>A) and 1 duplication variant (i.e. 16_100dup) have been reported to be causative for MOPD1 [27, 28, 36–41]. Six of the 11 SNVs (i.e. 30G>A, 50G>A, 50G>C, 51G>A, 53C>G and 55G>A) occurred within the critical canonical stem region (i.e. 28–33, 50–55) of a functionally indispensable element of U4atac, the 5′ stem-loop structure (Fig. 1b). In what follows, we shall term this critical canonical stem region the mutational hotspot region for MOPD1-causative variants.

We, therefore, speculated that variants in the *RNU4ATAC* gene, which would not have been detected by WES, might underlie the severe microcephaly in this family. Subsequent targeted testing of the *RNU4ATAC* gene by Sanger sequencing identified compound heterozygous variants, 29T>C (rs779143800) and 51G>A (rs188343279), in II:2. Carrier analysis confirmed that the two variants had been inherited from the mother and father, respectively (Fig. 1a). Although it was suspected that the affected foetus II:1 had also inherited these two variants, this could not be confirmed due to the non-availability of genetic material.

51G>A was among the first described MOPD1-causative variants [27, 28] and represents the most common MOPD1-causative variant so far reported. By contrast, 29T>C has not been previously reported in MOPD1 patients. It has however been reported at a very low frequency in normal populations; thus, it is present in heterozygous form in two individuals in the Genome Aggregation Database [42], corresponding to an allele frequency of 1.6×10^{-5}. Further, in common with all 11 previously reported MOPD1-causative *RNU4ATAC* SNVs, 29T>C affected one of the evolutionarily highly conserved positions of U4atac (Fig. 1b). Furthermore, and most importantly, 29T is located within the mutational hotspot region for MOPD1-causative variants (Fig. 1b). In this latter regard, our current understanding of the pathogenetic mechanism underlying the six known MOPD1-causing *RNU4ATAC* SNVs occurring within the mutational hotspot is that they abrogate U4atac snRNA function by disrupting the 5′ stem-loop structure [27, 28, 41]. Accordingly, we compared the potential effect of 29T>C on the 5′ stem-loop structure of U4atac with those of the aforementioned six known MOPD1-causative SNVs. To this end, wild-type and the seven mutant sequences spanning positions 20 to 58 (i.e. the sequence forming the 5′ stem-loop structure; Fig. 1b) of U4atac were separately subjected to Mfold analysis [43] under default conditions. All seven SNVs were predicted to significantly affect the 5′ stem-loop structure of U4atac as compared with the wild-type.

In particular, 29T>C was predicted to alter the secondary structure in the same way as the pathogenic nucleotide substitutions 30G>A, 50G>A and 53C>G; it was also predicted to alter the secondary structure in a similar way to the most common pathogenic 51G>A variant (Fig. 3). These observations, taken together, strongly suggest that 29T>C constitutes a novel causative variant for MOPD1.

Prenatal diagnosis of the third pregnancy

Prenatal diagnosis was performed on the third foetus (II:3) (Fig. 1a). Genomic DNA was prepared from amniotic fluid cells taken by ultrasound-mediated amniocentesis at 16 GW. However, neither the *RNU4ATAC* 51G>A variant nor the 29T>C variant was detected. Normal foetal growth was confirmed by continual ultrasound monitoring during the whole period of pregnancy (Fig. 2g, h). The third foetus was born healthy after 40 GW.

Discussion

In this study, we relate our experience of how the genetic cause was finally identified in a Chinese family presenting with two successive foetuses with severe microcephaly. In brief, negative findings from karyotyping, chromosomal microarray analysis and WES in foetus II:2 prompted us to consider the potential role of noncoding genes in causing microcephaly in the family. Consequently, targeted sequencing of the noncoding *RNU4ATAC* gene resulted in the identification of compound heterozygous variants, one being the most frequently reported MOPD1-causative 51G>A, the other being a novel 29T>A variant. Based upon the four lines of evidence, namely allele frequency in normal populations, evolutionary conservation, occurrence within a known mutational hotspot for MOPD1-causative variants and predicted effect on the 5′ stem-loop structure, we were able to conclude with confidence that the newly found 29T>A variant represents a new causative variant for MOPD1. Here, we should like to make two additional points. First, in the context of in silico analysis, many algorithms have been designed to predict the functional consequences of intronic or missense variants found in protein-coding genes. However, these tools are inappropriate for use with the *RNU4ATAC* variants discussed here, whose functional consequences depend upon their potential effect on RNA secondary structure. Currently, Mfold analysis is the gold standard for performing RNA secondary structure predictions. Second, stringent standards and guidelines have been proposed for investigating the causality of sequence variants in human genetic disease [44, 45]. Apart from the aforementioned four lines of evidence supporting causality of the detected compound heterozygous *RNU4ATAC* variants, we would like to add a new consideration. The genomic

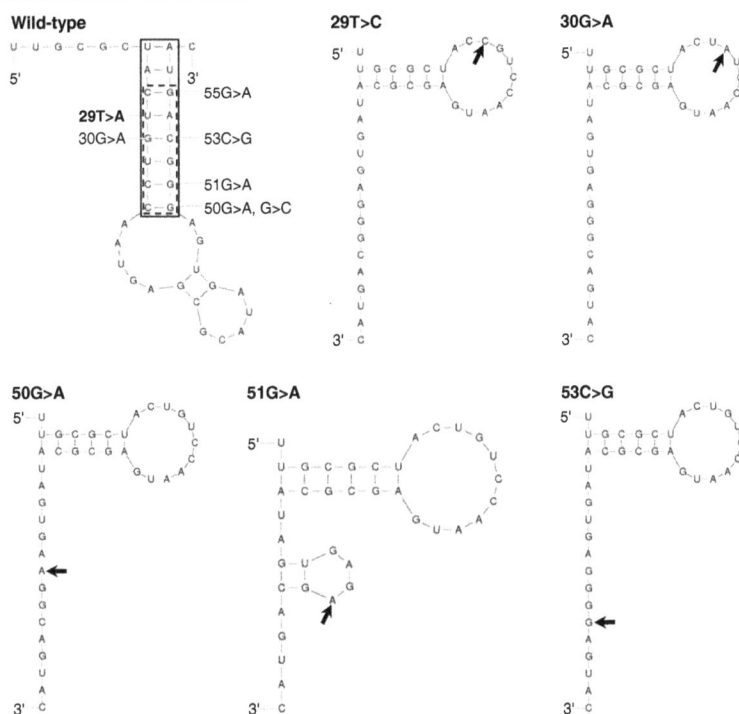

Fig. 3 Predicted effects of 29T>C, 30G>A, 50G>A, 51G>A and 53C>G on the 5′ stem-loop structure of U4atac. Secondary structures of the wild-type and corresponding mutant sequences between positions 20 and 58 of U4atac (see Fig. 2b) were predicted by Mfold under default parameters, with the lowest energy structures being shown. In the wild-type panel, the solid-boxed area indicates the canonical stem of the 5′ stem-loop structure in accordance with [53, 54]; the dotted-boxed area indicates the mutational hotspot region for MOPD1-causative variants; all six known MOPD1-causative SNVs as well as the newly found 29T>C variant occurring within the mutational hotspot region are also indicated. In the mutant panels, the respective variants are indicated by arrows

structure of *RNU4ATAC* is very simple, comprising only 130 nucleotides. All the so far reported MOPD1-causative *RNU4ATAC* variants were invariably located within the 130 nucleotides.

Most previous studies have reported homozygous or compound heterozygous *RNU4ATAC* variants in patients with a diagnosis, or suggestive diagnosis, of MOPD1 [27, 28, 36–40]. Only very recently have *RNU4ATAC* variants been described in foetuses [41]; all four foetuses (two of whom were twins) had severe microcephaly together with some other brain and skeletal abnormalities including corpus callosum agenesis, short limb, brachydactyly and ossification delay, suggestive of a diagnosis of MOPD1. By contrast, the two foetuses in the family under study here showed only severe microcephaly and corpus callosum agenesis. The identification of compound heterozygous *RNU4ATAC* variants in II:2, therefore, provided a definite diagnosis of the disease that could not otherwise have been made merely on the basis of clinical findings.

In a more general context, our study adds to the increasing appreciation that variants in noncoding RNA genes are an underestimated cause of human inherited disease. Here, we further emphasise this point by citing

a recent finding concerning the *RNU4ATAC* gene. Compound heterozygous *RNU4ATAC* variants have also been reported to cause Roifman syndrome (OMIM #300258) [10, 46], a rare congenital association of antibody deficiency, spondyloepiphyseal chondro-osseous dysplasia, retinal dystrophy, poor pre- and postnatal growth and cognitive delay, which is phenotypically quite different from MOPD1. It should be noted that Roifman syndrome-causative compound heterozygous *RNU4ATAC* variants comprise one variant that is located within MOPD1-implicated structural elements and one variant that is located outside of MOPD1-implicated structural elements [10, 46].

Conclusions

In a general context, our findings highlight one key limitation of WES, namely that it fails to detect disease causative variants within noncoding RNA genes. This provides support for a role for whole-genome sequencing as a first-tier genetic test in paediatric medicine [9]. This is also the first report of MOPD1-causative *RNU4ATAC* variants in the Chinese population and the first report of prenatal diagnosis and genetic counselling provided for a subsequent pregnancy once *RNU4ATAC* variants had been identified as a cause of MOPD1. Finally, the identification of a novel *RNU4ATAC* variant within the mutational hotspot for MOPD1-causative variants further strengthens the critical role of the 5′ stem-loop structure of U4atac in health and disease.

Methods

Karyotyping and chromosomal microarray analysis

Standard G-banding karyotyping was performed. The array experiments were performed using the high-resolution Affymetrix CytoScan HD microarray (Affymetrix Inc., Santa Clara, CA) in accordance with the manufacturer's protocols. The results were analysed using the Chromosome Analysis Suite software version 1.2.2; the reporting threshold of the copy number was set at 10 kb, with marker count at ≥ 50, as previously reported [47].

WES

Genomic DNA was fragmented randomly and then purified by means of the magnetic particle method. Sequences were captured by Agilent SureSelect version 4 (Agilent Technologies, Santa Clara, CA) according to the manufacturer's protocols. The DNA libraries, after enrichment and purification, were sequenced on the NextSeq500 sequencer according to the manufacturer's instructions (Illumina, San Diego). The sequencing reads were aligned to GRCh37.p10 using Burrows-Wheeler Aligner software (version 0.59) [48]. Local realignment and base quality recalibration of the Burrows-Wheeler aligned reads were then performed using the GATK IndelRealigner [49] and GATK BaseRecalibrator [50], respectively. SNVs and small indels were identified by the GATK UnifiedGenotyper [51]. Variants were annotated using the Consensus Coding Sequences Database at the National Centre for Biotechnology Information [52].

Targeted sequencing of the *RNU4ATAC* gene

Primer sequences and PCR conditions are available upon request.

RNA secondary structure prediction

This was performed by means of Mfold analysis under default conditions [43].

Abbreviations
BPD: Biparietal diameter; GW: Gestational weeks; HC: Head circumference; indels: Insertions or deletions; MOPD1: Microcephalic osteodysplastic primordial dwarfism type 1; snRNA: Small nuclear RNA; SNVs: Single-nucleotide variants; WES: Whole-exome sequencing

Acknowledgements
Not applicable.

Funding
This work was supported in part by the National Natural Science Foundation of China (nos. 81270705 and 81671464) and the Science and Technology Project of Guangdong, China (2017A020214013).

Authors' contributions
YW, YL and QF designed the study with the assistance of JMC. YW, XW and XB performed the genetic analysis. YL, LD, JZ, SD, QC and HX conducted the clinical and ultrasound evaluations. CF and JMC interpreted the pathogenic relevance of the novel *RNU4ATAC* variant. YW and JMC drafted the manuscript. DNC critically revised the manuscript. All authors analysed the data and approved the final manuscript.

Consent for publication
Written informed consent for the publication of their clinical details and/or clinical images was obtained from the parents. A copy of the consent form is available for review by the Editor of this journal.

Competing interests
The authors declare that they have no competing interests.

Author details
[1]Fetal Medicine Centre, Department of Obstetrics and Gynaecology, The First Affiliated Hospital of Sun Yat-Sen University, Guangzhou, China. [2]Department of Dermatology, Guangzhou Institute of Dermatology, Guangzhou, China. [3]Department of Ultrasonic Medicine, The First Affiliated Hospital of Sun Yat-Sen University, Guangzhou, China. [4]Guangzhou KingMed Center for Clinical Laboratory, Guangzhou, China. [5]Dongguan Women and Children's Hospital, Dongguan, China. [6]UMR1078 "Génétique, Génomique Fonctionnelle et Biotechnologies", INSERM, EFS - Bretagne, Université de Brest, CHRU Brest, Brest, France. [7]Institute of Medical Genetics, School of Medicine, Cardiff University, Cardiff, UK. [8]INSERM UMR1078, EFS, UBO, 22 avenue Camille Desmoulins, 29238 Brest, France.

References
1. Leviton A, Holmes LB, Allred EN, Vargas J. Methodologic issues in epidemiologic studies of congenital microcephaly. Early Hum Dev. 2002;69: 91–105.
2. Opitz JM, Holt MC. Microcephaly: general considerations and aids to nosology. J Craniofac Genet Dev Biol. 1990;10:175–204.
3. Duerinckx S, Abramowicz M. The genetics of congenitally small brains. Semin Cell Dev Biol. 2017. https://doi.org/10.1016/j.semcdb.2017.09.015. [Epub ahead of print]
4. The Human Phenotype Ontology database. http://compbio.charite.de/hpoweb/showterm?id=HP:0000252. Accessed 23 Nov 2017.
5. Boycott KM, Vanstone MR, Bulman DE, MacKenzie AE. Rare-disease genetics in the era of next-generation sequencing: discovery to translation. Nat Rev Genet. 2013;14:681–91.
6. Lee H, Deignan JL, Dorrani N, Strom SP, Kantarci S, Quintero-Rivera F, et al. Clinical exome sequencing for genetic identification of rare Mendelian disorders. JAMA. 2014;312:1880–7.
7. Posey JE, Harel T, Liu P, Rosenfeld JA, James RA, Coban Akdemir ZH, et al. Resolution of disease phenotypes resulting from multilocus genomic variation. N Engl J Med. 2017;376:21–31.
8. Ku CS, Cooper DN, Polychronakos C, Naidoo N, Wu M, Soong R. Exome sequencing: dual role as a discovery and diagnostic tool. Ann Neurol. 2012; 71:5–14.
9. Lionel AC, Costain G, Monfared N, Walker S, Reuter MS, Hosseini SM, et al. Improved diagnostic yield compared with targeted gene sequencing panels suggests a role for whole-genome sequencing as a first-tier genetic test. Genet Med. 2017. https://doi.org/10.1038/gim.2017.119. [Epub ahead of print]
10. Bogaert DJ, Dullaers M, Kuehn HS, Leroy BP, Niemela JE, De Wilde H, et al. Early-onset primary antibody deficiency resembling common variable immunodeficiency challenges the diagnosis of Wiedeman-Steiner and Roifman syndromes. Sci Rep. 2017;7:3702.
11. Online Mendelian Inheritance in Man. https://www.omim.org/. Accessed 23 Nov 2017.
12. Database of Genomic Variants; http://dgv.tcag.ca/dgv/app/home. Accessed 23 Nov 2017.
13. DatabasE of genomiC varIation and Phenotype in Humans using Ensembl Resources; https://decipher.sanger.ac.uk. Accessed 23 Nov 2017.

14. Abecasis GR, Altshuler D, Auton A, Brooks LD, Durbin RM, Gibbs RA, et al. A map of human genome variation from population-scale sequencing. Nature. 2010;467:1061–73.

15. The ESP5400 data of the National Heart, Lung and Blood Institute GO Exome Sequencing Project. http://evs.gs.washington.edu/EVS. Accessed 23 Nov 2017.

16. PolyPhen-2. http://genetics.bwh.harvard.edu/pph2/. Accessed 23 Nov 2017.

17. SIFT. http://sift.jcvi.org/www/SIFT_chr_coords_submit.html. Accessed 23 Nov 2017.

18. Mutation Taster. http://www.mutationtaster.org/. Accessed 23 Nov 2017.

19. de Pontual L, Yao E, Callier P, Faivre L, Drouin V, Cariou S, et al. Germline deletion of the miR-17 approximately 92 cluster causes skeletal and growth defects in humans. Nat Genet. 2011;43:1026–30.

20. Sharaidin HS, Knipe S, Bain N, Goel H. Clinical features associated with a 15.41 Mb deletion of chromosome 13q encompassing the MIR17HG locus. Clin Dysmorphol. 2013;22:68–70.

21. Valdes-Miranda JM, Soto-Alvarez JR, Toral-Lopez J, Gonzalez-Huerta L, Perez-Cabrera A, Gonzalez-Monfil G, et al. A novel microdeletion involving the 13q31.3-q32.1 region in a patient with normal intelligence. Eur J Med Genet. 2014;57:60–4.

22. Ganjavi H, Siu VM, Speevak M, MacDonald PA. A fourth case of Feingold syndrome type 2: psychiatric presentation and management. BMJ Case Rep. 2014. https://doi.org/10.1136/bcr-2014-207501.

23. Low KJ, Buxton CC, Newbury-Ecob RA. Tetralogy of Fallot, microcephaly, short stature and brachymesophalangy is associated with hemizygous loss of noncoding MIR17HG and coding GPC5. Clin Dysmorphol. 2015; 24:113–4.

24. Grote LE, Repnikova EA, Amudhavalli SM. Expanding the phenotype of feingold syndrome-2. Am J Med Genet A. 2015;167A:3219–25.

25. Sirchia F, Di Gregorio E, Restagno G, Grosso E, Pappi P, Talarico F, et al. A case of Feingold type 2 syndrome associated with keratoconus refines keratoconus type 7 locus on chromosome 13q. Eur J Med Genet. 2017;60: 224–7.

26. Hemmat M, Rumple MJ, Mahon LW, Strom CM, Anguiano A, Talai M, et al. Short stature, digit anomalies and dysmorphic facial features are associated with the duplication of miR-17 ~ 92 cluster. Mol Cytogenet. 2014;7:27.

27. Edery P, Marcaillou C, Sahbatou M, Labalme A, Chastang J, Touraine R, et al. Association of TALS developmental disorder with defect in minor splicing component U4atac snRNA. Science. 2011;332:240–3.

28. He H, Liyanarachchi S, Akagi K, Nagy R, Li J, Dietrich RC, et al. Mutations in U4atac snRNA, a component of the minor spliceosome, in the developmental disorder MOPD I. Science. 2011;332:238–40.

29. Taybi H, Linder D. Congenital familial dwarfism with cephaloskeletal dysplasia. Radiology. 1967;89:275–81.

30. Pierce MJ, Morse RP. The neurologic findings in Taybi-Linder syndrome (MOPD I/III): case report and review of the literature. Am J Med Genet A. 2012;158A:606–10.

31. Hall SL, Padgett RA. Requirement of U12 snRNA for in vivo splicing of a minor class of eukaryotic nuclear pre-mRNA introns. Science. 1996;271: 1716–8.

32. Tarn WY, Steitz JA. Highly diverged U4 and U6 small nuclear RNAs required for splicing rare AT-AC introns. Science. 1996;273:1824–32.

33. Sharp PA, Burge CB. Classification of introns: U2-type or U12-type. Cell. 1997; 91:875–9.

34. Nilsen TW. The spliceosome: the most complex macromolecular machine in the cell? BioEssays. 2003;25:1147–9.

35. Will CL, Luhrmann R. Splicing of a rare class of introns by the U12-dependent spliceosome. Biol Chem. 2005;386:713–24.

36. Kilic E, Yigit G, Utine GE, Wollnik B, Mihci E, Nur BG, et al. A novel mutation in RNU4ATAC in a patient with microcephalic osteodysplastic primordial dwarfism type I. Am J Med Genet A. 2015;167A:919–21.

37. Abdel-Salam GM, Abdel-Hamid MS, Issa M, Magdy A, El-Kotoury A, Amr K. Expanding the phenotypic and mutational spectrum in microcephalic osteodysplastic primordial dwarfism type I. Am J Med Genet A. 2012;158A: 1455–61.

38. Nagy R, Wang H, Albrecht B, Wieczorek D, Gillessen-Kaesbach G, Haan E, et al. Microcephalic osteodysplastic primordial dwarfism type I with biallelic mutations in the RNU4ATAC gene. Clin Genet. 2012;82:140–6.

39. Abdel-Salam GM, Miyake N, Eid MM, Abdel-Hamid MS, Hassan NA, Eid OM, et al. A homozygous mutation in RNU4ATAC as a cause of microcephalic osteodysplastic primordial dwarfism type I (MOPD I) with associated pigmentary disorder. Am J Med Genet A. 2011;155A:2885–96.

40. Kroigard AB, Jackson AP, Bicknell LS, Baple E, Brusgaard K, Hansen LK, et al. Two novel mutations in RNU4ATAC in two siblings with an atypical mild phenotype of microcephalic osteodysplastic primordial dwarfism type 1. Clin Dysmorphol. 2016;25:68–72.

41. Putoux A, Alqahtani A, Pinson L, Paulussen AD, Michel J, Besson A, et al. Refining the phenotypical and mutational spectrum of Taybi-Linder syndrome. Clin Genet. 2016;90:550–5.

42. The Genome Aggregation Database (gnomAD). http://gnomad. broadinstitute.org/. Accessed 23 Nov 2017.

43. Mfold. http://unafold.rna.albany.edu/?q=mfold/rna-folding-form. Accessed 23 Nov 2017.

44. MacArthur DG, Manolio TA, Dimmock DP, Rehm HL, Shendure J, Abecasis GR, et al. Guidelines for investigating causality of sequence variants in human disease. Nature. 2014;508:469–76.

45. Richards S, Aziz N, Bale S, Bick D, Das S, Gastier-Foster J, et al. Standards and guidelines for the interpretation of sequence variants: a joint consensus recommendation of the American College of Medical Genetics and Genomics and the Association for Molecular Pathology. Genet Med. 2015; 17:405–24.

46. Merico D, Roifman M, Braunschweig U, Yuen RK, Alexandrova R, Bates A, et al. Compound heterozygous mutations in the noncoding RNU4ATAC cause Roifman syndrome by disrupting minor intron splicing. Nat Commun. 2015;6:8718.

47. Wang Y, Su P, Hu B, Zhu W, Li Q, Yuan P, et al. Characterization of 26 deletion CNVs reveals the frequent occurrence of micro-mutations within the breakpoint-flanking regions and frequent repair of double-strand breaks by templated insertions derived from remote genomic regions. Hum Genet. 2015;134:589–603.

48. Li H, Durbin R. Fast and accurate short read alignment with Burrows-Wheeler transform. Bioinformatics. 2009;25:1754–60.

49. The GATK IndelRealigner. https://software.broadinstitute.org/gatk/ documentation/tooldocs/current/org_broadinstitute_gatk_tools_walkers_ indels_IndelRealigner.php. Accessed 23 Nov 2017.

50. The GATK BaseRecalibrator. https://software.broadinstitute.org/gatk/ documentation/tooldocs/current/org_broadinstitute_gatk_tools_walkers_ bqsr_BaseRecalibrator.php. Accessed 23 Nov 2017.

51. The GATK UnifiedGenotyper. https://software.broadinstitute.org/gatk/ documentation/tooldocs/current/org_broadinstitute_gatk_tools_walkers_ genotyper_UnifiedGenotyper.php. Accessed 23 Nov 2017.

52. The Consensus Coding Sequences Database at the National Centre for Biotechnology Information. https://www.ncbi.nlm.nih.gov/CCDS/. Accessed 23 Nov 2017.

53. Padgett RA, Shukla GC. A revised model for U4atac/U6atac snRNA base pairing. RNA. 2002;8:125–8.

54. Cojocaru V, Klement R, Jovin TM. Loss of G-A base pairs is insufficient for achieving a large opening of U4 snRNA K-turn motif. Nucleic Acids Res. 2005;33:3435–46.

15

Nonparametric approaches for population structure analysis

Luluah Alhusain* ⓘ and Alaaeldin M. Hafez

Abstract

The analysis of population structure has many applications in medical and population genetic research. Such analysis is used to provide clear insight into the underlying genetic population substructure and is a crucial prerequisite for any analysis of genetic data. The analysis involves grouping individuals into subpopulations based on shared genetic variations. The most widely used markers to study the variation of DNA sequences between populations are single nucleotide polymorphisms. Data preprocessing is a necessary step to assess the quality of the data and to determine which markers or individuals can reasonably be included in the analysis. After preprocessing, several methods can be utilized to uncover population substructure, which can be categorized into two broad approaches: parametric and nonparametric. Parametric approaches use statistical models to infer population structure and assign individuals into subpopulations. However, these approaches suffer from many drawbacks that make them impractical for large datasets. In contrast, nonparametric approaches do not suffer from these drawbacks, making them more viable than parametric approaches for analyzing large datasets. Consequently, nonparametric approaches are increasingly used to reveal population substructure. Thus, this paper reviews and discusses the nonparametric approaches that are available for population structure analysis along with some implications to resolve challenges.

Keywords: Population structure analysis, Clustering, Dimension reduction, Principal component analysis, Allele-sharing distance, Genetic data, Single nucleotide polymorphism, Population genetics

Background

Population structure analysis is a major area of interest within the field of genetics and bioinformatics. Population structure is the grouping of individuals into subpopulations based on observable characteristics, such as culture, language, geographical region, and physical appearance [1]. Since patterns of genetic variation exist among people, genetic research is concerned with characterizing the genetic variations of populations and summarizing the relationships between individuals from genetic data. Thus, the analysis of population structure involves the identification of shared genetic variations among individuals and, accordingly, the grouping of similar individuals into subpopulations.

The inference of population structure from genetic markers is very helpful in different applications, such as genome-wide association studies (GWAS) [2–8] and forensics [9]. In GWAS, case-control studies aim to scan a large portion of the genome to identify the responsible genes for different diseases via associations between a genetic marker and a disease. The presence of population structure might result in spurious associations between a marker and a disease, which occur when most of the samples in the case group are from a specific population. Subsequently, a marker appears significantly more frequently in the case than in the control group, so this marker is incorrectly considered to be associated with the disease. Consequently, inferring population structure is a prerequisite for association mapping studies to avoid making spurious correlations or missing genuine correlations, which would eventually reduce false positive rates. In forensics, identifying population substructure is a prerequisite for developing reference panels. Reference panels are composed of a set of genetic markers that can provide information on an individual's ancestry [10].

Populations are genetically structured into distinct subpopulations [11]. Thus, the main research question is how to assign n individuals using m genetic markers to K subpopulations. Therefore, research in population

* Correspondence: lalhusain@ksu.edu.sa
College of Computer and Information Sciences, King Saud University, Riyadh, Saudi Arabia

structure addresses the following problems: how to detect population structure, how to assign individuals to their corresponding subpopulation, how to determine the optimal number of subpopulations, how to reduce the number of genetic markers needed for inference of population structure, how to infer population structure at a fine scale, and finally, how to handle large genetic datasets [11–16].

Several methods can be utilized to uncover population substructure. In general, these methods can be categorized into two broad approaches: parametric and nonparametric. Parametric approaches use statistical models to infer population structure and assign individuals into subpopulations. However, these approaches suffer from many drawbacks that make them impractical for large datasets. Such drawbacks include an intensive computational cost, genetic assumptions that must be held, and sensitivity to sample size. In contrast, nonparametric approaches have the advantage of efficient computational cost and no modeling assumption requirements, making them more viable than parametric approaches for analyzing large datasets.

Advances in DNA sequencing technology have provided genome-wide single nucleotide polymorphisms (SNPs) that have enabled the study of genetic variation at an unprecedented resolution. Detailed characterization of genetic variations across all chromosomes is possible using thousands of markers spanning the entire genome. Consequently, nonparametric approaches are increasingly being used to reveal population structure because of their great advantage of efficiency in handling high-dimensional genetic datasets. Therefore, this paper reviews the literature on the topic of population structure analysis with an emphasis on nonparametric approaches. The purpose of this paper is to review the nonparametric methods available to infer population structure from genetic data. The paper comprises seven sections, including this background section. It begins by outlining the background information required to understand the genetic data used for the analysis, along with the data preprocessing. Then, an overview of the parametric and nonparametric approaches of population structure analysis is presented. Since nonparametric approaches are more viable than parametric approaches for analyzing large datasets, this paper is concentrated on the nonparametric approaches proposed to address the inference of population structure from genetic data. These approaches are categorized into dimension reduction-based methods and distance-based methods. Afterward, the paper discusses the literature on the selection of informative markers. Finally, the paper concludes with a comprehensive discussion of the literature. Figure 1 provides a general workflow for population structure analysis, where the input is the genetic dataset and the output is the population substructure as a set of subpopulations (i.e., clusters).

Genetic data
Data description

The most widely used markers to study the variation of DNA sequences are SNPs [17]. SNPs take the form of substitutions at a single base pair. An SNP occurs when a single nucleotide from a DNA sequence differs at the same position between individuals. Since SNPs arise in certain populations only, they are very useful to differentiate and analyze different populations. In practice, genotyping is an inexpensive process used to examine DNA samples to determine which alleles appear in particular loci. Therefore, genotyping produces a genotypic profile of an individual as an unordered set of alleles that appears at each locus. In this profile, the nucleotides are encoded as two alleles, allele (A) and allele (B). Therefore, three distinct genotypes can appear at a locus: wild-type homozygous (AA), homozygous (BB), and heterozygous (AB). Nevertheless, an SNP marker can be encoded as 0, 1, or 2 according to the number of reference alleles. Thus, it has the advantage of being handled as a numerical variable that represents the number of reference alleles.

Many datasets are available online to study population structure. These datasets consist of genotyped markers along with information about individuals, where the population label is the most required information for population structure analysis. The most well-known datasets are HapMap [18–20], 1000 Genomes Project [21], and Pan-Asian [22].

Data preprocessing

The preprocessing of genetic data is a necessary step to examine the quality of data and determine which markers or individuals can reasonably be included in the analysis [23]. First, the quality of the SNP markers is assessed, including the following:

- SNP call rate: SNP call rate is assessed to verify the amount of missing data for each marker. SNP call rate is the proportion of genotypes per marker with non-missing data. Usually, a threshold of 95% is used to remove these poorly genotyped SNPs. However, the threshold should be set carefully to avoid removing important markers.
- Hardy-Weinberg equilibrium (HWE): HWE [12] verifies the assumptions of Hardy-Weinberg. So, a statistical test is applied to determine whether a marker follows the Hardy-Weinberg equilibrium or not. If a marker deviates from the equilibrium, then

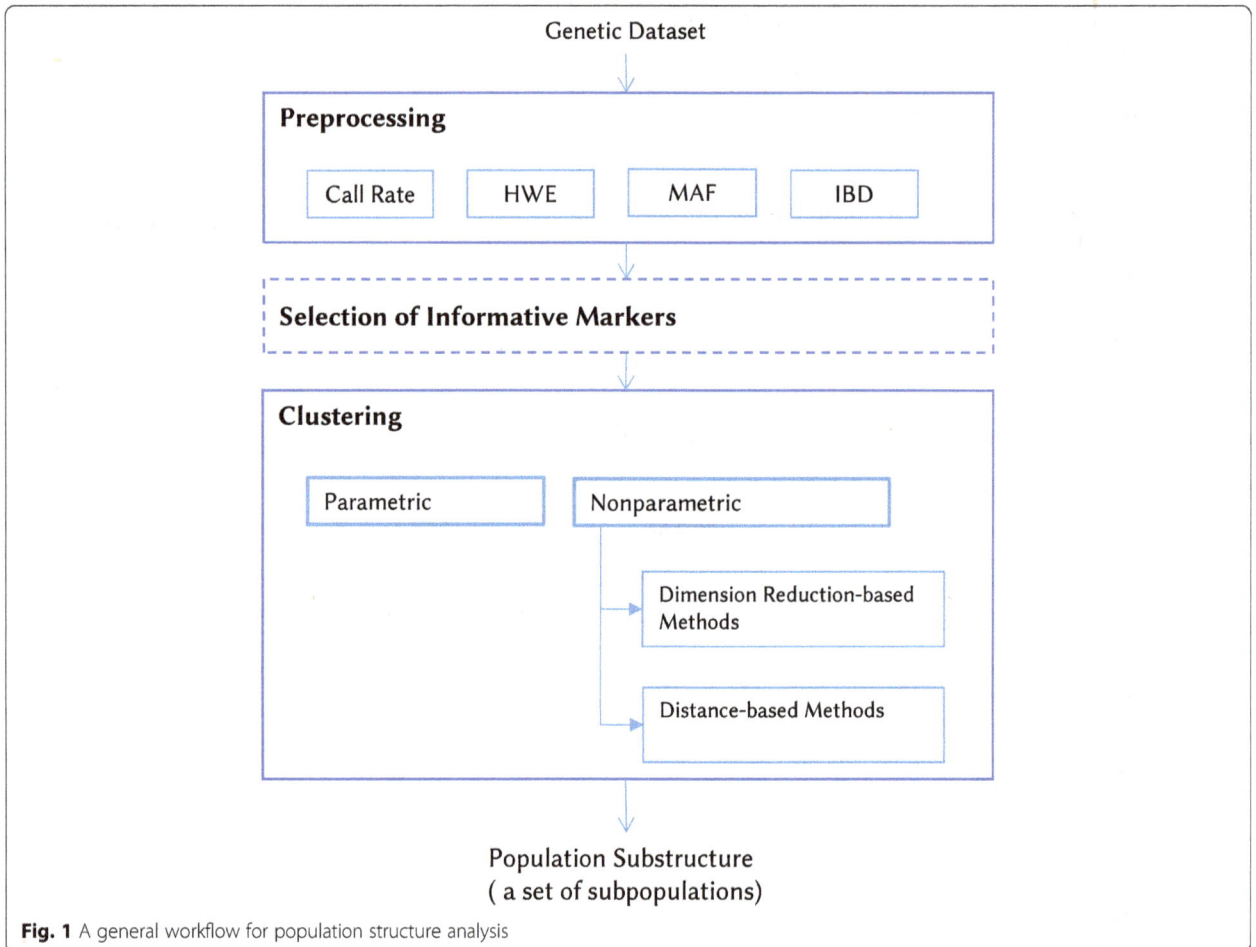

Fig. 1 A general workflow for population structure analysis

it may be because of genotyping errors; therefore, it should be excluded.

- Minor allele frequency (MAF): MAF denotes the frequency of a marker's less frequent allele in a given population. SNPs with low MAF should be excluded, and a threshold of 1–2% is typically applied.

For this assessment, PLINK [24] is typically used to prune SNPs with a minor allele frequency greater than 5%, a missing rate less than 5%, and a Hardy-Weinberg equilibrium (HWE) deviation p value of no less than 0.05.

Then, an assessment is performed to check the quality of the individuals, which includes the following:

- Individual call rate: Individual call rate refers to the proportion of genotypes per individual with non-missing data. The missingness rate should not exceed a certain threshold.
- Identity by descent (IBD): IBD [25] is calculated to assess which individuals are related. It indicates whether a pair of individuals has identical copies

of the same ancestral allele. The proportion of shared alleles between a pair of individuals determines the relation between them, such as identical twins, first-degree relatives (i.e., full siblings, parent–offspring), second-degree relatives (i.e., half-siblings, uncle/aunt, nephew/niece), and third-degree relatives (i.e., cousins). Related individuals are excluded. In practice, relatedness can be assessed using kinship coefficients estimated by KING [26]. The KING command can be used to filter out related individuals, where a threshold of a degree relationship can be specified.

Parametric approaches

Parametric approaches use statistical models to infer population structure and assign individuals into subpopulations. These models are used to estimate population parameters, such as allele frequency, for the population and to calculate the likelihood that an individual belongs to a specific subpopulation [12, 27]. Parametric approaches are based on several genetic assumptions about the data, including the Hardy-Weinberg equilibrium

(HWE) [12] for populations and the linkage equilibrium (LE) [28] between loci within each population.

Essentially, a parametric approach infers ancestral proportions for each individual and then groups individuals who have similar patterns of inferred ancestry [16]. The majority of parametric methods for population structure analysis apply Bayesian inference. Bayesian inference is applied to model the probability of observed genotypes given the individual ancestry proportions and population allele frequencies. These methods simultaneously assign individuals to populations and identify populations from genotype data based on the estimation of the allele frequencies for each population [13, 29, 30].

STRUCTURE is a widely used parametric method that relies on Bayesian MCMC [12, 29]. In particular, Markov chain Monte Carlo (MCMC) based on Gibbs sampling is implemented to estimate the posterior distribution of allele frequency given the probability of ancestral populations of individuals and allele frequencies for all populations. Similar to STRUCTURE, PARTITION [31], BAPS/BAPS2 [32, 33], and GENELAND [34] take the same modeling approach, which is based on an MCMC algorithm, to sample the posterior distribution. Moreover, FRAPPE [35] and ADMIXTURE [30, 36] adopt the same modeling approach but rely on maximizing the likelihood using an expectation-maximization (EM) algorithm instead of sampling the posterior distribution. In contrast, L-POP [27] implements a maximum likelihood approach based on latent class analysis (LSA), whereas PSMIX [37] uses the same approach via the implementation of a mixture model. Recently, fast STRUCTURE [38] was developed to improve the inference model underlying STRUCTURE using a variational Bayesian method. Variational methods optimize the computation of posterior distributions and accelerate the inference process.

Parametric approaches estimate the observed allele frequency for each population using statistical inference models that include some parameters and are based on probability distribution. Before running these methods, parameters must be set, such as the number of populations K, the most critical parameter. Accordingly, a parametric approach suffers from many drawbacks: First and most importantly, the intensive computational cost makes it impractical for large-scale datasets containing thousands of individuals and thousands of markers [39–41]. Second, parametric approaches are developed on the basis of the genetic assumptions of the Hardy-Weinberg equilibrium (HWE) and the linkage equilibrium (LE) between loci within each population. As a result, they can be very misleading when data assumptions cannot be verified or are invalidated [35, 40]. In specific, LE does not hold when a vast amount of genetic data are used [42]. Third, parametric methods depend on an estimation of allele frequency that is sensitive to sample size. Consequently, allele frequency is subject to high variations when using small samples representing each subpopulation [29, 42]. Lastly, parametric methods are not applicable to analyzing large and highly structured population datasets because of the limited number of K clusters that can be inferred [16].

Nonparametric approaches

Nonparametric approaches have been proposed to address the problem of analyzing population structure from genetic data in order to overcome the drawbacks of parametric approaches. Nonparametric approaches group individuals with similar genetic profiles together [16]. In 2006, Liu and Zhao [40] proposed a two-stage nonparametric strategy for analyzing population structure from genetic data with the goal of facilitating the clustering process of the high-dimensional space of genotype data. The first stage involves reducing the dimensionality of the genotypic dataset using multivariate analysis methods, such as singular value decomposition (SVD) and principal component analysis (PCA). The second stage involves applying clustering algorithms to identify population substructure from the reduced data. Another nonparametric strategy is to calculate the pairwise distances between individuals and then perform clustering. Both strategies have the advantage of identifying a population structure and assigning individuals to their corresponding subpopulation. Indeed, both strategies provide a framework for population structure analysis from genetic data where different methods can fit into that framework.

Nonparametric approaches have many advantages, including an efficient computational cost and no modeling assumption requirements. Nonparametric approaches have a more efficient computational cost compared to parametric approaches, making the former more viable for analyzing large datasets [15]. Also, nonparametric approaches do not make any assumption on genetic data, which is a great advantage over parametric approaches [43]. Therefore, when a large amount of genotype data is available, nonparametric approaches are preferred, as there is no need to verify the assumptions of Hardy-Weinberg and the linkage equilibrium [44]. Moreover, since these approaches are not dependent on estimating allele frequencies, they are unaffected when the number of individuals representing a subpopulation is small [42].

There are many nonparametric methods. Some methods use a dimension reduction technique to reduce the dimensions of genetic markers before conducting a clustering. Other methods consider computing dis/similarity matrices of the data where a clustering technique is applied. Thus, nonparametric methods can be categorized into dimension reduction-based methods and distance-based methods.

Dimension reduction-based methods

Dimension reduction-based methods are based on mapping high-dimensional genetic data to low-dimensional space and then applying clustering on the reduced dimensions. Principal component analysis (PCA) is the most cited dimension reduction method used to detect population structure based on genetic data [45, 46]. Typically, PCA's scatterplots are used to visualize population structure, where the most genetically isolated subpopulations appear as distinct clusters of individuals. Most importantly, PCA can be used to infer spatial population genetic variations [47].

EIGENSTRAT\smartpca [39, 41] is the most used PCA-based tool for detecting population structure. In EIGENSTRAT\smartpca, eigenanalysis is used to detect population substructure, such that eigenvalues and eigenvectors capture the amount and axes of variation among individuals, respectively. Thus, the principal components (PCs), or eigenvectors, serve as the new reduced dimensions. Similar to EIGENSTRAT\smartpca, PLINK [24] and SNPRelate [48] can be used to apply PCA on genetic datasets.

Principal components analysis

Given $x = (x_{i,l})_{\substack{1 \le i \le n \\ 1 \le l \le p}}$ is an $n \times p$ matrix, where n is the number of individuals and p is the number of SNPs. Each entry $x_{i,l}$ corresponds to the genotype of individual i for the marker l, coded as 0, 1, or 2 according to the number of reference alleles present at the locus l.

To perform a principal components analysis (PCA) on the matrix x, the data are first centered and normalized. The column means μ_l and the observed allele frequency of each marker p_l are computed as follows:

$$\mu_l = \frac{\sum_{i=1}^{n} x_{il}}{n}$$

$$p_l = \frac{1 + \sum_{i=1}^{n} x_{il}}{2 + 2n}$$

The new genotype matrix $\tilde{x}$ is defined, such that each entry is:

$$\tilde{x}_{il} = \frac{x_{il} - \mu_l}{\sqrt{p_l(1 - p_l)}}$$

Based on the $n \times n$ covariance matrix, a singular vector decomposition is computed as:

$$\frac{1}{p} \tilde{x} (\tilde{x})^T$$

Then, a set of principal components $(PC_1, PC_2, ..., PC_{n-1})$ are generated [41, 49].

A major issue with PCA applied to genetic data is how to determine the number of significant principal components, which is the number of principal components needed to sufficiently describe a structure of the population [13]. The EIGENSTRAT algorithm applies a variant of eigenanalysis to determine the significant principal components based on Tracy-Widom (TW) theory [50]. TW theory states that the distribution of the largest eigenvalue approximately follows the TW distribution when the dimension of a matrix is suitably large [51]. Hence, the TW distribution is used to determine the probability of population substructure.

Principal components can be used as the axes of variations to provide a graphical overview of the population structure. This graphical representation of the individuals can highlight outlier individuals, or those which seem to lie farther out than the others. Also, the set of significant principal components can be used to cluster individuals into genetically homogeneous subpopulations. For instance, the Gaussian mixture model or K-means algorithm can be applied to these principal components [52].

Clustering based on principal components

Different clustering algorithms can be applied to the principal components. Since the principal components are normally distributed, they fit well with the Gaussian mixture model (GMM) clustering. Therefore, the PCAclust algorithm [52] was proposed as three steps. The first step involves applying PCA to the genetic data to compute the principal components (PCs). Then, a set of significant PCs is selected using the TW statistic at a 5% level. Finally, the selected PCs are clustered using the GMM algorithm to group the individuals into populations.

Moreover, Lee et al. [52] have proposed using PCA for dimension reduction with three clustering algorithms: K-means [53], the mixture model [54], and spectral clustering [55]. They used Gap statistics [56] and the Bayesian information criterion (BIC) [57] to predict the optimal number of clusters. In their experiment, they showed that all three algorithms have comparable results. However, the different clustering algorithms showed different degrees of sensitivity to noisy and non-informative markers, which demonstrated the importance of selecting a proper set of informative markers.

Furthermore, iterative pruning PCA (ipPCA) was proposed to resolve the highly structured population that appears as a conglomerate in PCA space. ipPCA does this by iteratively applying PCA to decompose the structure of the population. The ipPCA method has two versions, TW-ipPCA [11] and EigenDev-ipPCA, [16], which differ in their termination tests. Recently, HiClust-ipPCA [58] was proposed as a variation of EigenDev-ipPCA wherein hierarchical clustering is used.

The PCA-based ipPCA method [11] has been proposed to address the overlapping problem that appears in PCA space when analyzing closely related subpopulations. The ipPCA method can detect population structure at a fine scale by iteratively bisecting individuals based on a termination test that checks whether a significant structure is present. In ipPCA, PCA is applied, and then a termination test is verified to decide whether to advance to clustering or to stop. Clustering is performed based on significant PCs. The number of significant PCs depends on the number of individuals in the dataset, such that later iterations require fewer PCs for clustering than earlier iterations. Therefore, the new bisected datasets will have fewer individuals. ipPCA iterates until all individuals have been assigned to homogeneous subpopulations. At the end, the number of subpopulations K is determined by counting all the terminal nodes or subpopulations. ipPCA uses a fuzzy C-mean algorithm to split the dataset into two parts. Indeed, the iterative pruning nature of ipPCA offers a logical way to present the degree of relatedness between subpopulations.

ipPCA has two different versions: TW-ipPCA [11] and EigenDev-ipPCA [16]. TW-ipPCA applies the TW test as a termination criterion [41]. TW, as previously mentioned, is implemented in the EIGENSTRAT/smartpca algorithm for detecting whether a significant structure is present in the dataset. TW-ipPCA suffers from type 1 error when the sample size is large, and subsequently, a group of individuals belonging to a single subpopulation would be assigned into separate subpopulations.

EigenDev-ipPCA was proposed to address the spurious cluster problem using a heuristic called EigenDev as a termination criterion [16]. EigenDev is inspired by the Eigenvalue Grads heuristic [59], which is applied in the signal processing domain. The EigenDev statistic is based on the eigenvalues of the data matrix; it has no hidden parameters and is more robust to type 1 error. The application of EigenDev to ipPCA improves the accuracy of individuals' assignments and the estimation of the number of subpopulations, especially when using huge and complex datasets. EigenDev-ipPCA reveals subpopulations that are subclusters of subpopulations generated by TW-ipPCA.

HiClust-ipPCA [58] is another variation of ipPCA that employs hierarchical clustering instead of fuzzy C-mean within the ipPCA framework. In addition, a PCA-based feature selection is applied as a data preprocessing step. In each iteration, PCA is applied to select the most informative markers. Then, PCA is applied to the selected markers to map them to a reduced space. Next, a hierarchical clustering with Ward's minimum variance is applied to cluster data into two groups. This process is iterated until satisfying a termination condition. The experiments illustrate that hierarchical clustering provides better clustering results than fuzzy C-mean and that the use of the feature selection technique is effective for reducing data dimensions and increasing computational efficiency.

Other dimension reduction methods

There are many alternatives to PCA, such as singular value decomposition (SVD) [60]. Liu and Zhao [40] used SVD for dimension reduction and density-based mean clustering (DBMC) for clustering. SVD is used because it is efficient for a large matrix of markers and individuals. DBMC was proposed as a variant of K-means that can determine the number of clusters automatically, because K-means requires the number of clusters to be given. The similarity between individuals is measured using Cosine similarity. The performance of DBMC was compared with K-means and the mixture model [40], and it was found that the mixture model and DBMC performed better than K-means. Another alternative of PCA is multi-dimensional scaling (MDS), which uses a similarity matrix between the individuals instead of the data matrix to create axes of variation [61].

Table 1 describes the nonparametric dimension reduction-based methods in terms of dimension reduction and/or proximity measure, clustering technique, and the package/tool if it is available.

Distance-based methods

Distance-based methods are based on computing the pairwise similarities/distances between individuals. The allele-sharing distance (ASD) [44, 62] is a measure proposed for determining the genetic proximity between each pair of individuals. Distance-based methods usually apply a clustering on the ASD matrix to infer population structure. For instance, allele-sharing distance and Ward's minimum variance hierarchical clustering (AWclust) [42, 44] applies an agglomerative hierarchical clustering to ASD, while Spectral Hierarchical clustering for the Inference of Population Structure (SHIPS) [43] uses divisive clustering. Furthermore, NETVIEW [63] reveals the hierarchy of population substructures based on a representation of the genetic data as a network of individuals connected by edges representing the ASD between each pair. Iterative neighbor-joining tree clustering (iNJclust) [64] performs a graph-based clustering on a neighbor-joining (NJ) tree. Table 2 describes the distance-based methods in terms of the proximity measure, clustering technique, and available package/tool.

Allele-sharing distance

For clustering genetic data, allele-sharing distance (ASD) is used to identify closely related and distantly related

Table 1 Dimension reduction-based methods of population structure analysis

Reference	Dimension reduction	Distance matrix	Clustering	Tool/package
Patterson at el. (2006) [41]	PCA (TW)	–	–	EIGENSTRAT/smartpca [82]: Perl
Liu at el. (2006) [40]	SVD	Cosine similarity	Density-based mean clustering (DBMC)	–
Lee at el.(2009) [52]	PCA (TW)	–	Spectral clustering (K-means, mixture model)	–
Intarapanich at el. (2009) [11]	PCA (TW)	Euclidean distance	Fuzzy C-means	TW-ipPCA [83]: MATLAB
Limpiti at el.(2011) [16]	PCA (EigenDev)	Euclidean distance	Fuzzy C-means	EigenDev-ipPCA [83]: MATLAB
Amornbunchornvej at el. (2012) [58]	PCA	ASD	Ward's clustering	–

pairs of individuals. ASD is similar to identity by state (IBS) metric [25].

Given $x = (x_{i,l})_{\substack{1 \le i \le n \\ 1 \le l \le p}}$ is a $n \times p$ matrix where n is the number of individuals and p the number of SNPs. Each entry $x_{i,\,l}$ corresponds to the genotype of individual i for the marker l. Then, the ASD between individuals i and j at locus l, denoted as $D_l(i,j)$, is defined as follows:

$$D_l(i,j) = \begin{cases} 0 & \text{if same genotype} \\ 1 & \text{if one common allele} \\ 2 & \text{if no common allele} \end{cases}$$

Therefore, the total distance between individuals i and j can be calculated as:

$$D(i,j) = \frac{1}{p}\sum_{l=1}^{p}(D_l(i,j)) \qquad \text{for each } i \text{ and } j \in [1,n]$$

or as

$$D(i,j) = \frac{1}{p}\sum_{l=1}^{p}(|x_{i,l}-x_{j,l}|) \qquad \text{for each } i \text{ and } j \in [1,n]$$

where $x_{i,\,l}$, $x_{j,\,l}$ are the individuals' genotypes, coded as 0, 1, or 2 according to the number of reference alleles present at the locus l. The closer the pair of individuals are, genetically, the smaller the value of $D(i,j)$.

Using the function $D(i,j)$ to quantify the distance between each pair of individuals i and j, a distance matrix can be formed by combining the information for all pairs of individuals. The distance matrix, $= (D_{i,j})_{\substack{1 \le i \le n \\ 1 \le j \le n}}$,

is a squared matrix of $n \times n$, where n is the number of individuals.

Based on ASD, a similarity measure can be inferred to measure the similarity between individuals i and j at locus l, denoted as $S_l(i,j)$, where:

$$S(i,j) = \frac{1}{p}\sum_{l=1}^{p}\left(2-|x_{il}-x_{jl}|\right) \qquad \text{for each } i \text{ and } j \in [1,n]$$

Clustering based on ASD

Distance-based clustering methods use the ASD matrix as an input to group individuals into populations. AWclust, SHIPS, NETVIEW, and iNJclust all distance-based clustering methods, are summarized in Table 2.

AWclust [42, 44] is a distance-based population structure exploration method. The first step of AWclust is to construct the ASD matrix between all pairs of individuals in the sample. The second step is to apply hierarchical clustering to infer clusters of individuals from the ASD matrix using Ward's minimum variance algorithm [65, 66]. AWclust uses gap statistics [56] to select the optimal number of subpopulations K. The employment of gap statistics is computationally intensive as it involves an iterative statistical inference process [67]. To deal with the slow speed of calculating gap statistics, AWclust limits the number of inferred K to be 16 at maximum [67]. The execution of AWclust slows down dramatically when using a larger number of SNPs due to the increase in the size of the ASD matrix [67]. Deejai et al. [67] found that AWclust performs well only with a small number of SNP markers and in individuals with low diversity (i.e., the number of inferred subpopulations K is small), and thus, it is not suitable for

Table 2 Distance-based methods of population structure analysis

Reference	Clustering	Tool/package
Gao at el.(2007) [44]	Ward's minimum variance algorithm	AWclust [84]: R package
Bouaziz et al. (2012) [43]	Spectral clustering (GMM)	SHIPS [85]: R package
Neuditschko at el.(2012) [63]	Super paramagnetic clustering (SPC)	NETVIEW [86]: MATLAB
Limpiti at el. (2014) [64]	Neighbor-joining (NJ) tree-based clustering	iNJclust [87]: C++

performing large-scale population genetic analysis. The application of AWclust on HapMap project phase 1 [18] provided good results. It successfully differentiated the four ethnic populations in the dataset: African, European, Han Chinese, and Japanese individuals [44].

SHIPS [43, 68], or Spectral Hierarchical clustering for the Inference of Population Structure, is a distance-based method for inferring the structure of populations from genetic data. SHIPS applies a divisive strategy of hierarchical clustering followed by a pruning procedure to investigate population structure progressively. SHIPS constructs a binary tree to represent the substructure of a population using spectral clustering. Spectral clustering is applied to a pairwise distance matrix to divide a population into two subpopulations, and this is iterated for each of the two subpopulations. ASD is used within SHIPS; however, SHIPS can be used with any similarity matrix. SHIPS applies a pruning procedure along with gap statistics to determine the optimal number of sub-populations. A pruning procedure provides all possible clustering results. Thus, it allows a fast calculation of the gap statistics that requires all the clustering results of specified numbers of clusters. Moreover, because calculating gap statistics is time consuming, SHIPS applies a version of gap statistics that is less precise but has better experimental performance in estimating the optimal K. Experiments have involved applying SHIPS on two datasets: HapMap project phase 3 [19] and Pan-Asian [22]. These experiments have shown that SHIPS can accurately assign individuals to clusters with relatively low computational cost and estimate the number of clusters as well [43, 68]. In addition, SHIPS is quite robust such that several applications of SHIPS algorithm on the same dataset produce the same clustering result.

NETVIEW [63] is an analysis pipeline that combines a network-based clustering method with a visualization tool to infer fine-scale population structure. NETVIEW is composed of three key steps: distance matrix calculation, network construction and clustering, and network-based visualization. NETVIEW first calculates the ASD matrix that represents the relationships between all individuals in the dataset. Then, the ASD matrix is used to construct a population network using super paramagnetic clustering (SPC) [69]. In this network, nodes represent individuals, edges represent the relationship between a pair of individuals, and the thickness of edges represents the genetic distance. SPC is based on computing the K-nearest neighborhood to produce a cluster relationship matrix and a hierarchical tree of clusters. Specifically, SPC is implemented as Sorting Points Into Neighborhood (SPIN) [69, 70], which employs the Potts Hamiltonian model [71] to identify the number and size of clusters, known as cluster stability. The problem with SPC is how to specify the number of the nearest

neighborhood an individual can have. Based on this number, NETVIEW produces clusters at optimal thresholds of genetic distance. The result of this algorithm provides a hierarchical clustering of individuals. However, NETVIEW uses a network-based visualization to present the population structure at a very fine scale, where highly interconnected individuals identify subpopulations. The empirical study in [63] involved applying NETVIEW on Human and Bovine HapMap datasets. The study demonstrated that NETVIEW could assign individuals to their corresponding subpopulations effectively and showed the genetic relatedness of individuals within their populations at a very fine scale.

iNJclust [64], or iterative Neighbor-Joining tree clustering, is an iterative application of graph-based clustering on a neighbor-joining (NJ) tree. The algorithm starts by computing the ASD matrix from the data. Then, an NJ tree is constructed based on the ASD matrix. Next, the algorithm performs a graph-based clustering to bisect the NJ tree into two subtrees. For each subtree, a new NJ tree is constructed based on the ASD matrix that contains only individuals within that subtree. The process of bisecting the NJ trees to create new subtrees is iterated until all subtrees become homogenous. The algorithm determines whether the cluster is homogeneous based on the fixation index. The fixation index (F_{ST}) is a measure of genetic population substructure used to examine the overall genetic divergence among subpopulations [72]. The construction of the NJ tree starts with all individuals as the leaf nodes. Then, the pair of nodes that are nearest to each other are merged. The merging process is repeated until all nodes are merged into the tree. The distance between nodes is measured using the minimum evolution criteria [73] based on the ASD. For NJ tree clustering, the NJ tree is split into two subtrees by cutting the edge between the two nodes with the longest length. iNJclust assigns the individuals into populations and estimates the optimal number of populations. The clustering result of iNJclust is a binary tree, where each leaf node represents a population of a set of individuals, and the tree structure represents the relationships between populations. The experimental results of applying iNJclust on real and simulated data have indicated that iNJclust yields a reasonable estimation of the number of populations, a robust assignment of individuals, and a meaningful representation of relationships among populations with the binary tree [64].

Selection of informative markers

Given that a large number of genetic markers can be used to infer population structure, reducing the number of markers is often desirable for efficient structure identification. In such settings, selecting ancestry informative markers (AIMs) aims to identify the minimum set of markers required to derive population structure and to reduce the genotyping cost. Selecting informative markers can be accomplished by using supervised or unsupervised methods. Supervised methods rely on prior knowledge of the ancestry of the individuals.

Informativeness for assignment (I_n) [74] is a supervised measure that computes mutual information based on allele frequencies and relies on self-reported ancestry information from individuals. In contrast, PCAIM [15] is an unsupervised algorithm proposed to identify a set of informative markers that captures the structure of a population. It does not demand prior information about the ancestry/origin of individuals. The PCAIM algorithm applies PCA to determine markers that are correlated with the significant principal components and then assigns a score to each marker. Then, the algorithm returns the top scoring markers that correlate well with the top few eigenvectors. The algorithm is efficient in selecting the informative markers. It is computationally fast and suitable for large datasets.

The performance of I_n and PCAIM in selecting informative markers has been evaluated in [15] and was found to attain comparable results; in addition, a considerable overlap was found between the selected markers. The overlapping was expected since PCAIM ranks markers based on how well they can reproduce the structure of the dataset, whereas I_n determines which markers are most likely to be associated with major clusters in the dataset. Therefore, PCAIM selects either the same markers or markers that are in high linkage disequilibrium (LD) with markers selected using the I_n measure.

The selection of informative markers could potentially suffer from redundant markers. Typically, redundancy exists due to the correlation among markers that are in the LD region. To select a minimal set of informative markers, a redundancy removal step should be applied after the initial markers selection step to avoid redundancy and determine the final set of AIMs.

In the literature, two different methods have been proposed to filter out redundant markers. The first method deals with the problem as a Column Subset Selection Problem, which is a well-known problem in linear algebra [75]. In [75], the algorithm Greedy QR [76, 77] is employed to select the minimally correlated subset of markers. The algorithm essentially works as an iterative process to pick up the uncorrelated markers. This algorithm has an implementation in MATLAB, and it can run efficiently in a shorter amount of time using thousands of markers. On the other hand, the redundancy removal problem can be resolved via the clustering technique. In particular, a clustering-based strategy was employed in [14] to minimize the number of markers to the most informative and uncorrelated ones,

which was inspired by [78] in data analysis. In simpler terms, the strategy applies a clustering technique to cluster markers into K clusters and then returns one representative marker for each cluster. In [14], the Cluto toolkit [79] was used with default parameters for clustering using a cosine similarity matrix. The advantage of applying clustering to identify redundant markers is that it returns K lists of markers. Within each list, the markers are interchangeable, thus providing some flexibility in choosing any informative marker that falls into the same cluster. In contrast, the first method just returns one set of non-redundant markers. Although the two approaches of redundancy removal had comparable performance, clustering was slightly more accurate but was five times slower than the first method [14].

Discussion

Nonparametric approaches are increasingly being used to reveal population structure because of their great advantages of efficiency in handling high-dimensional genetic datasets [74]. Due to the high dimensionality of genetic data, it is imperative to reduce the dimensions of the data before clustering. In the literature of population structure analysis, PCA is employed as a dimension reduction technique for two purposes. The first purpose is feature extraction, where PCA is applied to transform the data to low-dimensional space where clustering will be performed. The second purpose is feature selection, where PCA is applied to select the informative genetic markers. To accomplish this, PCA is applied to a covariance matrix of genetic markers, and then the genetic markers that are well correlated with significant principal components are selected.

PCA is considered computationally efficient and performs well in detecting the genetic structure of populations. However, it is also argued that PCA not be efficient when used with correlated markers that naturally arise in any genetic data, especially in densely genotyped data. The problem is that a large number of redundant and correlated markers may mask the real structure of data. In practice, with large genotype data, there are linked markers due to linkage disequilibrium (LD) [28], which is considered dependent and redundant, and this may seriously distort the results of PCA. Moreover, dimension reduction methods, like PCA, consider the complete markers of the dataset to produce only one subspace, in which the clustering can then be performed. However, an issue would arise when the correlation between markers or the relevance of markers are significant for some clusters (i.e., populations) but not for complete datasets. Consequently, this issue can be resolved by subspace clustering. Subspace clustering computes multiple subspaces, where a different set of features is selected for each subspace. Then, individuals are clustered differently in each subspace according to the relevance of markers to describe those individuals. Subspace clustering may be a significant solution, inferring the population structure at a very fine scale.

Many distance-based methods have been developed to resolve the problem of clustering individuals into subpopulations. These methods have utilized different clustering techniques that required a matrix of pairwise distance/similarity between individuals. Allele-sharing distance (ASD) is widely used for this purpose. In [80], it is shown that the ASD between individuals from different subpopulations is always larger than that of individuals from the same subpopulations. Moreover, calculating the ASD for many SNP markers allows differentiation of the populations through the accumulated effect of SNP loci. However, distance assessment using ASD between individuals becomes increasingly meaningless as dimensionality increases. As with increasing the number of SNPs, the distances of the individual to its similar individuals and dissimilar individuals tend to be almost the same. Individuals appear almost alike because of correlated SNPs, which are considered "redundant," while ASD treats each marker independently. Therefore, the identification of correlated markers might improve the inference of population structure from high-dimensional genetic data. Filtering those markers before calculating ASD could contribute to more accurate clustering results, as achieved within HiClust-ipPCA [58].

The clustering techniques used to identify the population genetic substructure can be categorized into partitional clustering and hierarchical clustering. Partitional clustering produces a flat clustering which divides the data into a pre-specified number of clusters K (e.g., K-means [81], DBMS [40], Lee's [52]). In contrast, hierarchical clustering produces a hierarchy of clusters (e.g., AWclust [44], SHIPS [43], NETVIEW [63], ipPCA [11, 16, 58], iNJclust [64]). Hierarchical clustering is preferable over partitional clustering in the context of population structure analysis. This is because it produces multiple nested partitions instead of one partition, which allows the choice of different partitions according to the desired level of similarity. Most importantly, a fine-scale population substructure can be obtained using hierarchical clustering because of the clustering's ability to capture data at different levels of granularity.

A major challenge in population structure analysis is the estimation of the optimal number of subpopulations (i.e., clusters). Gap statistics [56] have often been applied to determine the optimal number of clusters. However, gap statistics is computationally intensive and impractical for highly structured genetic datasets that comprise a large number of clusters. Some clustering methods can implicitly determine the optimal number of clusters—for instance, ipPCA [11, 16, 58], where the number of clusters is represented by the number of leaf nodes of the

binary tree constructed by iterative applications of PCA. However, determining the number of populations as a single number is not practical and may have no biological meaning when there are hierarchical levels of population structure (i.e., subpopulations within populations). Furthermore, the researcher must be able to control the level of granularity to uncover the substructure of the population. Overall, these provide insights into the importance of presenting the clustering result as a hierarchy whereby the researcher can visually determine the optimal level of separation from the number of major clusters in the dendrogram. The dendrogram serves as a visual means for both understanding the structure of the data and selecting a reasonable number of clusters.

Conclusion

The analysis of population structure is used to obtain a clear insight into the underlying genetic population substructure and is a crucial prerequisite for any analysis of genetic data, such as genome-wide association studies, to eventually reduce false positive rates, and for forensics to develop reference panels that provide information on an individual's ancestry. Single nucleotide polymorphisms (SNPs) are the most widely used markers to study the variation of DNA sequences between populations. Data preprocessing is a necessary step to assess the quality of the data before analysis, including the assessment of the call rates of both SNPs and individuals, minor allele frequency, and relatedness between individuals, where a threshold is set to eliminate SNPs/individuals that do not meet that threshold. Additionally, the selection of ancestry informative markers (AIMs), which are the minimal set of markers required to derive population structure, is considered important in preprocessing to improve the accuracy of clustering results.

After preprocessing, several analysis methods, including parametric and nonparametric, are used. Parametric approaches are impractical for large datasets because of their intensive computational cost, genetic assumptions that must be held, and sensitivity to sample size. In contrast, nonparametric approaches have the advantage of efficient computational cost with no modeling assumption requirements, making them more viable than parametric approaches for analyzing large datasets. Nonparametric approaches can be categorized into dimension reduction-based and distance-based methods. On the one hand, dimension reduction techniques are used to reduce the dimensions of genetic markers before conducting a clustering. The most used dimension reduction technique is principal components analysis (PCA), as it is implemented in EIGENSTRAT\smartpca. On the other hand, distance-based methods include computing dis/similarity matrices of the data where the clustering method is applied, such as AWclust, SHIPS, NETVIEW, and iNJclust. In these methods, similarity is measured using allele-sharing distance (ASD). ASD is a measure to determine how genetically close each pair of individuals is.

All in all, as evident in the challenges introduced by the ever-growing sizes and complexity of genetic datasets, accurate and efficient analysis methods are increasingly desirable to take full advantage of these available genetic datasets.

Abbreviations
AIMs: Ancestry informative markers; ASD: Allele-sharing distance; AWclust: Allele-sharing distance and Ward's minimum variance hierarchical clustering; DBMC: Density-based mean clustering; GWAS: Genome-wide association studies; HWE: Hardy-Weinberg equilibrium; iNJclust: Iterative neighbor-joining tree clustering; ipPCA: Iterative pruning PCA; LD: Linkage disequilibrium; MDS: Multi-dimensional scaling; NJ: Neighbor-joining; PCA: Principal component analysis; SHIPS: Spectral Hierarchical clustering for the Inference of Population Structure; SVD: Singular value decomposition; TW: Tracy-Widom

Acknowledgements
This research project was supported by a grant from the "Research Center of the Female Scientific and Medical Colleges," Deanship of Scientific Research, King Saud University.

Authors' contributions
LA performed the literature review, analyzed the findings, and wrote the manuscript. AH assisted with the analysis and reviewed the manuscript. Both authors read and approved the final manuscript.

Competing interests
The authors declare that they have no competing interests.

References
1. Lawson DJ, Falush D. Population identification using genetic data. Annu Rev Genomics Hum Genet. 2012;13:337–61.
2. Pritchard JK, Donnelly P. Case-control studies of association in structured or admixed populations. Theor Popul Biol. 2001;60:227–37.
3. Hoggart CJ, Parra EJ, Shriver MD, Bonilla C, Kittles RA, Clayton DG, McKeigue PM. Control of confounding of genetic associations in stratified populations. Am J Hum Genet. 2003;72:1492–504.
4. Marchini J, Cardon LR, Phillips MS, Donnelly P. The effects of human population structure on large genetic association studies. Nat Genet. 2004; 36:512–7.
5. Helgason A, Yngvadóttir B, Hrafnkelsson B, Gulcher J, Stefánsson K. An Icelandic example of the impact of population structure on association studies. Nat Genet. 2005;37:90–5.
6. Ziv E, Burchard EG. Human population structure and genetic association studies. Pharmacogenomics. 2003;4:431–41.
7. Freedman ML, Reich D, Penney KL, McDonald GJ, Mignault AA, Patterson N, Gabriel SB, Topol EJ, Smoller JW, Pato CN. Assessing the impact of population stratification on genetic association studies. Nat Genet. 2004;36:388–93.
8. Price AL, Zaitlen NA, Reich D, Patterson N. New approaches to population stratification in genome-wide association studies. Nat Rev Genet. 2010;11:459.

9. Kidd KK, Pakstis AJ, Speed WC, Grigorenko EL, Kajuna SL, Karoma NJ, Kungulilo S, Kim J-J, Lu R-B, Odunsi A. Developing a SNP panel for forensic identification of individuals. Forensic Sci Int. 2006;164:20–32.
10. Kidd KK, Speed WC, Pakstis AJ, Furtado MR, Fang R, Madbouly A, Maiers M, Middha M, Friedlaender FR, Kidd JR. Progress toward an efficient panel of SNPs for ancestry inference. Forensic Sci Int Genet. 2014;10:23–32.
11. Intarapanich A, Shaw PJ, Assawamakin A, Wangkumhang P, Ngamphiw C, Chaichoompu K, Piriyapongsa J, Tongsima S. Iterative pruning PCA improves resolution of highly structured populations. BMC bioinformatics. 2009;10:382.
12. Pritchard JK, Stephens M, Donnelly P. Inference of population structure using multilocus genotype data. Genetics. 2000;155:945–59.
13. Liu Y, Nyunoya T, Leng S, Belinsky SA, Tesfaigzi Y, Bruse S. Softwares and methods for estimating genetic ancestry in human populations. Hum Genomics. 2013;7(1):1.
14. Paschou P, Lewis J, Javed A, Drineas P. Ancestry informative markers for fine-scale individual assignment to worldwide populations. J Med Genet. 2010;47:835–47.
15. Paschou P, Ziv E, Burchard EG, Choudhry S, Rodriguez-Cintron W, Mahoney MW, Drineas P. PCA-correlated SNPs for structure identification in worldwide human populations. PLoS Genet. 2007;3:e160.
16. Limpiti T, Intarapanich A, Assawamakin A, Shaw PJ, Wangkumhang P, Piriyapongsa J, Ngamphiw C, Tongsima S. Study of large and highly stratified population datasets by combining iterative pruning principal component analysis and structure. BMC bioinformatics. 2011;12:255.
17. Brookes AJ. The essence of SNPs. Gene. 1999;234:177–86.
18. The International HapMap C. A haplotype map of the human genome. Nature. 2005;437:1299–320.
19. Pemberton TJ, Wang C, Li JZ, Rosenberg NA. Inference of unexpected genetic relatedness among individuals in HapMap phase III. Am J Hum Genet. 2010;87:457–64.
20. Consortium IH. A second generation human haplotype map of over 3.1 million SNPs. Nature. 2007;449:851.
21. Consortium GP. An integrated map of genetic variation from 1,092 human genomes. Nature. 2012;491:56.
22. Ngamphiw C, Assawamakin A, Xu S, Shaw PJ, Yang JO, Ghang H, Bhak J, Liu E, Tongsima S, Consortium HP-AS. PanSNPdb: the Pan-Asian SNP genotyping database. PLoS One. 2011;6:e21451.
23. Laurie CC, Doheny KF, Mirel DB, Pugh EW, Bierut LJ, Bhangale T, Boehm F, Caporaso NE, Cornelis MC, Edenberg HJ. Quality control and quality assurance in genotypic data for genome-wide association studies. Genet Epidemiol. 2010;34:591–602.
24. Purcell S, Neale B, Todd-Brown K, Thomas L, Ferreira Manuel AR, Bender D, Maller J, Sklar P, de Bakker Paul IW, Daly Mark J, Sham Pak C. PLINK: a tool set for whole-genome association and population-based linkage analyses. Am J Hum Genet. 2007;81:559–75.
25. Stevens EL, Heckenberg G, Roberson ED, Baugher JD, Downey TJ, Pevsner J. Inference of relationships in population data using identity-by-descent and identity-by-state. PLoS Genet. 2011;7:e1002287.
26. Manichaikul A, Mychaleckyj JC, Rich SS, Daly K, Sale M, Chen W-M. Robust relationship inference in genome-wide association studies. Bioinformatics. 2010;26:2867–13.
27. Purcell S, Sham P. Properties of structured association approaches to detecting population stratification. Hum Hered. 2005;58:93–107.
28. Reich DE, Cargill M, Bolk S, Ireland J, Sabeti PC, Richter DJ, Lavery T, Kouyoumjian R, Farhadian SF, Ward R, Lander ES. Linkage disequilibrium in the human genome. Nature. 2001;411:199–204.
29. Porras-Hurtado L, Ruiz Y, Santos C, Phillips C, Carracedo Á, Lareu MV. An overview of STRUCTURE: applications, parameter settings, and supporting software. Front Genet. 2013;4:98.
30. Alexander DH, Lange K. Enhancements to the ADMIXTURE algorithm for individual ancestry estimation. BMC bioinformatics. 2011;12:246.
31. Dawson KJ, Belkhir K. A Bayesian approach to the identification of panmictic populations and the assignment of individuals. Genet Res. 2001;78:59–77.
32. Corander J, Waldmann P, Sillanpää MJ. Bayesian analysis of genetic differentiation between populations. Genetics. 2003;163:367–74.
33. Corander J, Waldmann P, Marttinen P, Sillanpää MJ. BAPS 2: enhanced possibilities for the analysis of genetic population structure. Bioinformatics. 2004;20:2363–9.
34. Guillot G, Mortier F, Estoup A. GENELAND: a computer package for landscape genetics. Mol Ecol Notes. 2005;5:712–5.

35. Tang H, Peng J, Wang P, Risch NJ. Estimation of individual admixture: analytical and study design considerations. Genet Epidemiol. 2005;28:289–301.

36. Alexander DH, Novembre J, Lange K. Fast model-based estimation of ancestry in unrelated individuals. Genome Res. 2009;19:1655–64.

37. Wu B, Liu N, Zhao H. PSMIX: an R package for population structure inference via maximum likelihood method. BMC bioinformatics. 2006;7:317.

38. Raj A, Stephens M, Pritchard JK. fastSTRUCTURE: variational inference of population structure in large SNP datasets. Genetics. 2014;197(2):573–89.

39. Price AL, Patterson NJ, Plenge RM, Weinblatt ME, Shadick NA, Reich D. Principal components analysis corrects for stratification in genome-wide association studies. Nat Genet. 2006;38:904–9.

40. Liu N, Zhao H. A non-parametric approach to population structure inference using multilocus genotypes. Human genomics. 2006;2:353.

41. Patterson N, Price AL, Reich D. Population structure and eigenanalysis. PLoS Genet. 2006;2:e190.

42. Gao X, Starmer JD. AWclust: point-and-click software for non-parametric population structure analysis. BMC bioinformatics. 2008;9:77.

43. Bouaziz M, Paccard C, Guedj M, Ambroise C. SHIPS: spectral hierarchical clustering for the inference of population structure in genetic studies. PLoS One. 2012;7:e45685.

44. Gao X, Starmer J. Human population structure detection via multilocus genotype clustering. BMC Genet. 2007;8:34.

45. Bryc K, Auton A, Nelson MR, Oksenberg JR, Hauser SL, Williams S, Froment A, Bodo J-M, Wambebe C, Tishkoff SA. Genome-wide patterns of population structure and admixture in West Africans and African Americans. Proc Natl Acad Sci. 2010;107:786–91.

46. Bryc K, Velez C, Karafet T, Moreno-Estrada A, Reynolds A, Auton A, Hammer M, Bustamante CD, Ostrer H. Genome-wide patterns of population structure and admixture among Hispanic/Latino populations. Proc Natl Acad Sci. 2010;107:8954–61.

47. Novembre J, Stephens M. Interpreting principal component analyses of spatial population genetic variation. Nat Genet. 2008;40:646–9.

48. Zheng X, Levine D, Shen J, Gogarten SM, Laurie C, Weir BS. A high-performance computing toolset for relatedness and principal component analysis of SNP data. Bioinformatics. 2012;28:3326–8.

49. McVean G. A genealogical interpretation of principal components analysis. PLoS Genet. 2009;5:e1000686.

50. Tracy CA, Widom H. Level-spacing distributions and the airy kernel. Commun Math Phys. 1994;159:151–74.

51. Johnstone IM. On the distribution of the largest eigenvalue in principal components analysis. Ann Stat. 2001;29(2):295–327.

52. Lee C, Abdool A, Huang C-H: PCA-based population structure inference with generic clustering algorithms. BMC bioinformatics 2009, 10:S73.

53. Hartigan JA, Wong MA. Algorithm AS 136: a k-means clustering algorithm. Appl Stat. 1979:100–8.

54. Fraley C, Raftery AE. Enhanced model-based clustering, density estimation, and discriminant analysis software: MCLUST. J Classif. 2003;20:263–86.

55. Ng AY, Jordan MI, Weiss Y. On spectral clustering: analysis and an algorithm. In: Proceedings of advances in neural information processing systems. Cambridge: MIT Press; 2001. p. 849–56.

56. Tibshirani R, Walther G, Hastie T. Estimating the number of clusters in a data set via the gap statistic. J Royal Stat Soc Series B (Statistical Methodology). 2001;63:411–23.

57. Schwarz G. Estimating the dimension of a model. Ann Stat. 1978;6:461–4.

58. Amornbunchornvej C, Limpiti T, Assawamakin A, Intarapanich A, Tongsima S: Improved iterative pruning principal component analysis with graph-theoretic hierarchical clustering. In 9th international conference on electrical engineering/electronics, computer, telecommunications and information technology; 16–18 2012. 2012: 1–4.

59. Luo J, Zhang Z: Using eigenvalue grads method to estimate the number of signal source. In 2000 5th International Conference on Signal Processing Proceedings; Beijing. IEEE; 2000: 223–225.

60. Wall ME, Rechtsteiner A, Rocha LM. Singular value decomposition and principal component analysis, A practical approach to microarray data analysis; 2003. p. 91.

61. Li M, Reilly C, Hanson T. A semiparametric test to detect associations between quantitative traits and candidate genes in structured populations. Bioinformatics. 2008;24:2356–62.

62. Mountain JL, Cavalli-Sforza LL. Inference of human evolution through cladistic analysis of nuclear DNA restriction polymorphisms. Proc Natl Acad Sci. 1994;91(14):6515–19.

63. Neuditschko M, Khatkar MS, Raadsma HW. NetView: a high-definition network-visualization approach to detect fine-scale population structures from genome-wide patterns of variation. PLoS One. 2012;7:e48375.

64. Limpiti T, Amornbunchornvej C, Intarapanich A, Assawamakin A, Tongsima S. iNJclust: iterative neighbor-joining tree clustering framework for inferring population structure. IEEE/ACM Trans Comput Biol Bioinformatics. 2014;11:903–14.

65. Ward Jr JH. Hierarchical grouping to optimize an objective function. J Am Stat Assoc. 1963;58:236–44.

66. Ward Jr JH, Hook ME. Application of an hierarchial grouping procedure to a problem of grouping profiles. Educ Psychol Meas. 1963;23(1):69–81.

67. Deejai P, Assawamakin A, Wangkumhang P, Poomputsa K, Tongsima S: On assigning individuals from cryptic population structures to optimal predicted subpopulations: an empirical evaluation of non-parametric population structure analysis techniques. In Computational Systems-Biology and Bioinformatics. Berlin: Springer; 2010. p. 58–70.

68. Bouaziz M: SHIPS: spectral hierarchical clustering for the inference of population structure. In Annals of Human Genetics; NJ,USA. WILEY-BLACKWELL; 2012: 413–413.

69. Blatt M, Wiseman S, Domany E. Superparamagnetic clustering of data. Phys Rev Lett. 1996;76:3251.

70. Tsafrir D, Tsafrir I, Ein-Dor L, Zuk O, Notterman DA, Domany E. Sorting points into neighborhoods (SPIN): data analysis and visualization by ordering distance matrices. Bioinformatics. 2005;21:2301–8.

71. Tetko IV, Facius A, Ruepp A, Mewes H-W. Super paramagnetic clustering of protein sequences. BMC Bioinformatics. 2005;6:82.

72. Holsinger KE, Weir BS. Genetics in geographically structured populations: defining, estimating and interpreting F ST. Nat Rev Genet. 2009;10:639.

73. Gascuel O, Steel M. Neighbor-joining revealed. Mol Biol Evol. 2006;23:1997–2000.

74. Rosenberg NA, Li LM, Ward R, Pritchard JK. Informativeness of genetic markers for inference of ancestry. Am J Hum Genet. 2003;73:1402–22.

75. Paschou P, Drineas P, Lewis J, Nievergelt CM, Nickerson DA, Smith JD, Ridker PM, Chasman DI, Krauss RM, Ziv E. Tracing sub-structure in the European American population with PCA-informative markers. PLoS Genet. 2008;4(7): e1000114.

76. Golub G. Numerical methods for solving linear least squares problems. Numer Math. 1965;7:206–16.

77. Gu M, Eisenstat SC. Efficient algorithms for computing a strong rank-revealing QR factorization. SIAM J Sci Comput. 1996;17:848–69.

78. Boutsidis C, Sun J, Anerousis N: Clustered subset selection and its applications on it service metrics. In Proceedings of the 17th ACM conference on Information and knowledge management. ACM; 2008: 599–608.

79. Zhao Y, Karypis G: Evaluation of hierarchical clustering algorithms for document datasets. In Proceedings of the eleventh international conference on Information and knowledge management. ACM; 2002: 515–524.

80. Gao X, Martin ER. Using allele sharing distance for detecting human population stratification. Hum Hered. 2009;68:182–91.

81. Jombart T, Devillard S, Balloux F. Discriminant analysis of principal components: a new method for the analysis of genetically structured populations. BMC Genet. 2010;11:94.

82. EIGENSTRAT/smartpca [http://www.hsph.harvard.edu/alkes-price/software/]. Accessed 20 Jan 2018.

83. ipPCA [http://www4a.biotec.or.th/GI/tools/ippca]. Accessed 20 Jan 2018.

84. AWclust [http://awclust.sourceforge.net/]. Accessed 20 Jan 2018.

85. SHIPS [http://www.math-evry.cnrs.fr/logiciels/ships]. Accessed 20 Apr 2018.

86. NETVIEW [http://sydney.edu.au/vetscience/reprogen/netview/]. Accessed 20 Jan 2018.

87. iNJclust [http://www4a.biotec.or.th/GI/tools/injclust]. Accessed 20 Jan 2018.

16

Associations between hypertension and the peroxisome proliferator-activated receptor-δ (PPARD) gene rs7770619 C>T polymorphism in a Korean population

Minjoo Kim[1], Minkyung Kim[1], Hye Jin Yoo[2], Jayoung Shon[2,3] and Jong Ho Lee[1,2,3]*

Abstract

Background: Oxidative stress is associated with the increased risk of hypertension (HTN). This cross-sectional study is aimed to identify the association between the peroxisome proliferator-activated receptor-δ (*PPARD*) polymorphism and plasma malondialdehyde (MDA), an oxidative stress marker which is related to HTN development, and to determine whether *PPARD* gene is a candidate gene for HTN.

Results: One thousand seven hundred ninety-three individuals with normal blood pressure (BP) and HTN were included in this cross-sectional study. The Korean Chip was used to obtain genotype data. Through the analysis, the ten most strongly associated single-nucleotide polymorphisms (SNPs) were nominated for an MDA-related SNP. Among them, the rs7770619 polymorphism was identified in the *PPARD* gene. The CT genotype of the *PPARD* rs7770619 C>T polymorphism was associated with a lower risk of HTN before and after adjustments for age, sex, body mass index, smoking, and drinking. Significant associations were observed between plasma MDA and the *PPARD* rs7770619 C>T polymorphism and between systolic BP and the *PPARD* rs7770619 SNP in the controls. The CT controls showed significantly lower systolic BP and plasma MDA than the CC controls. Additionally, in both controls and HTN patients, the CT subjects showed significantly lower serum glucose and higher adiponectin levels than the CC subjects. Furthermore, the CT subjects showed significantly higher serum free fatty acid levels than the CC subjects among the HTN patients.

Conclusion: This is a new finding that the *PPARD* rs7770619 C>T SNP is a novel candidate variant for HTN based on the association between *PPARD* and plasma MDA in a Korean population.

Keywords: Peroxisome proliferator-activated receptor-δ gene, Genetic polymorphism, Blood pressure, Hypertension, Malondialdehyde

Background

Oxidative stress is defined as a sustained increase in the levels of reactive oxygen species (ROS), such as hydrogen peroxide, superoxide anion radicals, and other free radicals. Lipids have been reported as one of the primary targets of ROS. Lipid peroxidation produces highly reactive aldehydes, including malondialdehyde (MDA), which has been reported as a primary biomarker of free radical-mediated lipid damage and oxidative stress [1]. Increased MDA levels as a marker of oxidative stress were higher in hypertensive patients than in normotensive individuals [2, 3]. Additionally, a positive correlation between serum MDA levels and systolic and diastolic blood pressure (BP) has been reported [4].

Hypertension (HTN) is a multifactorial disorder involving both genetic and environmental factors [5]. Therefore, genetic factors affecting oxidative stress may include a common genetic basis of susceptibility to HTN. Although some studies have focused on the association between peroxisome proliferator-activated receptor

---

* Correspondence: jhleeb@yonsei.ac.kr
[1]Research Center for Silver Science, Institute of Symbiotic Life-TECH, Yonsei University, Seoul 03722, Korea
[2]Department of Food and Nutrition, Brain Korea 21 PLUS Project, College of Human Ecology, Yonsei University, 50 Yonsei-ro, Seodaemun-gu, Seoul 03722, Korea
Full list of author information is available at the end of the article

(PPAR) and HTN [6–8], the association between PPAR-δ (PPARD) and HTN has not been extensively studied previously. Indeed, PPARD has been suggested to regulate BP by modulating risk factors of HTN, including obesity and fatty acid catabolism [9]. Since a close relationship was observed between BP and MDA levels in HTN [2, 3], the MDA-related single-nucleotide polymorphisms (SNPs) analyzed with the Korean Chip (K-CHIP) could also be novel SNPs associated with HTN risk. The K-CHIP is a customized chip optimized for genetic studies on diseases and complex traits in the Korean population. Therefore, the objective of this study was to determine whether the *PPARD* gene is a candidate gene for HTN by identifying any association between *PPARD* and MDA, which is increased in HTN [2, 3].

Methods

Study population

All individuals who visited the Health Service Center (HSC) at National Health Insurance Corporation Ilsan Hospital, Goyang, Korea, for their routine checkups (from January 2010 to March 2015) were potential study subjects for this research. Based on the data screened from HSC, men and women aged over 20 years (adult subjects) with nondiabetic normotension (systolic BP < 140 mmHg and diastolic BP < 90 mmHg) or HTN (systolic BP ≥ 140 mmHg or diastolic BP ≥ 90 mmHg) were asked to participate in this study and were given detailed explanation regarding the study, and then, individuals who agreed to take part in the study were recruited. These potential subjects were referred to the Department of Family Medicine, and their health and BP were reexamined. Finally, individuals who met the study criteria were included ($n = 2167$). The exclusion criteria were a current diagnosis or history of cardiovascular disease, liver disease, renal disease, pancreatitis, or cancer; pregnancy or lactation; and regular use of any medication except for HTN treatments. The inclusion criteria were men and women adults (aged over 20 years), nondiabetic (fasting glucose < 126 mg/dL and no use of glucose-lowering medication), and individuals who do not correspond to the exclusion criteria (Additional file 1: Figure S1). The aim of the study was carefully explained to all participants, who provided their written informed consent. The Institutional Review Board of Yonsei University and the National Health Insurance Corporation Ilsan Hospital approved the study protocol, which complied with the Declaration of Helsinki.

Blood sample collection

Venous blood samples were collected following an overnight fast for at least 12 h. The fasting blood specimens were collected in EDTA-treated tubes and serum tubes (BD Vacutainer; Becton, Dickinson and Company,

Franklin Lakes, NJ, USA) and were then centrifuged (1200 rpm, 20 min, 4 °C) to obtain plasma and serum. The plasma and serum sample aliquots were stored at −80 °C prior to analysis.

BP measurement

Systolic and diastolic BP were measured using a random-zero sphygmomanometer (HM-1101, Hico Medical Co., Ltd., Chiba, Japan) with appropriately sized cuffs after a rest period of at least 20 min in a seated position. BP was measured three times in both arms. The differences among the three systolic BP measurements were always less than 2 mmHg. Participants were instructed not to smoke or drink alcohol for at least 30 min before each BP measurement.

Clinical and biochemical assessments

Body weight (UM0703581; Tanita, Tokyo, Japan) and height (GL-150; G-tech International, Uijeongbu, Korea) were measured after subjects removed their shoes, and the body mass index (BMI) was calculated (kg/m^2).

The serum fasting triglyceride (TG) and total cholesterol (TC) levels were measured enzymatically using TG and CHOL Kits (Roche, Mannheim, Germany), respectively. Serum fasting high-density lipoprotein (HDL)-cholesterol was measured by a selective inhibition method with an HDL-C Plus Kit (Roche, Mannheim, Germany). The resulting color reactions of the assays were monitored using a Hitachi 7600 autoanalyzer (Hitachi, Tokyo, Japan). Low-density lipoprotein (LDL)-cholesterol values were obtained indirectly using the Friedewald formula: LDL-cholesterol = TC − [HDL-cholesterol + (TG/5)]. Serum fasting free fatty acid was measured with enzymatic assays using an NEFA-M Kit (Shinyang Diagnostics, Gyeonggi, Korea), and the resulting color reactions of the assays were monitored with a Hitachi 7600 autoanalyzer (Hitachi, Tokyo, Japan).

The serum fasting glucose level was measured by a hexokinase method using a GLU Kit (Roche, Mannheim, Germany). The serum fasting insulin was measured by an immunoradiometric assay using an Insulin IRMA Kit (DIAsource, Louvain, Belgium). The resulting color reaction was monitored with a Hitachi 7600 autoanalyzer (Hitachi, Tokyo, Japan) and an SR-300 system (Stratec, Birkenfeld, Germany), respectively. To calculate insulin resistance (IR), the equation for homeostatic model assessment (HOMA) was used: HOMA-IR = [fasting insulin (μIU/mL) × fasting glucose (mg/dL)]/405. Plasma adiponectin was measured via an enzyme immunoassay using a Human Adiponectin ELISA Kit (B-Bridge International Inc., San Jose, CA, USA), and the resulting color reaction was monitored with a Victor2 (PerkinElmer Life Sciences, Turku, Finland).

Serum high-sensitivity C-reactive protein (hs-CRP) levels were measured using a CRP Kit (Roche, Mannheim, Germany), and the resulting colorimetric reaction was monitored with a Hitachi 7600 autoanalyzer (Hitachi, Tokyo, Japan). Plasma MDA was measured from thiobarbituric acid reactive substances (TBARS) using a TBARS Assay Kit (ZeptoMetrix Co., Buffalo, NY, USA).

Affymetrix Axiom™ KORV1.0-96 Array hybridization and SNP selection

The detailed information for this protocol is described in our previous study [10]. A total of 2167 samples were genotyped according to the manufacturer's protocol, which recommended the Axiom® 2.0 Reagent Kit (Affymetrix Axiom® 2.0 Assay User Guide; Affymetrix, Santa Clara, CA, USA). The genotype data were produced using the K-CHIP, which was available through the K-CHIP consortium. The K-CHIP was designed by the Center for Genome Science at the Korea National Institute of Health (4845-301, 3000-3031).

Samples that revealed the following features were excluded during the quality control process: sex inconsistency, markers with a high missing rate (> 5%), individuals with a high missing rate (> 10%), minor allele frequency < 0.01, and a significant deviation from Hardy-Weinberg equilibrium (HWE) ($p < 0.001$). In addition, SNPs were excluded if they were related to each other in linkage disequilibrium. Consequently, among a total of 833,535 SNPs on the arrays and 2167 samples, 395,787 SNPs and 2158 samples remained, and they were used in subsequent association analyses.

Statistical analysis

HWE and association assessments between SNPs and MDA using linear regression analysis were performed in PLINK version 1.07 (http://zzz.bwh.harvard.edu/plink); for issues of multiple comparisons between SNPs and MDA, false discovery rate (FDR) correction was used. Descriptive statistical analyses were conducted using SPSS version 23.0 (IBM, Chicago, IL, USA). Logarithmic transformation was used for the skewed variables, and a two-tailed p value of < 0.05 was considered statistically significant. An independent t test was performed on the continuous variables. Sex distribution, smoking and drinking status, and genotype frequency were tested using the chi-squared test. The association of HTN with a genotype was calculated using the odds ratio (OR) [95% confidence interval (CI)] of a logistic regression model with an adjustment for confounding factors.

Results

Through the subsequent analysis using 395787 SNPs and 2158 samples, the ten SNPs that were most strongly associated with plasma MDA were nominated

(Additional file 1: Table S1). Among them, one SNP, rs7770619, was identified in the *PPARD* gene. Therefore, we conducted an association analysis of *PPARD* rs7770619 polymorphism. Among 2158 subjects, 313 and 52 subjects did not have data of plasma MDA and rs7770619, respectively; thus, a total of 1793 subjects who had both plasma MDA and rs7770619 data were finally included in the final analysis (Additional file 1: Figure S1).

The clinical and biochemical characteristics of the normotensive controls ($n = 1359$) and HTN patients ($n = 434$) are shown in Table 1. HTN patients included those who use antihypertensive medication (35.9%). Thus, we subdivided HTN patients into two groups: those not treated with antihypertensive drugs (HTN without treatment, $n = 278$) and those treated with antihypertensive drugs (HTN with treatment, $n = 156$). Compared with normotensive controls, both HTN subgroups were older and heavier. After adjusting for age, sex, BMI, smoking, and drinking, the patients in both HTN subgroups showed higher systolic and diastolic BP and plasma MDA than normotensive controls. Serum TG was higher in HTN patients without treatment than in normotensive controls. The HTN with treatment subgroup showed lower TC and LDL-cholesterol and higher glucose than normotensive controls (Table 1).

Distribution of the *PPARD* rs7770619 C>T polymorphism

The observed and expected frequencies of the *PPARD* rs7770619 C>T polymorphism were in HWE in the entire population and in the control and patient groups. The relative *PPARD* rs7770619 C>T genotypes in the HTN patients differed significantly from those in the normotensive controls (Table 2). There was no homozygous mutation TT genotype in either the normotensive controls or the HTN patients. Frequencies of the T allele of the *PPARD* rs7770619 C>T polymorphism in the HTN patients (0.012) were significantly lower than those in the normotensive controls (0.028) ($p = 0.007$) (Table 2).

The presence of the CT genotype of the *PPARD* rs7770619 C>T SNP was associated with a lower risk of HTN [OR 0.404 (95% CI 0.207–0.788), $p = 0.008$] (Table 3). The significance of the association remained after adjusting for age, sex, BMI, smoking, and drinking [OR 0.478 (95% CI 0.238–0.960), $p = 0.038$], and the p value of Hosmer-Lemeshow goodness-of-fit test was 0.700 for this model implying that our model is well-fitted.

Associations of plasma MDA, systolic BP, serum glucose, free fatty acids, and adiponectin with the *PPARD* rs7770619 C>T polymorphism

A significant association was observed between plasma MDA and the *PPARD* rs7770619 C>T polymorphism in the normotensive controls. The CT carriers showed

Table 1 Clinical and biochemical characteristics in normotensive controls and HTN patients

	Normotensive controls ($n = 1359$)	HTN patients ($n = 434$)		
		Total ($n = 434$)	HTN without treatment ($n = 278$)	HTN with treatment ($n = 156$)
Age (year)	48.4 ± 0.28	$53.8 \pm 0.52^{***}$	$51.5 \pm 0.66^{***}$	$57.9 \pm 0.73^{***}$
Male/female, n (%)	499 (36.7)/860 (63.3)	229 (52.8)/205 (47.2)***	153 (55.0)/125 (45.0)***	76 (48.7)/80 (51.3)**
Current smoker, n (%)	193 (14.2)	74 (17.1)	55 (19.8)*	19 (12.2)
Current drinker, n (%)	824 (60.6)	259 (59.7)	177 (63.7)	82 (52.6)
BMI (kg/m^2)	23.8 ± 0.08	$25.4 \pm 0.15^{***}$	$25.3 \pm 0.19^{***}$	$25.5 \pm 0.23^{***}$
Systolic BP (mmHg)	116.5 ± 0.31	$139.2 \pm 0.75^{\dagger\dagger\dagger}$	$144.4 \pm 0.75^{\dagger\dagger\dagger}$	$129.9 \pm 1.31^{\dagger\dagger\dagger}$
Diastolic BP (mmHg)	72.9 ± 0.23	$88.3 \pm 0.49^{\dagger\dagger\dagger}$	$92.6 \pm 0.46^{\dagger\dagger\dagger}$	$80.6 \pm 0.77^{\dagger\dagger\dagger}$
Triglyceride (mg/dL)^{f}	119.2 ± 1.88	$148.3 \pm 4.24^{\dagger}$	$152.3 \pm 5.62^{\dagger\dagger}$	141.2 ± 6.20
Total cholesterol (mg/dL)^{f}	199.4 ± 0.98	$199.3 \pm 1.73^{\dagger}$	202.9 ± 2.14	$192.8 \pm 2.86^{\dagger\dagger\dagger}$
HDL-cholesterol (mg/dL)^{f}	54.4 ± 0.37	51.2 ± 0.60	51.1 ± 0.74	51.5 ± 1.05
LDL-cholesterol (mg/dL)^{f}	122.0 ± 0.90	$119.2 \pm 1.58^{\dagger\dagger\dagger}$	122.3 ± 1.98	$113.7 \pm 2.55^{\dagger\dagger\dagger}$
Glucose (mg/dL)^{f}	95.4 ± 0.55	$102.5 \pm 1.10^{\dagger}$	100.9 ± 1.39	$105.3 \pm 1.80^{\dagger\dagger}$
Insulin (μIU/mL)^{f}	8.92 ± 0.12	9.59 ± 0.28	9.83 ± 0.38	9.11 ± 0.36
Free fatty acids (μEq/L)^{f}	552.8 ± 6.65	567.9 ± 12.6	560.9 ± 16.0	581.8 ± 20.4
HOMA-IR^{f}	2.09 ± 0.03	2.46 ± 0.10	2.54 ± 0.14	2.30 ± 0.10
hs-CRP (mg/dL)^{f}	1.28 ± 0.08	1.60 ± 0.15	1.59 ± 0.16	1.61 ± 0.28
Adiponectin (ng/mL)^{f}	6.49 ± 0.10	5.98 ± 0.17	5.94 ± 0.22	6.06 ± 0.27
Malondialdehyde (nmol/mL)^{f}	8.89 ± 0.09	9.78 ± 0.27	$9.33 \pm 0.35^{\dagger\dagger}$	$10.6 \pm 0.43^{\dagger\dagger}$

Mean ± SE. ^{f}Tested following logarithmic transformation. $^{*}p < 0.05$, $^{**}p < 0.01$, and $^{***}p < 0.001$ derived from an independent t test between normotensive controls and each subgroup of hypertensive patients. $^{\dagger}p < 0.05$, $^{\dagger\dagger}p < 0.01$, and $^{\dagger\dagger\dagger}p < 0.001$ derived after adjusting for age, sex, BMI, smoking, and drinking

significantly lower MDA than the CC carriers (CC 8.98 ± 0.09 nmol/mL, CT 7.31 ± 0.20 nmol/mL; $p < 0.001$) (Fig. 1). Similarly, in the HTN patients, the CT carriers showed lower MDA than the CC carriers, but the difference was not statistically significant. Systolic BP and the *PPARD* rs7770619 C>T polymorphism were significantly associated in the normotensive controls (CC 116.7 ± 0.32 mmHg, CT 113.0 ± 1.36 mmHg; $p = 0.007$), and there was a trend toward an association between systolic BP and the *PPARD* rs7770619 C>T polymorphism in the HTN patients (CC 139.4 ± 0.76 mmHg, CT 130.9 ± 3.43 mmHg; $p = 0.090$). In the normotensive controls, the CT carriers showed significantly lower systolic BP than the CC carriers (Fig. 1). Additionally, in both normotensive controls and HTN patients, the CT carriers showed lower glucose and higher adiponectin than the CC carriers. In the HTN patients, compared with the CC carriers,

the CT carriers showed higher free fatty acid. In the normotensive controls, there was a trend toward an increase of free fatty acids in the CT carriers compared with the CC carriers (Fig. 1).

Discussion

The major finding of the present study was that the frequency of the *PPARD* rs7770619 CT genotype was significantly lower in patients with HTN than in the normotensive controls, suggesting that there was an association between the *PPARD* rs7770619 C>T SNP and HTN. This observation correlated with recent findings that the *PPARD* polymorphism has a key role for HTN development [9]. The significance of the present observations was established by the identification of the human polymorphism in the *PPARD* locus with altered BP and plasma MDA levels, which are reliable oxidative stress markers in HTN [2, 3].

Table 2 Frequencies of the *PPARD* rs7770619 genotypes in the normotensive controls and the HTN patients

PPARD rs7770619	Normotensive controls ($n = 1359$)		HTN patients ($n = 434$)		p values
	n	%	n	%	
CC	1284	94.5	424	97.7	0.006
CT	75	5.5	10	2.3	
T allele frequency	75	2.8	10	1.2	0.007

A chi-squared test was used to calculate the p values

Table 3 Unadjusted and adjusted odds ratios (ORs) for all the HTN patients according to the *PPARD* rs7770619 genotypes

PPARD rs7770619	HTN patients (*n* = 434) OR (95% CI)	*p* values
Model 1		
C[‡] compared with T	0.411 (0.211 to 0.798)	0.009
CC[‡] compared with CT	0.404 (0.207 to 0.788)	0.008
Model 2		
C[‡] compared with T	0.486 (0.243 to 0.970)	0.041
CC[‡] compared with CT	0.478 (0.238 to 0.960)	0.038

Model 1: unadjusted; Model 2: adjusted for age, sex, BMI, smoking, and drinking. *CI* confidence interval. [‡]Reference

Patients with HTN tend to have several conditions that accelerate the atherogenic process, including an increase in free radicals. ROS are the most important free radicals in the human body and cause increased oxidative stress and tissue injury under pathological conditions [11, 12]. Several studies have reported evidence for enhanced ROS production and decreases in the antioxidant reserves in the plasma and tissues of hypertensive animals and humans [13, 14]. MDA is produced during the attack of ROS upon membrane lipoproteins and polyunsaturated fatty acids. Kashyap et al. [15] have reported increased MDA levels in hypertensive subjects compared with those in normotensive subjects and suggested that elevated lipid peroxidation reflected increased oxidative stress in patients with HTN. In a recent study, similar results were obtained in terms of MDA [3]. The authors found significantly higher MDA levels in the essential HTN group than in the control group. Similarly, this study also showed higher MDA levels in the HTN group regardless of whether they were taking antihypertensive medication than in the control group.

In the present study, subjects with the *PPARD* rs7770619 CT genotype showed significantly lower systolic BP than those with the CC genotype in the normotensive controls. Additionally, the significantly lower

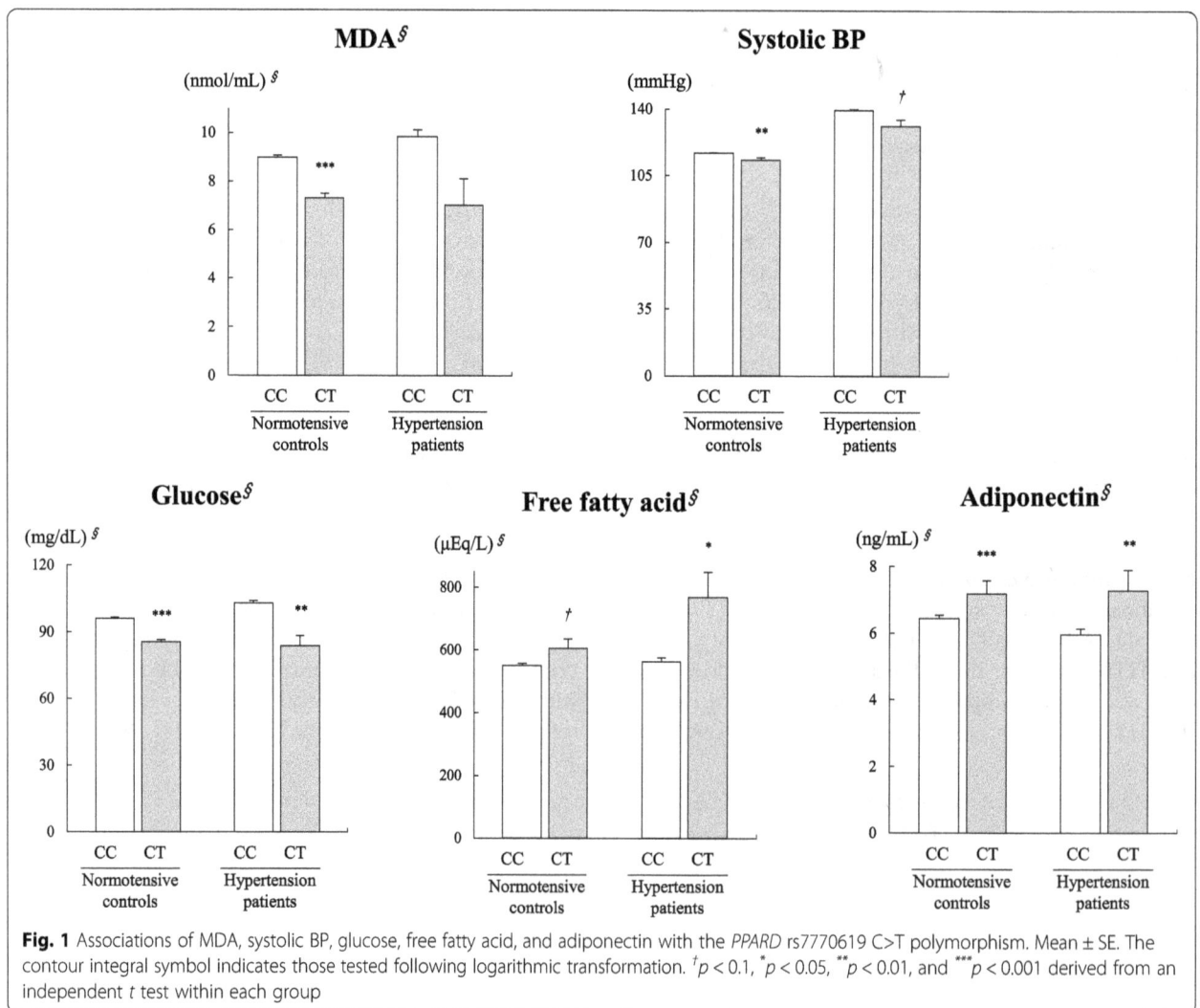

Fig. 1 Associations of MDA, systolic BP, glucose, free fatty acid, and adiponectin with the *PPARD* rs7770619 C>T polymorphism. Mean ± SE. The contour integral symbol indicates those tested following logarithmic transformation. $^{†}p < 0.1$, $^{*}p < 0.05$, $^{**}p < 0.01$, and $^{***}p < 0.001$ derived from an independent *t* test within each group

concentrations of plasma MDA in the subjects carrying the *PPARD* rs7770619 CT genotype than in the subjects with the CC genotype in the normotensive controls and the tendency toward a decrease in the HTN patients were found in this study. Recently, Li et al. [9] have observed associations between rs2016520 and the rs9794 minor allele of *PPARD* and decreased risk of HTN and additional interactions between these two SNPs. Although the *PPARD* rs7770619 SNP is not equivalent to the SNPs identified in other studies [9], the results of this study indicate that the *PPARD* rs7770619 SNP may represent a critical locus that negatively influences HTN and oxidative stress.

The *PPARD* rs7770619 C>T SNP is considered functional since serum glucose and the *PPARD* rs7770619 C>T polymorphism were significantly associated in both normotensive controls and HTN patients in this study. *PPARD* activation in the liver appears to decrease hepatic glucose output, which contributes to improved glucose control [16]. *PPARD* also appears to have a role in the regulation of fatty acid oxidation in several tissues, including skeletal muscle and adipose tissue [17]. It has been suggested that the mechanisms of action of this gene involve redistribution of the non-esterified fatty acid (NEFA) flux. The increasing oxidative capability draws the NEFA to the muscle to be preferentially oxidized rather than stored in adipose tissue, which leads to a decrease in adipocyte size, enhanced lipolysis, and increased adiponectin secretion [18]. In this study, subjects with the *PPARD* rs7770619 CT genotype showed significantly higher adiponectin concentrations than those with the CC genotype in both the normotensive controls and the HTN patients. Additionally, HTN patients with the CT genotype showed an increase in serum free fatty acid compared with subjects with the CC genotype.

The rs7770619 SNP, showing the features above in this study, is an intron mutation located at chromosome 6:35382265 (GRC38p.12 assembly of the human genome). Introns are involved in regulation of alternative splicing and gene expression [19]; therefore, the rs7770619 C>T polymorphism "may" affect the *PPARD* gene splicing, *PPARD* splice variants generation, and/or PPARD function. So far, however, there is a lack of studies on rs7770619, and studies about an impact of rs7770619 on *PPARD* gene splicing do not exist. Thus, at the present stage of knowledge and information, it is difficult to prove that the intron mutation of rs7770619 "really" influences splicing of *PPARD* gene. Moreover, to the best of our knowledge, no literatures that investigated altered risk of HTN, oxidative stress, or MDA levels according to *PPARD* splice variants exist. Our study only approached the association between *PPARD* rs7770619 C>T polymorphism and the risk of HTN in the Korean cohort; thus, the exact mechanism on

development of HTN through the SNP is still unknown. Therefore, attempts to verify the underlying mechanism is needed in the future.

According to the dbSNP (http://www.ncbi.nlm.nih.gov/snp), MAF for rs7770619 SNP is 7.5% in the 1000 Genomes Project (phase 3) and 10.5% in TOPMED. In the present study, MAF of rs7770619 was 2.4% in the whole participants (2.8% and 1.2% for normotensive controls and HTN patients, respectively); somewhat low MAF was shown in the Korean cohort compared to the world's average. Indeed, MAFs for rs7770619 vary based on ethnic groups. Many African populations have high rs7770619 MAF (> 20%), whereas in Asian populations (especially East Asians such as Korean, Chinese, and Japanese), rs7770619 MAF is generally low; based on the 1000 Genomes Project (phase3) (https://www.ncbi.nlm.nih.gov/variation/tools/1000genomes), in the population of Han Chinese in Beijing (HCB population) and Japanese in Tokyo (JPT population), MAF is 1.5% and 0.5%, respectively. Therefore, considering ethnic origin, MAF for rs7770619 observed in our cohort is not exclusively low. Many studies have reported the ethnic-dependent differences of HTN prevalence; studies have shown that black people are under higher risk of HTN than white people [20–23]. Moreover, Cappuccio et al. [24] showed that the South Asian population had two- to three-fold higher prevalence of HTN than white people, and Holly et al. [23] also showed significant higher odds ratio of HTN in the Chinese population compared with the white population after adjusting confounding factors. In dbSNP, the Caucasian population (CAUC1) shows 8.1% of MAF for rs7770619. Since our results demonstrated that rs7770619 minor T allele is associated with lower risk of HTN, higher risk of HTN in Asian populations whose MAF for rs7770619 is lower than white people makes sense; however, higher prevalence of HTN was also observed in the black population, thus, our results do not entirely correspond with other researches in terms of ethnicity. HTN can be caused by various risks such as genetic, metabolic, and environmental (e.g., dietary habit and physical activity) factors. A complex interaction of these risk factors induces HTN; thus, risk of HTN according to ethnic groups cannot be explained by only one polymorphism. At least, in a Korean population, rs7770619 was found as a novel SNP for HTN risk via a connection with oxidative stress (MDA). In addition, rs7770619 C>T polymorphism has not been studied with regard to oxidative stress and HTN, and this is the first study to identify association between rs7770619 SNP and HTN risk. Therefore, based on the results of our study, gradual expansion of the HTN study in respect of rs7770619 polymorphism is needed.

There are several limitations. First, when interpreting the findings of this study, it should be considered that our results share the limitations of cross-sectional observational studies becuase we evaluated only an association rather than prospective prediction. Second, in terms of conventional genome-wide association study (GWAS), this study did not share the exact analysis method of conventional GWAS, because (1) imputation could not be conducted due to the limit of equipment's calculation capacity and (2) when we selected SNPs, we used FDR correction instead of Bonferroni correction, widely used in conventional GWAS. Bonferroni correction has a very conservative threshold ($p < 5 \times 10^{-8}$) so that it can cause decrease of statistical power [25, 26]. Recently, indeed, not only Bonferroni correction but also FDR correction has been used in common practice [27, 28]. Third, unexpectedly, low MAF generated a too small sample size of individuals with rs7770619 CT genotype; thus, in Fig. 1, a subset analysis with the HTN without treatment group (for control of HTN medication effects) according to the rs7770619 genotypes could not be performed due to a problem of statistical power. Lastly, we specifically focused on a representative group of Korean subjects. Therefore, our results cannot be generalized to other ethnicities, age groups, or geographical groups. Taken together, further study needs to be conducted by exact manner of GWAS with a cohort having a much larger sample size to confirm replication of the results observed in this study. Despite these limitations, our results show interesting associations between the *PPARD* rs7770619 CT genotype, a decreased risk of HTN, and decreased oxidative stress. These results suggest that the *PPARD* rs7770619 C>T SNP is a novel candidate gene for HTN through the association between *PPARD* and MDA, a biomarker of oxidative stress.

Conclusion

According to the individuals' *PPARD* rs7770619 genotype, the risk of HTN development is predictable via association between *PPARD* rs7770619 SNPs and MDA. Reducing MDA in subjects with *PPARD* rs7770619 CC genotype is necessary to decrease the risk of HTN development. Therefore, by analyzing personal genetic background, MDA, the oxidative stress marker, can be considered as a therapeutic target of HTN in a Korean population.

Abbreviations
BP: Blood pressure; CI: Confidence interval; FDR: False discovery rate; GWAS: Genome-wide association study; HDL: High-density lipoprotein; HOMA: Homeostatic model assessment; hs-CRP: High-sensitivity C-reactive protein; HTN: Hypertension; HWE: Hardy-Weinberg equilibrium; IR: Insulin resistance; K-CHIP: Korean Chip; LDL: Low-density lipoprotein; MDA: Malondialdehyde; NEFA: Non-esterified fatty acid; OR: Odds ratio; PPAR: Peroxisome proliferator-activated receptor; PPARD: Peroxisome proliferator-activated receptor-δ; ROS: Reactive oxygen species; SNP: Single-nucleotide polymorphism; TC: Total cholesterol; TG: Triglyceride

Acknowledgements
The genotype data were generated using the Korean Chip (K-CHIP), which is available through the K-CHIP consortium. The K-CHIP was designed by the Center for Genome Science at the Korea National Institute of Health, Korea (4845-301, 3000-3031). We appreciate Dr. Sang-Hyun Lee, who belongs to the Department of Family Practice, National Health Insurance Corporation, Ilsan Hospital, Goyang, Korea, for collaborating on the acquisition of the valuable data.

Funding
This study was funded by the Mid-career Researcher Program (NRF-2016R1A2B4011662) of the Ministry of Science and ICT through the National Research Foundation of Korea, Republic of Korea.

Authors' contributions
All authors contributed to the conception and design of the study. MJK and JHL were responsible for the analysis and interpretation of the data and preparation of the manuscript. MKK and HJY took part in the analysis of the data and preparation of the manuscript. JS carried out the acquisition and analysis of the data. All authors contributed to the critical revisions of the paper and have approved the study for publication.

Competing interests
The authors declare that they have no competing interests.

Author details
[1]Research Center for Silver Science, Institute of Symbiotic Life-TECH, Yonsei University, Seoul 03722, Korea. [2]Department of Food and Nutrition, Brain Korea 21 PLUS Project, College of Human Ecology, Yonsei University, 50 Yonsei-ro, Seodaemun-gu, Seoul 03722, Korea. [3]Department of Food and Nutrition, National Leading Research Laboratory of Clinical Nutrigenetics/Nutrigenomics, College of Human Ecology, Yonsei University, Seoul 03722, Korea.

References
1. Tiwari BK, Pandey KB, Abidi AB, Rizvi SI. Markers of oxidative stress during diabetes mellitus. J Biomark. 2013;2013:378790.
2. Armas-Padilla MC, Armas-Hernández MJ, Sosa-Canache B, Cammarata R, Pacheco B, Guerrero J, et al. Nitric oxide and malondialdehyde in human hypertension. Am J Ther. 2007;14:172–6.
3. Gönenç A, Hacışevki A, Tavil Y, Çengel A, Torun M. Oxidative stress in patients with essential hypertension: a comparison of dippers and non-dippers. Eur J Intern Med. 2013;24:139–44.
4. Uzun H, Karter Y, Aydin S, Curgunlu A, Simşek G, Yücel R, et al. Oxidative stress in white coat hypertension; role of paraoxonase. J Hum Hypertens. 2004;18:523–8.
5. Yagil Y, Yagil C. The search for the genetic basis of hypertension. Curr Opin Nephrol Hypertens. 2005;14:141–7.
6. Zhu Q, Guo Z, Hu X, Wu M, Chen Q, Luo W, et al. Haplotype analysis of *PPARγ* C681G and intron CT variants. Positive association with essential hypertension. Herz. 2014;39:264–70.
7. Gu SJ, Guo ZR, Wu M, Ding Y, Luo WS. Association of peroxisome proliferator-activated receptor γ polymorphisms and haplotypes with essential hypertension. Genet Test Mol Biomarkers. 2013;17:418–23.
8. Usuda D, Kanda T. Peroxisome proliferator-activated receptors for hypertension. World J Cardiol. 2014;6:744–54.
9. Li Y, Sun G. Case-control study on association of peroxisome proliferator-activated receptor-δ and SNP-SNP interactions with essential hypertension in Chinese Han population. Funct Integr Genomics. 2016;16:95–100.
10. Kim M, Kim M, Yoo HJ, Yun R, Lee S-H, Lee JH. Estrogen-related receptor γ gene (*ESRRG*) rs1890552 A>G polymorphism in a Korean population: association with urinary prostaglandin-$F_{2\alpha}$ concentration and impaired fasting glucose or newly diagnosed type 2 diabetes. Diabetes Metab. 2017; 43:385–8.
11. Ischiropoulos H, Beckman JS. Oxidative stress and nitration in neurodegeneration: cause, effect, or association? J Clin Invest. 2003;111:163–9.

12. Schopfer FJ, Baker PR, Freeman BA. NO-dependent protein nitration: a cell signaling event or an oxidative inflammatory response? Trends Biochem Sci. 2003;28:646–54.
13. Dhalla NS, Temsah RM, Netticadan T. Role of oxidative stress in cardiovascular diseases. J Hypertens. 2000;18:655–73.
14. Touyz RM, Schiffrin EL. Reactive oxygen species in vascular biology: implications in hypertension. Histochem Cell Biol. 2004;122:339–52.
15. Kashyap MK, Yadav V, Sherawat BS, Jain S, Kumari S, Khullar M, et al. Different antioxidants status, total antioxidant power and free radicals in essential hypertension. Mol Cell Biochem. 2005;277:89–99.
16. Villegas R, Williams S, Gao Y, Cai Q, Li H, Elasy T, et al. Peroxisome proliferator-activated receptor delta (PPARD) genetic variation and type 2 diabetes in middle-aged Chinese women. Ann Hum Genet. 2011;75:621–9.
17. Fredenrich A, Grimaldi PA. Roles of peroxisome proliferator-activated receptor delta in skeletal muscle function and adaptation. Curr Opin Clin Nutr Metab Care. 2004;7:377–81.
18. Schulze MB, Hu FB. Primary prevention of diabetes: what can be done and how much can be prevented? Annu Rev Public Health. 2005;26:445–67.
19. Jo BS, Choi SS. Introns: the functional benefits of introns in genomes. Genomics Inform. 2015;13:112–8.
20. Ortega LM, Sedki E, Nayer A. Hypertension in the African American population: a succinct look at its epidemiology, pathogenesis, and therapy. Nefrologia. 2015;35:139–45.
21. Lackland DT. Racial differences in hypertension: implications for high blood pressure management. Am J Med Sci. 2014;348:135–8.
22. Brown MJ. Hypertension and ethnic group. BMJ. 2006;332:833–6.
23. Kramer H, Han C, Post W, Goff D, Diez-Roux A, Cooper R, et al. Racial/ethnic differences in hypertension and hypertension treatment and control in the multi-ethnic study of atherosclerosis (MESA). Am J Hypertens. 2004;17:963–70.
24. Cappuccio FP, Cook DG, Atkinson RW, Strazzullo P. Prevalence, detection, and management of cardiovascular risk factors in different ethnic groups in south London. Heart. 1997;78:555–63.
25. Nakagawa S. A farewell to Bonferroni: the problems of low statistical power and publication bias. Behav Ecol. 2004;15:1044–5.
26. Benjamini Y, Hochberg Y. Controlling the false discovery rate: a practical and powerful approach to multiple testing. J R Statist Soc B. 1995;57:289–300.
27. Li S, Xie L, Du M, Xu K, Zhu L, Chu H, et al. Association study of genetic variants in estrogen metabolic pathway genes and colorectal cancer risk and survival. Arch Toxicol. 2018. https://doi.org/10.1007/s00204-018-2195-y.
28. Duan B, Hu J, Liu H, Wang Y, Li H, Liu S, et al. Genetic variants in the platelet-derived growth factor subunit B gene associated with pancreatic cancer risk. Int J Cancer. 2018;142:1322–31.

Associations of high-altitude polycythemia with polymorphisms in PIK3CD and COL4A3 in Tibetan populations

Xiaowei Fan[1,2†], Lifeng Ma[1,2†], Zhiying Zhang[1,2†], Yi Li[3,5], Meng Hao[3], Zhipeng Zhao[1,2], Yiduo Zhao[1,2], Fang Liu[1,2], Lijun Liu[1,2], Xingguang Luo[4], Peng Cai[1,2], Yansong Li[1,2] and Longli Kang[1,2*] (iD)

Abstract

Background: High-altitude polycythemia (HAPC) is a chronic high-altitude disease that can lead to an increase in the production of red blood cells in the people who live in the plateau, a hypoxia environment, for a long time. The most frequent symptoms of HAPC include headache, dizziness, breathlessness, sleep disorders, and dilation of veins. Although chronic hypoxia is the main cause of HAPC, the fundamental pathophysiologic process and related molecular mechanisms responsible for its development remain largely unclear yet.

Aim/methods: This study aimed to explore the related hereditary factors of HAPC in the Chinese Han and Tibetan populations. A total of 140 patients (70 Han and 70 Tibetan) with HAPC and 60 healthy control subjects (30 Han and 30 Tibetan) were recruited for a case-control association study. To explore the genetic basis of HAPC, we investigated the association between HAPC and both phosphatidylinositol-4,5-bisphosphonate 3-kinase, catalytic subunit delta gene (*PIK3CD*) and collagen type IV α3 chain gene (*COL4A3*) in Chinese Han and Tibetan populations.

Results/conclusion: Using the unconditional logistic regression analysis and the false discovery rate (FDR) calculation, we found that eight SNPs in *PIK3CD* and one SNP in *COL4A3* were associated with HAPC in the Tibetan population. However, in the Han population, we did not find any significant association. Our study suggested that polymorphisms in the *PIK3CD* and *COL4A3* were correlated with susceptibility to HAPC in the Tibetan population.

Keywords: High-altitude polycythemia, PIK3CD, COL4A3

Introduction

High-altitude polycythemia (HAPC) is a chronic high-altitude disease, characterized by excessive erythrocytosis. The clinical HAPC is diagnosed by a hemoglobin concentration ≥ 19 g/dL for females and ≥ 21 g/dL for males, according to the criteria established in the VI World Congress on Mountain Medicine and High-altitude Physiology in 2004 [1]. More than 140 million people are living at high altitudes above 2500 m worldwide, majorly in the Andes, Ethiopian Highlands, and Qinghai-Tibet Plateau

[2]. The Qinghai-Tibet Plateau is the highest plateau in the world, which covers a large area with low oxygen in natural environment, and millions of people are living and working in this region. It is well known that the body's hemoglobin concentration increases due to the hypoxic environment of high altitude, and therefore, this response is crucial for people who adapt to live at high altitudes. Some studies show that a number of populations suffer from chronic mountain sickness because they stay long at high altitudes [3]. HAPC mainly leads to a significant increase in blood viscosity, causing damage to microcirculatory and immune response disturbances such as vascular thrombosis, extensive organ damage, and sleep disorders [4, 5]. It is reported that the prevalence of HAPC in the Qinghai-Tibet Plateau is around 5 to 18% [1], and the prevalence of HAPC increases with the altitude. As the construction of the Qinghai-Tibet Railway has been

* Correspondence: longli_kang@163.com
†Xiaowei Fan, Lifeng Ma and Zhiying Zhang contributed equally to this work.
[1]Key Laboratory for Molecular Genetic Mechanisms and Intervention Research on High Altitude Disease of Tibet Autonomous Region, School of Medicine, Xizang Minzu University, Xianyang 712082, Shaanxi, China
[2]Key Laboratory of High Altitude Environment and Genes Related to Diseases of Tibet Autonomous Region, School of Medicine, Xizang Minzu University, Xianyang 712082, Shaanxi, China

completed, a number of Han populations migrate to Tibet. The incidence of HAPC among immigrants is significantly higher than the high-altitude natives [6]. As the Tibetan population keeps genetic adaptations, they can easily adapt to the high-altitude hypoxia environment, for example, showing lower hemoglobin levels and lower hematocrit. Many studies have noted that there are some significant differences in the genomes between immigrants and high-altitude natives, which indicates that genetic factors may contribute to the development of HAPC, although the molecular mechanisms and pathogenesis are still under study. In our study, we aimed to investigate the associations between susceptibility to HAPC and two new candidate genes that are related to the oxygen metabolism in red blood cells but have not been reported before.

The first candidate, *PIK3CD*, encodes the p110δ catalytic subunit of phosphoinositide 3-kinaseδ (PI3Kδ), a member of a big family of metalloenzymes. PI3Kδ is a heterodimer comprising the p110δ and p85 family regulatory subunit and expressed predominantly in leukocytes. Therefore, it plays an important role in the proliferation, survival, and activation of leukocytes [7–9]. The expression pattern and functions of *PIK3CD* are very important in PI3K/Akt pathway. Recently, research studies revealed that PI3K/Akt mediated the stabilization of HIF-1α (hypoxia-inducible factors-1α) [10], and it was also involved in the increase of HIF-1α protein level [11]. Meanwhile, HIF-1α plays an important role in transcriptionally upregulating erythropoietin (EPO) in hypoxia and affecting the amount of red blood cells [12].

The second candidate, *COL4A3*, encodes a subunit of type IV collagen that is a structural protein of the alveolar extracellular matrix (ECM) and mostly found in the kidney, lung, and basement membranes. It is located at 2q35-q3 and mainly contains 51 exons [13]. Type IV collagen is involved in various physiological conditions, including aging, diabetes, kidney disease, scarring, and pulmonary fibrosis [14]. The ECM is important to the structure and function of cell types. It contributes to many processes, such as cellular proliferation, differentiation, migration, and apoptosis [15].

Results

The demographics of HAPC patients and controls are shown in Table 1. The basic characteristics of candidate SNPs in the Han and Tibetan subjects are summarized in Table 2 (Fig. 1) and Table 3 (Fig. 2). We analyzed the associations between SNPs and HAPC using unconditional logistic regression analysis. In the Han population, rs72633866 ($P1 = 0.033$ before adjustment and $P2 = 0.014$ after adjustment for age), rs9430220 ($P1 = 0.081$

Table 1 Demographics of the control individuals and patients with high-altitude polycythemia

Variables	Han		Tibetan	
	Case ($n = 70$)	Control ($n = 30$)	Case ($n = 70$)	Control ($n = 30$)
Male	35	15	35	15
Female	35	15	35	15

and $P2 = 0.029$), rs199962152 ($P1 = 0.024$ and $P2 = 0.034$), and rs10864435 ($P1 = 0.013$ and $P2 = 0.002$) in *PIK3CD* were significantly associated with HAPC. In the Tibetan subjects, rs2230735 ($P1 = 0.008$ and $P2 = 0.008$), rs28730671 ($P1 = 0.007$ and $P2 = 0.007$), rs111888887 ($P1 = 0.034$ and $P2 = 0.034$), rs28730674 ($P1 = 0.007$ and $P2 = 0.007$), rs371870925 ($P1 = 0.007$ and $P2 = 0.007$), rs199962152 ($P1 = 0.045$ and $P2 = 0.040$), rs77571929 ($P1 = 0.005$ and $P2 = 0.005$), rs117226273 ($P1 = 0.007$ and $P2 = 0.007$), rs28730676 ($P1 = 0.007$ and $P2 = 0.007$), and rs28730677 ($P1 = 0.007$ and $P2 = 0.007$) in *PIK3CD* were significantly associated with HAPC. Furthermore, rs34505188 ($P1 = 0.028$ and $P2 = 0.028$), rs11677877 ($P1 = 0.013$ and $P2 = 0.013$), rs34019152 ($P1 = 0.018$ and $P2 = 0.018$), and rs28381984 ($P1 = 0.001$ and $P2 = 0.001$) in *COL4A3* were associated with HAPC.

After using FDR to correct for multiple comparisons, in the Tibetan subjects, we found that rs2230735 (OR = 0.844, 95% CI = 0.337–2.079, $P = 0.046$), rs28730671 (OR = 0.821, 95% CI = 0.324–2.035, $P = 0.046$), rs28730674 (OR = 0.821, 95% CI = 0.324–2.035, $P = 0.046$), rs371870925 (OR = 0.812, 95% CI = 0.320–2.017, $P = 0.046$), rs77571929 (OR = 0.814, 95% CI = 0.320–2.024, $P = 0.046$), rs117226273 (OR = 0.821, 95% CI = 0.324–2.035, $P = 0.046$), rs28730676 (OR = 0.821, 95% CI = 0.324–2.035, $P = 0.046$), rs28730676 (OR = 0.821, 95% CI = 0.324–2.035, $P = 0.046$), and rs28730677 (OR = 0.821, 95% CI = 0.324–2.035, $P = 0.046$) in *PIK3CD* were significantly associated with HAPC. Furthermore, rs28381984 (OR = 0.761, 95% CI = 0.294–1.928, $P = 0.035$) in *COL4A3* was associated with HAPC in the Tibetan population. But in the Han population, we did not find any significant association. In addition, using haplotype analysis, two blocks were detected among the *PIK3CD* SNPs (Fig. 3): block 1 contains rs7518602, rs7516138, rs7516214, and rs11805716 and block 2 contains rs79190623, rs72633866, rs2230735, rs182137610, rs188191807, rs28730671, rs111888887, rs9430220, rs28730674, rs371870925, rs199962152, rs77571929, rs117226273, rs28730676, rs10864435, and rs28730677. Two blocks were detected among the *COL4A3* SNPs too (Fig. 4): block 1 contains rs10178458 and rs6436669 and block 2 contains rs55703767, rs10205042, rs34505188, rs11677877, and rs34019152. These SNPs within the same genes showed strong linkage in-between.

Table 2 Basic information of candidate SNPs in Han subjects

SNP_ID	Gene	Alleles A/B	Case (N)			Control (N)			OR (95% CI)	P	P1	P2
			AA	AB	BB	AA	AB	BB				
rs7518602	PIK3CD	C/T	4	16	50	0	6	23	3.899 (1.515–10.290)	0.961	0.261	0.425
rs7516138	PIK3CD	G/A	4	19	47	0	11	18	4.150 (1.615–10.989)	0.993	0.959	0.800
rs7516214	PIK3CD	G/A	4	19	46	0	11	18	4.070 (1.582–10.786)	0.993	0.924	0.829
rs11805716	PIK3CD	T/C	13	17	32	3	9	14	3.610 (1.335–10.014)	0.961	0.516	0.488
rs11806839	PIK3CD	G/C	7	11	12	2	6	6	2.800 (0.640–12.570)	0.993	0.627	0.616
rs79190623	PIK3CD	C/T	63	7	0	23	6	0	3.952 (1.533–10.448)	0.766	0.160	0.245
rs72633866	PIK3CD	G/A	64	4	0	22	6	0	5.107 (1.892–14.498)	0.236	0.033	0.014
rs2230735	PIK3CD	A/G	63	7	0	23	6	0	3.952 (1.533–10.448)	0.766	0.160	0.245
rs182137610	PIK3CD	A/C	63	7	0	24	5	0	3.961 (1.540–10.444)	0.961	0.321	0.510
rs188191807	PIK3CD	G/A	57	3	0	23	3	0	4.280 (1.564–12.125)	0.961	0.287	0.486
rs28730671	PIK3CD	C/T	65	5	0	23	6	0	3.808 (1.463–10.129)	0.509	0.061	0.129
rs111888887	PIK3CD	T/C	65	5	0	24	5	0	3.816 (1.471–10.117)	0.821	0.141	0.312
rs9430220	PIK3CD	T/C	37	28	4	9	18	2	5.044 (1.871–14.464)	0.245	0.081	0.029
rs28730674	PIK3CD	A/G	65	5	0	23	6	0	3.808 (1.463–10.129)	0.509	0.061	0.129
rs371870925	PIK3CD	T/C	65	3	0	24	5	0	4.016 (1.520–10.884)	0.509	0.050	0.112
rs199962152	PIK3CD	A/G	63	2	0	22	5	0	3.065 (1.118–8.502)	0.245	0.024	0.034
rs77571929	PIK3CD	T/C	64	5	0	23	6	0	3.737 (1.439–9.970)	0.509	0.065	0.132
rs117226273	PIK3CD	G/T	65	5	0	23	6	0	3.808 (1.439–10.129)	0.509	0.061	0.129
rs28730676	PIK3CD	T/C	65	5	0	23	6	0	3.808 (1.436–10.129)	0.509	0.061	0.129
rs10864435	PIK3CD	C/T	63	7	0	20	9	0	6.247 (2.235–18.787)	0.051	0.013	0.002
rs28730677	PIK3CD	G/A	65	5	0	24	5	0	3.816 (1.471–10.117)	0.821	0.141	0.312
rs10178458	COL4A3	T/C	0	9	61	0	4	25	4.116 (1.069–10.826)	0.993	0.900	0.803
rs6436669	COL4A3	A/G	0	9	61	0	4	25	4.116 (1.069–10.826)	0.993	0.900	0.803
rs80109666	COL4A3	G/A	49	19	2	22	7	0	3.984 (1.545–10.552)	0.993	0.433	0.707
rs55703767	COL4A3	G/T	53	15	2	21	7	1	4.088 (1.600–10.732)	0.993	0.732	0.762
rs10205042	COL4A3	C/T	2	15	53	1	7	21	4.073 (1.592–10.704)	0.993	0.732	0.869
rs34505188	COL4A3	G/A	46	19	5	18	11	0	4.214 (1.637–11.199)	0.993	0.787	0.587
rs11677877	COL4A3	A/G	46	19	5	17	12	0	4.187 (1.625–11.138)	0.993	0.997	0.725
rs34019152	COL4A3	G/A	46	19	5	18	11	0	4.214 (1.637–11.199)	0.993	0.787	0.587
rs28381984	COL4A3	C/T	27	32	11	9	16	4	4.160 (1.623–10.984)	0.993	0.711	0.599

SNP single-nucleotide polymorphism, *OR* odds ratio, *95% CI* 95% confidence interval, *P value* FDR-calculated *P* value, *P1* *P* value calculated by unconditional logistic regression analysis, *P2* *P* value adjusted for age

Discussion

Tibet covers a vast area with a harsh hypoxic natural environment. According to a report in 2006, approximately 12 million people permanently settled down in this region. This number constantly increases every year; the increase mainly comes from the Han population that are emigrating from plain areas [16]. HAPC is a serious disease that threatens the health of people in the plateau area, especially those who have emigrated from a low-altitude area. In the past, a large number of patients with HAPC have been investigated with a focus on the pathophysiologic mechanisms of this disease. Nevertheless, a lot of questions remain to be elucidated. With the completion of the human genome project, much research has been shifted to human genetic variation, which was one of NIH's Roadmap Initiatives for 2008 [17]. As we all know, in Tibet, in order to adapt to altitude hypoxia, the body increases the hemoglobin concentration to increase the efficiency of carrying oxygen, and this response is crucial for the Han population who adapt to live at high altitudes. Compared with the Han people, the Tibetan population keeps genetic adaptations; they can easily adapt to the high-altitude hypoxia environment. Several studies have indicated that natural selection associated with high-altitude adaptation appears to act on genes in the hypoxic response pathway

Gene	RS	OR (95% CI)		P value
PIK3CD	rs7518602	3.90 (1.52,10.29)		0.84
PIK3CD	rs7516138	4.15 (1.61,10.99)		0.8571
PIK3CD	rs7516214	4.07 (1.58,10.79)		0.8571
PIK3CD	rs11805716	3.61 (1.33,10.01)		0.84
PIK3CD	rs11806839	2.80 (0.64,12.57)		0.84
PIK3CD	rs79190623	3.95 (1.53,10.45)		0.6125
PIK3CD	rs72633866	5.11 (1.89,14.50)		0.2125
PIK3CD	rs2230735	3.95 (1.53,10.45)		0.6125
PIK3CD	rs182137610	3.96 (1.54,10.44)		0.84
PIK3CD	rs188191807	4.28 (1.56,12.12)		0.84
PIK3CD	rs28730671	3.81 (1.46,10.13)		0.3971
PIK3CD	rs111888887	3.82 (1.47,10.12)		0.6683
PIK3CD	rs9430220	5.04 (1.87,14.46)		0.2568
PIK3CD	rs28730674	3.81 (1.46,10.13)		0.3971
PIK3CD	rs371870925	4.02 (1.52,10.88)		0.3971
PIK3CD	rs199962152	3.07 (1.12,8.50)		0.2568
PIK3CD	rs77571929	3.75 (1.44,9.97)		0.3971
PIK3CD	rs117226273	3.81 (1.46,10.13)		0.3971
PIK3CD	rs28730676	3.81 (1.46,10.13)		0.3971
PIK3CD	rs10864435	6.25 (2.23,18.79)		0.0612
PIK3CD	rs28730677	3.82 (1.47,10.12)		0.6683
COL4A3	rs10178458	4.12 (1.61,10.83)		0.8571
COL4A3	rs6436669	4.12 (1.61,10.83)		0.8571
COL4A3	rs80109666	3.98 (1.54,10.55)		0.8571
COL4A3	rs55703767	4.09 (1.60,10.73)		0.8571
COL4A3	rs10205042	4.07 (1.59,10.70)		0.8686
COL4A3	rs34505188	4.21 (1.64,11.20)		0.84
COL4A3	rs11677877	4.19 (1.62,11.14)		0.8571
COL4A3	rs34019152	4.21 (1.64,11.20)		0.84
COL4A3	rs28381984	4.16 (1.62,10.98)		0.84

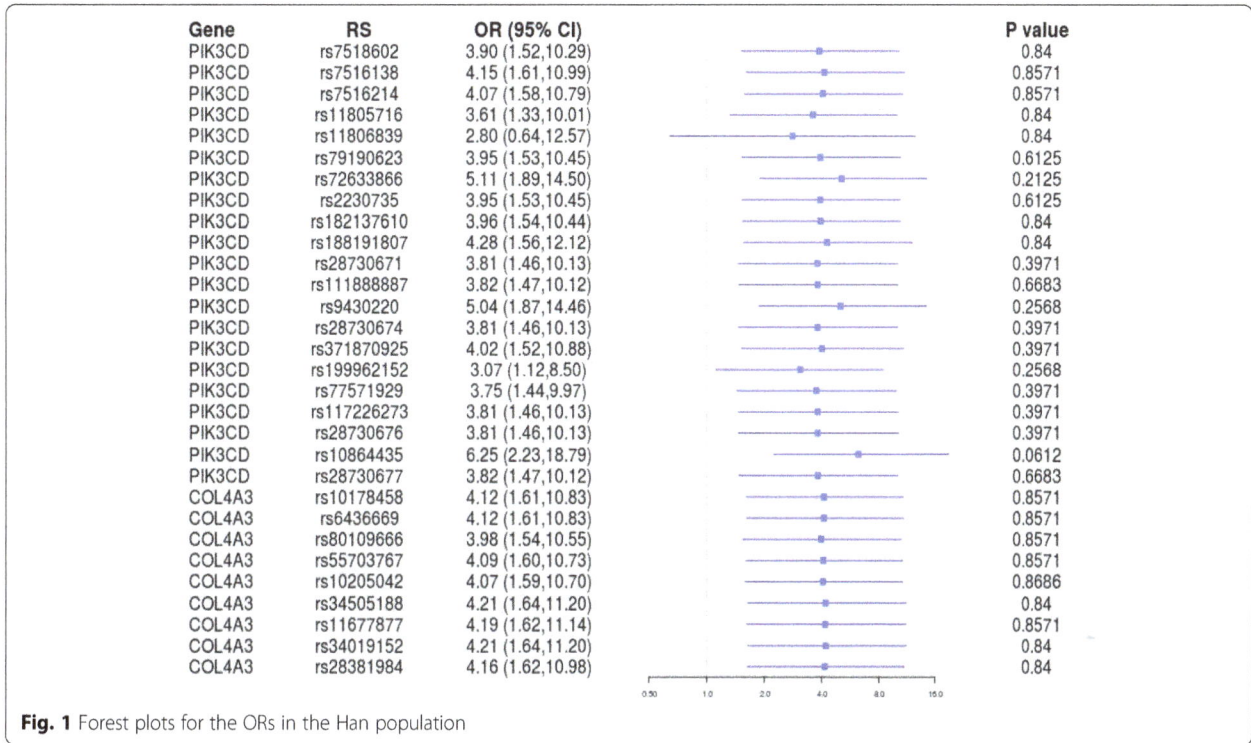

Fig. 1 Forest plots for the ORs in the Han population

to regulate erythrocyte production, possibly to prevent or reduce erythrocyte growth [18–22]. Recently, significant progress has been made in the study of the genetic basis of HAPC in Tibetans and Han, and some studies have confirmed that many genes are associated with HAPC. Namely, integrin subunit alpha 6 (ITGA6), erb-b2 receptor tyrosine kinase 4 (ERBB4), EPH receptor A2 (EPHA2), angiotensinogen (AGT), and endothelial PAS domain protein 1 (EPAS1) have been reported to play important roles in HAPC in Tibetans and Han [23–26]. Of these genes, EPHA2 can affect erythrocyte production by regulating EPO production and EPAS1 has been implicated as making the greatest contribution to genetic adaptation to high altitude and to the low Hb concentrations observed in the Tibetan population [18, 21]. Some research have shown that genetic variants selected for adaptation at extreme environmental conditions not only increase cancer risk later on age but may also be the downregulation of erythropoiesis in Tibetans in high altitude [22, 27, 28]. In our study, we found rs2230735, rs28730671, rs28730674, rs371870925, rs775 71929, rs117226273, rs28730676, rs28730677, and rs283 81984 in *PIK3CD* and *COL4A3* were significantly associated with decreased HAPC risk in the Tibetan population. However, in the Han population, we did not find any significant association. Inspired by the development of genome research and the genetic findings of high-altitude natives, we consider that genetic factors may be involved in the formation of such kind of disease.

Our study revealed an association of HAPC with SNPs in *PIK3CD* and *COL4A3* in the Tibetan population.

PIK3CD gene encodes P110δ catalytic subunit that is expressed predominantly in leukocytes and plays a vital role in the phosphoinositide 3-kinase (PI3K)/Akt signaling pathway. According to recent reports, p110δ contributes to the activation of Akt and cell proliferation in primary AML (acute myeloid leukemia) cells [29, 30]. PI3K signaling contributes to many processes, including cell cycle progression, proliferation and differentiation, survival, and migration [8, 9, 31]. PI3K/Akt pathways are critical to HIF-1α transcriptional activity in hypoxia. HIF-1, which is basically a heterodimer transcription factor, composed of HIF-1α and HIF-1β subunits, serves as a central regulator of metabolic adaptation to low oxygen [32]. The HIF-1α subunit is stabilized under hypoxia, translocating to the nucleus, forming a heterodimer with HIF-1β, and transactivating its target genes including EPO. HIF-1α is a factor that was originally thought to be bound to the 3′ enhancer region of the EPO genes, controlling for 100–200 genes that are involved in angiogenesis, glycolysis, and erythropoiesis [33]. The main organ for EPO production is the liver during the fetus stage, whereas it becomes the kidney after birth. However, there is small amount of expression in the other organs of the body, such as the brain, spleen, lungs, testis, and placenta. Further, it is a necessary glycoprotein, which does not only promote the maturation of red blood cell from erythroid progenitors but also mediates

Table 3 Basic information of candidate SNPs in Tibetan subjects

SNP_ID	Gene	Alleles A/B	Case (N)			Control (N)			OR (95% CI)	P	P1	P2
			AA	AB	BB	AA	AB	BB				
rs7518602	PIK3CD	C/T	5	29	36	1	10	19	1.043 (0.438–2.490)	0.612	0.240	0.238
rs7516138	PIK3CD	G/A	11	26	33	3	13	14	1.015 (0.428–2.411)	0.992	0.735	0.733
rs7516214	PIK3CD	G/A	11	26	32	3	13	14	0.992 (0.416–2.366)	0.982	0.688	0.691
rs11805716	PIK3CD	T/C	15	27	17	6	8	13	1.061 (0.416–2.718)	0.555	0.207	0.206
rs11806839	PIK3CD	G/C	5	7	4	2	9	3	1.205 (0.270–5.393)	0.893	0.593	0.612
rs79190623	PIK3CD	C/T	58	10	1	18	11	1	0.862 (0.349–2.106)	0.081	0.018	0.018
rs72633866	PIK3CD	G/A	57	8	1	25	2	0	1.071 (0.434–2.656)	0.791	0.366	0.367
rs2230735	PIK3CD	A/G	59	10	1	17	12	1	0.844 (0.337–2.079)	0.046	0.008	0.008
rs182137610	PIK3CD	A/C	57	10	1	18	9	1	0.764 (0.301–1.897)	0.150	0.050	0.045
rs188191807	PIK3CD	G/A	53	5	1	22	5	0	0.685 (0.266–1.726)	0.791	0.454	0.396
rs28730671	PIK3CD	C/T	63	6	1	19	10	1	0.821 (0.324–2.035)	0.046	0.007	0.007
rs111888887	PIK3CD	T/C	61	6	1	21	8	1	0.906 (0.368–2.205)	0.130	0.034	0.034
rs9430220	PIK3CD	T/C	29	34	5	12	14	4	0.940 (0.395–2.234)	0.851	0.536	0.535
rs28730674	PIK3CD	A/G	63	6	1	19	10	1	0.821 (0.324–2.035)	0.046	0.007	0.007
rs371870925	PIK3CD	T/C	64	4	1	20	9	1	0.812 (0.320–2.017)	0.046	0.007	0.007
rs199962152	PIK3CD	A/G	58	4	1	19	6	1	0.685 (0.256–1.766)	0.146	0.045	0.040
rs77571929	PIK3CD	T/C	63	5	1	19	10	1	0.814 (0.320–2.024)	0.046	0.005	0.005
rs117226273	PIK3CD	G/T	63	6	1	19	10	1	0.821 (0.324–2.035)	0.046	0.007	0.007
rs28730676	PIK3CD	T/C	63	6	1	19	10	1	0.821 (0.324–2.035)	0.046	0.007	0.007
rs10864435	PIK3CD	C/T	55	15	0	25	4	1	1.011 (0.423–2.417)	0.992	0.879	0.878
rs28730677	PIK3CD	G/A	63	6	1	19	10	1	0.821 (0.324–2.035)	0.046	0.007	0.007
rs10178458	COL4A3	T/C	4	21	45	4	7	19	1.026 (0.432–2.442)	0.851	0.539	0.537
rs6436669	COL4A3	A/G	4	21	45	4	7	19	1.026 (0.432–2.442)	0.851	0.539	0.537
rs80109666	COL4A3	G/A	60	9	1	28	2	0	1.070 (0.449–2.559)	0.655	0.271	0.267
rs55703767	COL4A3	G/T	47	21	2	18	8	4	0.990 (0.415–2.360)	0.524	0.184	0.184
rs10205042	COL4A3	C/T	0	23	47	4	8	18	1.017 (0.425–2.437)	0.305	0.102	0.102
rs34505188	COL4A3	G/A	42	20	8	10	14	6	0.948 (0.390–2.292)	0.115	0.028	0.028
rs11677877	COL4A3	A/G	42	21	7	10	13	7	0.941 (0.385–2.290)	0.070	0.013	0.013
rs34019152	COL4A3	G/A	42	20	8	10	13	7	0.928 (0.380–2.254)	0.081	0.018	0.018
rs28381984	COL4A3	C/T	24	39	7	23	6	1	0.761 (0.294–1.928)	0.035	0.001	0.001

The abbreviations were the same as Table 2

erythropoiesis. It is identified as an inducer of erythropoiesis and can promote excessive cell production. In addition, previous studies showed that inhibited PI3K/Akt signaling pathway led to decreased hematopoietic stem cell (HSC) proliferation. This suggests that such kind of pathway is important for HSC proliferation. HSC is involved in the formation of HAPC, expansion of the population, and enforcement of erythroid lineage-committed differentiation [34]. Therefore, we speculate that PIKCD may affect the generation of EPO and the decrease of HSC appreciation through the PI3K/Akt signaling pathway. In this study, we show here for the first time that the *PIK3CD* gene plays a crucial role in the production of erythrocyte, so *PIK3CD* has a significant influence on the formation of HAPC.

COL4A3 is an important risk gene for HAPC that is also linked to many diseases such as the Alport syndrome, focal segmental glomerulosclerosis, and type 2 diabetes [35–37]. Furthermore, it is important to the structure and function of various cell types and contributes to a variety of processes. Although the functional effects of the polymorphisms have not yet been elucidated fully, our current results show that the variants may have an effect on *COL4A3* expression or activity. Therefore, it may play an important role in modulating the susceptibility to HAPC. By searching the KEGG pathway database, we found that

Gene	RS	OR (95% CI)		P value
PIK3CD	rs7518602	1.04 (0.44,2.49)		0.3572
PIK3CD	rs7516138	1.01 (0.43,2.41)		0.7587
PIK3CD	rs7516214	0.99 (0.42,2.37)		0.7406
PIK3CD	rs11805716	1.06 (0.42,2.72)		0.3247
PIK3CD	rs11806839	1.20 (0.27,5.39)		0.6799
PIK3CD	rs79190623	0.86 (0.35,2.11)		0.045
PIK3CD	rs72633866	1.07 (0.43,2.66)		0.4999
PIK3CD	rs2230735	0.84 (0.34,2.08)		0.0257
PIK3CD	rs182137610	0.76 (0.30,1.90)		0.0834
PIK3CD	rs188191807	0.68 (0.27,1.73)		0.5159
PIK3CD	rs28730671	0.82 (0.32,2.03)		0.0255
PIK3CD	rs111888887	0.91 (0.37,2.20)		0.072
PIK3CD	rs9430220	0.94 (0.40,2.23)		0.6196
PIK3CD	rs28730674	0.82 (0.32,2.03)		0.0255
PIK3CD	rs371870925	0.81 (0.32,2.02)		0.0255
PIK3CD	rs199962152	0.68 (0.26,1.77)		0.081
PIK3CD	rs77571929	0.81 (0.32,2.02)		0.0255
PIK3CD	rs117226273	0.82 (0.32,2.03)		0.0255
PIK3CD	rs28730676	0.82 (0.32,2.03)		0.0255
PIK3CD	rs10864435	1.01 (0.42,2.42)		0.8777
PIK3CD	rs28730677	0.82 (0.32,2.03)		0.0255
COL4A3	rs10178458	1.03 (0.43,2.44)		0.6196
COL4A3	rs6436669	1.03 (0.43,2.44)		0.6196
COL4A3	rs80109666	1.07 (0.45,2.56)		0.3813
COL4A3	rs55703767	0.99 (0.42,2.36)		0.3074
COL4A3	rs10205042	1.02 (0.43,2.44)		0.1794
COL4A3	rs34505188	0.95 (0.39,2.29)		0.0638
COL4A3	rs11677877	0.94 (0.38,2.29)		0.039
COL4A3	rs34019152	0.93 (0.38,2.25)		0.045
COL4A3	rs28381984	0.76 (0.29,1.93)		0.0193

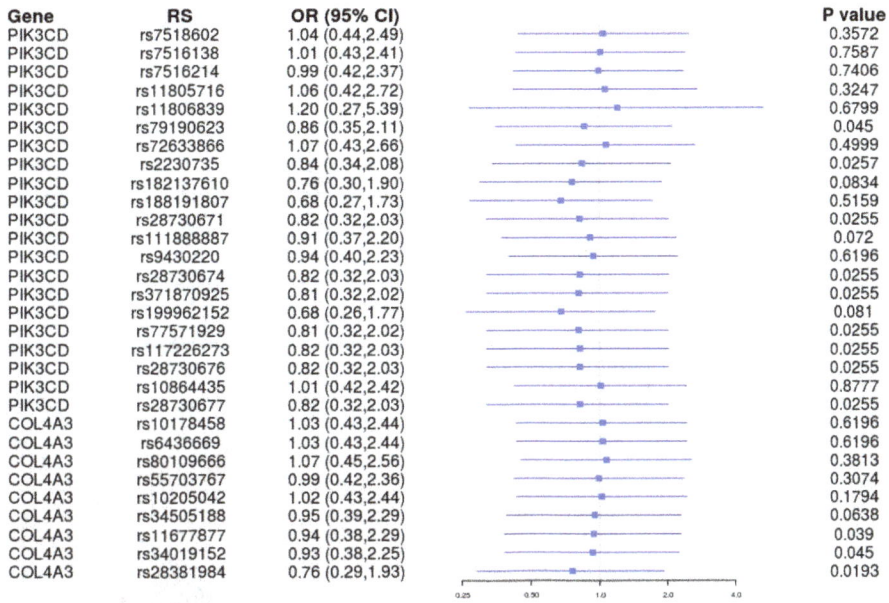

Fig. 2 Forest plots for the ORs in the Tibetan population

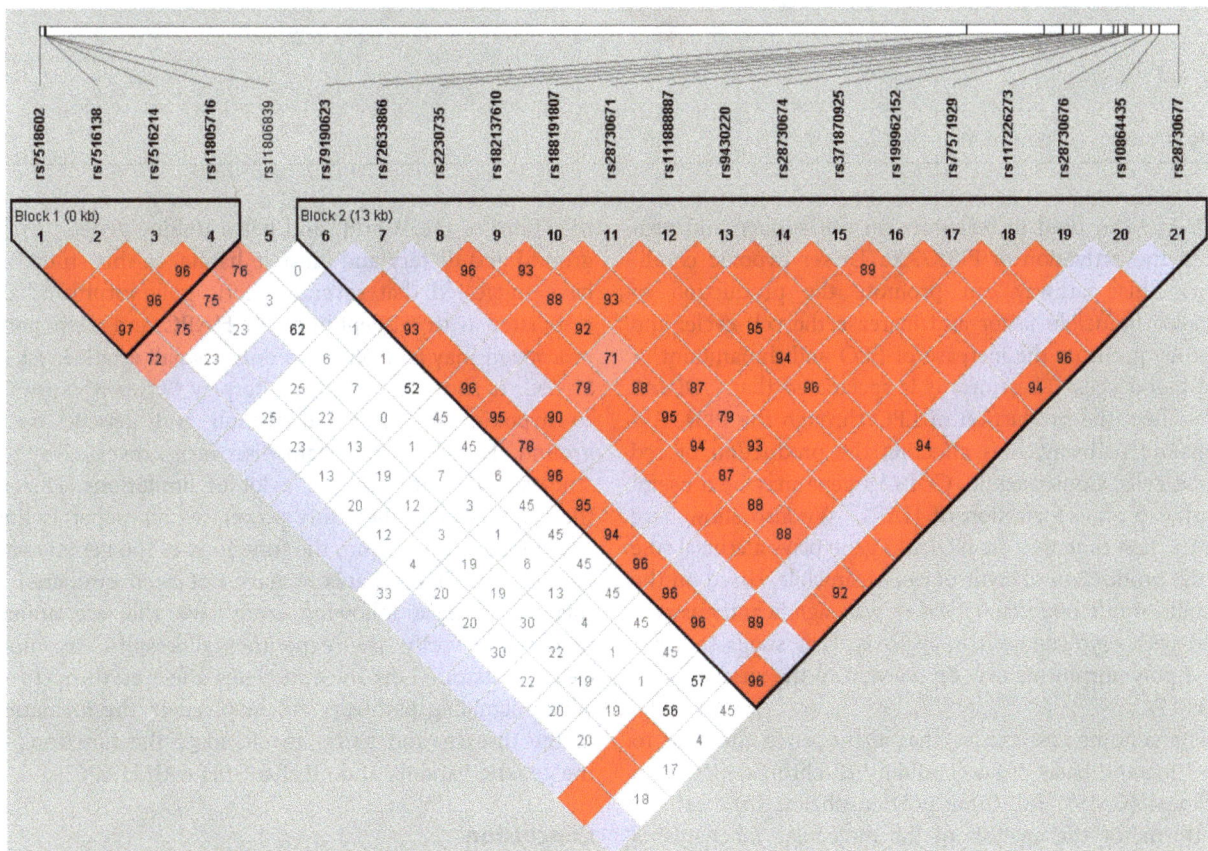

Fig. 3 Haplotype block map for the 15 *PIK3CD* SNPs

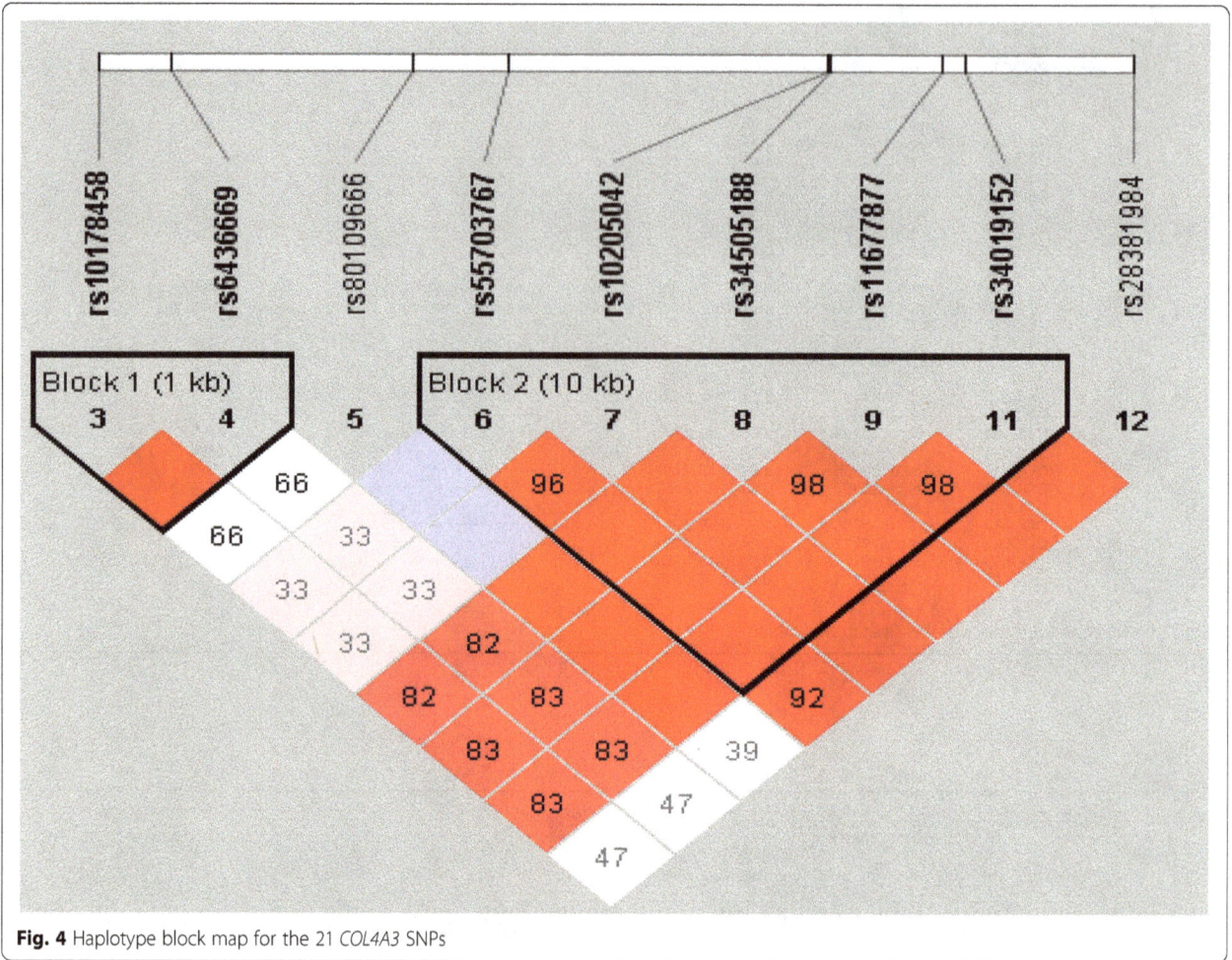

Fig. 4 Haplotype block map for the 21 *COL4A3* SNPs

COL4A3 can bind to receptors on cell surface and promote the activation of PI3K/Akt. Under hypoxic conditions, this pathway can promote the production of hypoxia-inducible factor and increase the cell cycle, and thereby promote the increase of EPO and the amount of red blood cells. Therefore, it is speculated that *COL4A3* may affect the production of EPO through the PI3K/Akt signaling pathway, thus affecting the production of red blood cells. Consequently, *COL4A3* gene may be a useful marker for the formation of HAPC. We also show here for the first time that the *COL4A3* gene plays a crucial role in the production of erythrocyte. Meanwhile, based on the results of our research, *COL4A3* was significantly associated with erythropoiesis in hypoxia. It is suggested that gene polymorphisms may be relevant to the susceptibility to HAPC.

The genome research era has also opened the road to studying the basis of susceptibility to chronic mountain sickness (CMS) [38]. Gene polymorphisms have set the platform for the analysis of the molecular mechanisms of adaptation to life at high altitudes [39]. Tibet, an average elevation above 4000 m, is commonly regarded as

the "Roof of the World" and has a unique genetic background and dietary and lifestyle habits. In this study, we have suggested that several genetic polymorphisms are associated with susceptibility to HAPC and each polymorphism may contribute to only a small relative risk of HAPC. It shows a complex interplay between exposure to hypoxic environmental stimuli and genetic background. There are important discoveries revealed by the studies, but there are still a lot of limitations. Due to these limitations, the study power of this paper is limited. On the other hand, the functions of the genetic variants and their mechanisms have not been evaluated in this study. In a following study, we will use animal models to verify the experimental results, to more clearly illustrate how the two genes affect erythrocytosis, which signaling pathways are involved in the formation of the disease, and to try to elucidate the functions of the genetic variants and mechanisms with HAPC.

Conclusion

We analyzed SNPs in *PIK3CD* and *COL4A3* and identified a relationship between genetic polymorphisms and

HAPC in the Tibetan people. This study sets out to improve the quality of life of people living in the Qinghai-Tibet Plateau, determines paramount insights into the etiology of HAPC, and may provide more guidance for such people with regard to prolonged and healthy living. However, additional genetic risk factors and functional investigations should be identified in order to further confirm our results.

Materials and methods
Study populations
A total of 140 patients (70 Han and 70 Tibetan) with HAPC and 60 healthy control subjects (30 Han and 30 Tibetan) were recruited for a case-control association study. The 200 subjects who participated in this research had resided at an altitude of above 4000 m, and these samples were collected from the General Hospital in Tibet Military Region and the second People's Hospital of Tibet Autonomous Region. Written informed consent was obtained from each individual. Patients met the diagnostic criteria for HAPC, i.e., males with hemoglobin ≥ 21 g/dL or females with hemoglobin ≥ 19 g/dL, and had no high-altitude cerebral edema and chronic respiratory disorders or secondary polycythemia due to hypoxemia caused by certain chronic diseases. Moreover, subjects have no endocrinological, nutritional, and metabolic diseases. Healthy individuals were randomly selected as controls. The experimental protocol was established by the Ethics Committee of the Xizang Minzu University.

Epidemiological and clinical data
We used a standardized epidemiological questionnaire to collect demographic and clinical data, including information on gender, age, residential region, ethnicity, family history of cancer, and education status. Furthermore, the patient information was collected through physicians or from medical chart review. All participants signed informed consent, and 5 ml of peripheral blood was taken from each participant in this study.

Selection of SNPs and methods of genotyping
Thirty SNPs from *PIK3CD* and *COL4A3* were chosen for analysis in this study, including 21 SNPs in *PIK3CD* and 9 SNPs in *COL4A3* with minor allele frequency (MAF) > 0.05 in the Asian population HapMap database, and SNP genotyping was performed utilizing Illumina sequencing platform for exon sequencing of PIKCD and COL4A3. Because the genetic background of Han and Tibetan populations has not been compared yet, we selected these two candidate genes based on their relations to the oxygen metabolism in red blood cells, which were related to high-altitude adaptation in the Chinese Han and Tibetan populations.

Statistical analysis
The data were analyzed using an R program, Haploview, and Excel. Unconditional logistic regression analysis was used to calculate odds ratios (ORs), 95% confidence intervals (CIs), and P values for comparisons between cases and controls. Multiple comparisons were corrected using FDR, and FDR-corrected $P < 0.05$ was considered to indicate a significant difference.

Abbreviations
95%CI: 95% confidence intervals; AGT: Angiotensinogen; AML: Acute myeloid leukemia; CMS: Chronic mountain sickness; COL4A3: Collagen type IV α3 chain gene; ECM: Extracellular matrix; EPAS1: Endothelial PAS domain protein 1; EPHA2: EPH receptor A2; ERBB4: Erb-b2 receptor tyrosine kinase 4; FDR: False discovery rate; HAPC: High-altitude polycythemia; HIF-1α: Hypoxia-inducible factors-1α; HSC: Hematopoietic stem cell; ITGA6: Integrin subunit alpha 6; OR: Odds ratio; PIK3CD: Catalytic subunit delta gene

Acknowledgements
We are grateful to those who collected samples in the Tibetan Plateau, and we thank those who participated in the study and worked hard on the research. This work was supported by the National Natural Science Foundation of China (No. 31460286; 31660307; 31260252; 31330038), the Natural Science Foundation of Xizang (Tibet) Autonomous Region (No. Z2014A09G2-3), the Innovation Support Program for Young Teachers of Tibet Autonomous Region (No. QCZ2016-27; QCZ2016-29; QCZ2016-34), and the Science and Technology Department Project of Tibet Autonomous Region (No. 2016ZR-MQ-06; 2015ZR- 13-19).

Funding
This work was supported by the National Natural Science Foundation of China (No. 31460286; 31660307; 31260252; 31330038), the Natural Science Foundation of Xizang (Tibet) Autonomous Region (No. Z2014A09G2-3), the Innovation Support Program for Young Teachers of Tibet Autonomous Region (No. QCZ2016-27; QCZ2016-29; QCZ2016-34), and the Science and Technology Department Project of Tibet Autonomous Region (No. 2016ZR-MQ-06; 2015ZR- 13-19).

Authors' contributions
LK conceived and designed the study, supervised the project, and drafted the manuscript. XF participated in the design of study and data analysis and helped to draft the manuscript. LM, ZZhang, YL, MH, LL, YZ, FL, PC, and YL contributed to the sample collection and experiments. XL helped in the English language editing. ZZhao contributed to sample collection and experiments. All authors have read and approved the final manuscript.

Consent for publication
Not applicable

Competing interests
The authors declare that they have no competing interests.

Author details
[1]Key Laboratory for Molecular Genetic Mechanisms and Intervention Research on High Altitude Disease of Tibet Autonomous Region, School of Medicine, Xizang Minzu University, Xianyang 712082, Shaanxi, China. [2]Key Laboratory of High Altitude Environment and Genes Related to Diseases of Tibet Autonomous Region, School of Medicine, Xizang Minzu University, Xianyang 712082, Shaanxi, China. [3]Ministry of Education Key Laboratory of

Contemporary Anthropology, Collaborative Innovation Center for Genetics and Development, School of Life Sciences, Fudan University, Shanghai 200433, China. [4]Division of Human Genetics, Department of Psychiatry, Yale University School of Medicine, New Haven, CT 06510, USA. [5]Six Industrial Research Institute, Fudan University, Shanghai 200433, China.

References

1. Leon-Velarde F, et al. Consensus statement on chronic and subacute high altitude diseases. High Alt Med Biol. 2005;6(2):147–57.

2. Otten, E.J., High altitude: an exploration of human adaptation.: edited by Hornbein TF and Schoene RB. New York, Marcel Dekker, Inc. 2001, 982 pages, $235. J Emerg Med, 2003. 25(3): p. 345-346.

3. Reeves JT, Leon-Velarde F. Chronic mountain sickness: recent studies of the relationship between hemoglobin concentration and oxygen transport. High Altitude Medicine & Biology. 2004;5(2):147.

4. Guan W, et al. Sleep disturbances in long-term immigrants with chronic mountain sickness: a comparison with healthy immigrants at high altitude. Respir Physiol Neurobiol. 2015;206:4–10.

5. Jiang C, et al. Gene expression profiling of high altitude polycythemia in Han Chinese migrating to the Qinghai-Tibetan plateau. Mol Med Rep. 2012; 5(1):287–93.

6. Wu TY. Chronic mountain sickness on the Qinghai-Tibetan plateau. Chin Med. 2005;118(2):161–8.

7. Vanhaesebroeck B, et al. P110delta, a novel phosphoinositide 3-kinase in leukocytes. Proc Natl Acad Sci U S A. 1997;94(9):4330.

8. Chantry D, et al. p110delta, a novel phosphatidylinositol 3-kinase catalytic subunit that associates with p85 and is expressed predominantly in leukocytes. J Biol Chem. 1997;272(31):19236.

9. Kok K, Geering B, Vanhaesebroeck B. Regulation of phosphoinositide 3-kinase expression in health and disease. Trends Biochem Sci. 2009;34(3):115.

10. Gaber T, et al. Hypoxia inducible factor (HIF) in rheumatology: low O2! See what HIF can do! Ann Rheum Dis. 2005;64(7):971.

11. Dayan F, et al. A dialogue between the hypoxia-inducible factor and the tumor microenvironment. Cancer Microenvironmen. 2008;1(1):53–68.

12. Tanaka T, Nangaku M. Recent advances and clinical application of erythropoietin and erythropoiesis-stimulating agents. Exp Cell Res. 2012; 318(9):1068–73.

13. Stabuc-Silih M, et al. Polymorphisms in COL4A3 and COL4A4 genes associated with keratoconus. Mol Vis. 2009;15(300-01):2848–60.

14. Osman OS, et al. A novel method to assess collagen architecture in skin. Bmc Bioinformatics. 2013;14(1):1–10.

15. Khan T, et al. Metabolic dysregulation and adipose tissue fibrosis: role of collagen VI. Mol Cell Biol. 2009;29(6):1575–91.

16. Wu T, Kayser B. High altitude adaptation in Tibetans. High Alt Med Biol. 2006;7(3):193.

17. Pennisi, E., Research funding. Are epigeneticists ready for big science? Science, 2008. 319(5867): p. 1177.

18. Yi X, et al. Sequencing of fifty human exomes reveals adaptation to high altitude. Science. 2010;329(5987):75–8.

19. Crawford JE, et al. Natural selection on genes related to cardiovascular health in high-altitude adapted Andeans. Am J Hum Genet. 2017;101(5): 752–67.

20. Simonson TS, et al. Genetic evidence for high-altitude adaptation in Tibet. Science. 2010;329(5987):72–5.

21. Beall CM, et al. Natural selection on EPAS1 (HIF2alpha) associated with low hemoglobin concentration in Tibetan highlanders. Proc Natl Acad Sci U S A. 2010;107(25):11459–64.

22. Bigham A, et al. Identifying signatures of natural selection in Tibetan and Andean populations using dense genome scan data. PLoS Genet. 2010;6(9): e1001116.

23. Xu J, et al. EPAS1 gene polymorphisms are associated with high altitude polycythemia in Tibetans at the Qinghai-Tibetan Plateau. Wilderness Environ Med. 2015;26(3):288–94.

24. Chen Y, et al. An EPAS1 haplotype is associated with high altitude polycythemia in male Han Chinese at the Qinghai-Tibetan plateau. Wilderness Environ Med. 2014;25(4):392–400.

25. Zhao Y, et al. Associations of high altitude polycythemia with polymorphisms in EPAS1, ITGA6 and ERBB4 in Chinese Han and Tibetan populations. Oncotarget. 2017;8(49):86736.

26. Liu L, et al. Associations of high altitude polycythemia with polymorphisms in EPHA2 and AGT in Chinese Han and Tibetan populations. Oncotarget. 2017;8(32):53234–43.

27. Macinnis MJ, Koehle MS, Rupert JL. Evidence for a genetic basis for altitude illness: 2010 update. High Alt Med Biol. 2010;11(4):349–68.

28. Voskarides K. Combination of 247 genome-wide association studies reveals high cancer risk as a result of evolutionary adaptation. Mol Biol Evol. 2017; 35(2):473–85.

29. Billottet C, et al. A selective inhibitor of the p110|[delta]| isoform of PI 3-kinase inhibits AML cell proliferation and survival and increases the cytotoxic effects of VP16. Oncogene. 2006;25(50):6648.

30. Sujobert P, et al. Essential role for the p110delta isoform in phosphoinositide 3-kinase activation and cell proliferation in acute myeloid leukemia. Blood. 2005;106(3):1063.

31. Vanhaesebroeck B, et al. p110δ, a novel phosphoinositide 3-kinase in leukocytes. Proc Natl Acad Sci U S A. 1997;94(9):4330.

32. Zhou J, et al. PI3K/Akt is required for heat shock proteins to protect hypoxia-inducible factor 1α from pVHL-independent degradation. J Biol Chem. 2004;279(14):13506–13.

33. Tanaka T, Nangaku M. Recent advances and clinical application of erythropoietin and erythropoiesis-stimulating agents. Exp Cell Res. 2012; 318(9):1068.

34. Li P, et al. Regulation of bone marrow hematopoietic stem cell is involved in high-altitude erythrocytosis. Exp Hematol. 2011;39(1):37–46.

35. Jingyuan X, et al. COL4A3 mutations cause focal segmental glomerulosclerosis. J Mol Cell Biol. 2014;6(6):498–505.

36. Guo L, et al. Mutation analysis of COL4A3 and COL4A4 genes in a Chinese autosomal-dominant Alport syndrome family. J Genet. 2017;96(2):389.

37. Saravani S, et al. Association of COL4A3 (rs55703767), MMP-9 (rs17576) and TIMP-1 (rs6609533) gene polymorphisms with susceptibility to type 2 diabetes. Biomed Rep. 2017;6(3):329–34.

38. Thomas PK, et al. Neurological manifestations in chronic mountain sickness: the burning feet-burning hands syndrome. J Neurol Neurosurg Psychiatry. 2000;69(4):447–52.

39. Cohen J. DNA duplications and deletions help determine health. Science. 2007;317(5843):1315–7.

Architecture of polymorphisms in the human genome reveals functionally important and positively selected variants in immune response and drug transporter genes

Yu Jin[1,3], Jingbo Wang[2], Maulana Bachtiar[2,3], Samuel S. Chong[4] and Caroline G. L. Lee[1,2,3,5*] (iD)

Abstract

Background: Genetic polymorphisms can contribute to phenotypic differences amongst individuals, including disease risk and drug response. Characterization of genetic polymorphisms that modulate gene expression and/or protein function may facilitate the identification of the causal variants. Here, we present the architecture of genetic polymorphisms in the human genome focusing on those predicted to be potentially functional/under natural selection and the pathways that they reside.

Results: In the human genome, polymorphisms that directly affect protein sequences and potentially affect function are the most constrained variants with the lowest single-nucleotide variant (SNV) density, least population differentiation and most significant enrichment of rare alleles. SNVs which potentially alter various regulatory sites, e.g. splicing regulatory elements, are also generally under negative selection.

Interestingly, genes that regulate the expression of transcription/splicing factors and histones are conserved as a higher proportion of these genes is non-polymorphic, contain ultra-conserved elements (UCEs) and/or has no non-synonymous SNVs (nsSNVs)/coding INDELs. On the other hand, major histocompatibility complex (*MHC*) genes are the most polymorphic with SNVs potentially affecting the binding of transcription/splicing factors and microRNAs (miRNA) exhibiting recent positive selection (RPS). The drug transporter genes carry the most number of potentially deleterious nsSNVs and exhibit signatures of RPS and/or population differentiation. These observations suggest that genes that interact with the environment are highly polymorphic and targeted by RPS.

Conclusions: In conclusion, selective constraints are observed in coding regions, master regulator genes, and potentially functional SNVs. In contrast, genes that modulate response to the environment are highly polymorphic and under positive selection.

Keywords: Single-nucleotide variant, Natural selection, Potentially functional SNV, Immune response genes, Drug transporters

* Correspondence: bchleec@nus.edu.sg
[1]NUS Graduate School for Integrative Sciences and Engineering, National University of Singapore, Singapore 117456, Singapore
[2]Department of Biochemistry, National University of Singapore, Singapore 119077, Singapore
Full list of author information is available at the end of the article

Background

Genetic polymorphisms may contribute to the differences in disease risks and drug responses amongst different individuals. Different forms of genetic variants are found in the human genome. Single-nucleotide variants (SNVs) account for more than 90% of genomic variants and are the major form of genetic polymorphisms [1].

Some polymorphisms can affect phenotype. These polymorphisms are likely to alter gene expression or protein function leading to modulation of cellular function and influencing disease risk or drug response. However, to identify the single or a group of causal variants for a particular phenotype from a pool of more than 100 million polymorphisms is like 'finding a needle in a haystack' and remains a great challenge since not all genetic variants are functionally important.

While non-synonymous SNVs (nsSNVs) have been extensively investigated as they are the most likely to modulate phenotypes via changing the amino acid composition of proteins, synonymous SNVs (sSNVs) and non-coding variants can also account for phenotypic differences since these variants can affect mRNA stability and transcriptional or translational efficiency and have been associated with gene expression levels in various cell lines and tissues [2–10]. While it may not be feasible to experimentally test every single polymorphism for its function, a variety of bioinformatics tools is now available. These tools can reasonably predict the potential functions of genetic variants, including the likelihood of nsSNVs to disrupt protein structures and/or functions [11–19], SNVs that potentially modify splicing [20, 21] or transcription [22], and SNVs in 3′ untranslated regions (3′UTRs) with potential to alter miRNA target sites [23–25]. There are also comprehensive web tools for predicting various potential functions of both regulatory and coding SNVs, e.g. pfSNP [26] and PupaSNP finder [27]. They can facilitate our understanding of how polymorphisms can lead to phenotype change and help us prioritize the potentially functional SNVs (pfSNVs) for further investigation.

In addition to the above-mentioned predictive bioinformatics tools, signatures of natural selections can also facilitate the identification of causal variants since variants under natural selection are likely to be functionally significant. Patterns of population differentiation were employed to identify 174 candidate gene loci showing signatures of purifying or positive selection [28]. 'Long-range haplotype' methods have been employed to identify a list of targets under recent positive selection (RPS) [29]. Another study utilizing HapMap Phase II data found that negative selection preferentially targets non-synonymous sites, while both non-synonymous and 5′ untranslated regions (5′UTRs) show an excess of highly differentiated SNVs, suggesting the evidence of

positive selection as well. The authors also reported that variants under selective pressures (either positive or negative) occur more frequently in disease-related genes and are more likely to contribute to disease phenotypes [30].

Although previous reports examined the association of SNVs in regulatory regions with natural selection, these studies were limited. They either merely focussed on only one class of regulatory SNVs (e.g. SNVs within miRBS) [31], on SNVs residing in non-coding regions [32] or within regulatory elements [33] without predicting whether these SNVs alter function (e.g. if a SNV will abolish or create a regulatory site).

In this study, we present the architecture of all genetic polymorphisms of the human genome, focusing on SNVs that are potentially functional and/or positively selected and the pathways that they reside.

Results

Polymorphisms are most constrained in coding regions

Of the > 14 million polymorphisms in the human genome validated in the dbSNV database (Build 131), 38% of the polymorphisms are within the protein-coding genes while 62% resides in the intergenic regions. More than 95% of the variants within human genes reside within introns (Fig. 1a). Coding polymorphisms constitute ~ 3% of the total polymorphisms within genes, of which 2.55% are SNVs while 0.35% are short insertion/deletions (INDELs) (Fig. 1a). Upon normalization against the length of each genic region, coding regions contain the lowest average densities of both SNVs and INDELs (Fig. 1b). Notably, frame-shift INDELs (i.e. length of INDELs is not in multiples of three) are significantly under-represented in the coding regions compared to non-coding regions in the human genes (p value < 0.001 by Fisher's exact test, Fig. 1c). These data suggest that both SNVs and INDELs are selectively constrained within coding sequences, especially the INDELs with potential to cause frame-shift.

To further investigate the regions within genes that may be most subjected to negative selection pressure, the derived allele frequencies (DAFs) of SNVs in different regions are further compared using allele frequency data of the International HapMap Project individuals. As evident in Fig. 1d, coding regions (red) contain a higher percentage of rare SNVs, defined as having DAF < 0.05 in all the three population groups, namely African, East Asian and European populations. nsSNVs (brown) within the coding region are also enriched with rare alleles compared to sSNVs (orange) (Fig. 1e). As negative selection increases the fraction of rare alleles [34], our results from the analysis of allele frequency data again suggest that coding SNVs, especially nsSNVs, tend to be targeted by negative selection.

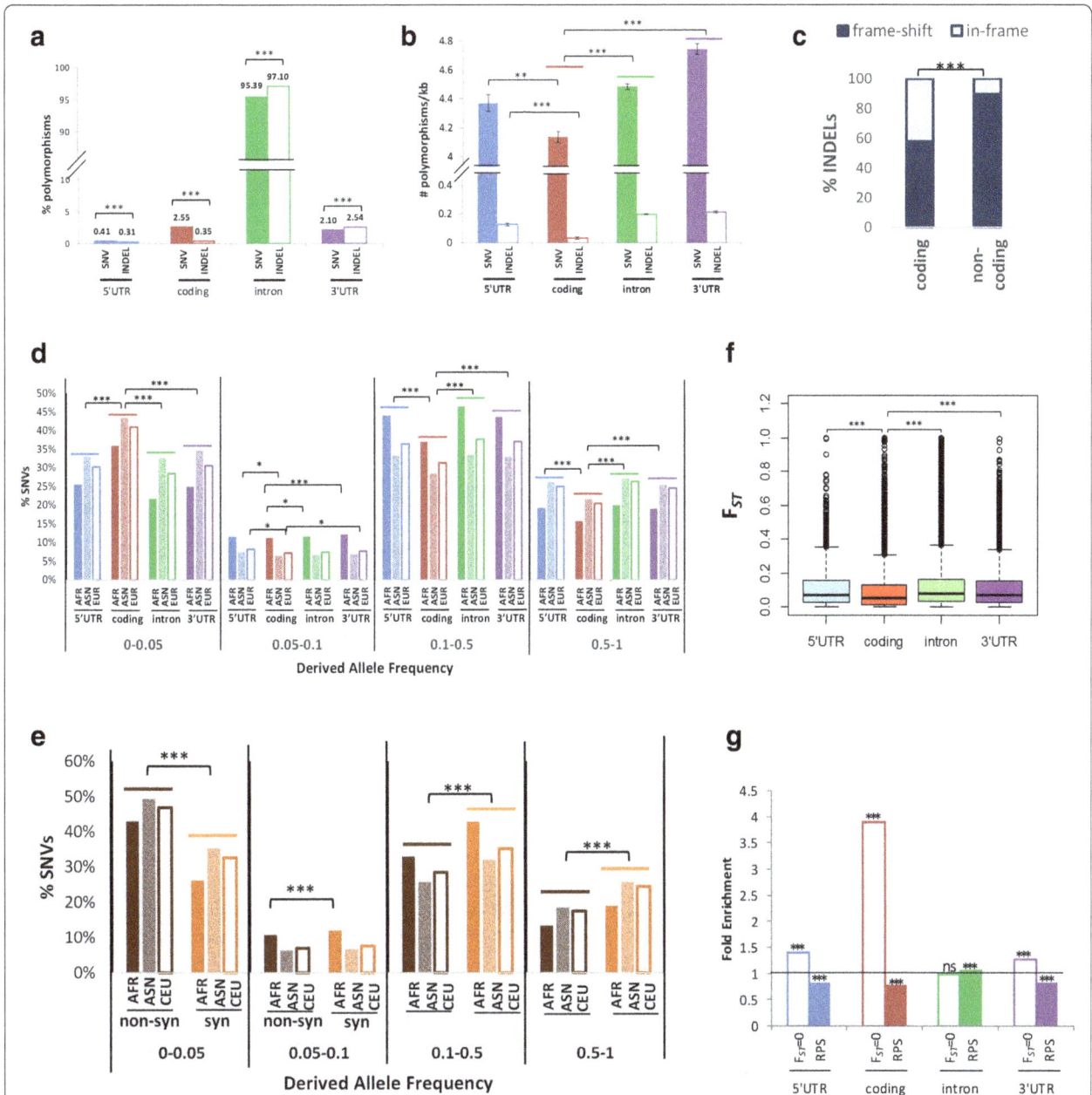

Fig. 1 Architecture of polymorphisms in the human genome. **a** Percentage of polymorphisms in different regions (5'UTRs, coding regions, introns, 3'UTRs) of the human genes. **b** Average SNV and INDEL densities (# polymorphisms/kb) in the different regions (5'UTR, coding region, intron and 3'UTR) of a gene in the human genome. Error bars represent the standard errors of the mean SNV and INDEL densities. **c** Percentage of frame-shift and in-frame INDELs in coding and non-coding regions in human genes. Frame-shift INDELS are defined as INDELs whose lengths are not in multiples of three while in-frame INDELs are those whose lengths are in multiples of three. **d** Percentage of SNVs with different DAFs in the four genic regions, as measured in HapMap individuals from African, Asian and European populations. **e** Percentage of synonymous and non-synonymous variants with different DAFs, as measured in HapMap individuals from African, Asian and European population groups. **f** Distribution of F_{ST} statistics in four genic regions. SNVs in coding regions show significantly lower median F_{ST} compared to the other non-coding regions. **g** Fold enrichment of SNVs showing signatures of negative selection ($F_{ST} = 0$) (open bar) or RPS (shaded bar) in the genic regions. Fold enrichment is determined by the percentage of SNVs with $F_{ST} = 0$ or under RPS in a specific region (e.g. coding region) divided by the percentage of all genotyped SNVs in that region. Coding, coding region; non-syn, non-synonymous; syn, synonymous. AFR, African; ASN, Asian; EUR, European. ***$p < 0.001$, **$p < 0.01$, *$p < 0.05$; ns, not significant

Signatures of natural selection are also examined through determining population differentiation using the F_{ST} statistics [28] across the different population groups (African, East Asian and European) since high F_{ST} is associated with a positive selection [34], while low F_{ST} is associated with a negative selection [30]. As shown in

Fig. 1f, coding SNVs have lower median F_{ST} than SNVs in other regions including 5′UTRs, 3′UTRs and introns (Bonferroni corrected p values < 0.001 by Mann-Whitney test). In fact, zero-F_{ST} SNVs are significantly over-represented in coding exons (Bonferroni corrected p value < 0.001 by Fisher's exact test) (Fig. 1g, non-shaded bars). Patterns of RPS are examined using linkage disequilibrium (LD) and haplotype-based methods. As shown in Fig. 1g (shaded bars), exonic regions, i.e. 5′ UTRs, coding regions and 3′UTRs, are significantly less enriched with RPS SNVs (Bonferroni corrected p values < 0.001 by Fisher's exact test), while introns are more enriched with RPS SNVs (Bonferroni corrected p value < 0.001 by Fisher's exact test).

Taken together, coding regions are generally under strong negative selection pressures as they show the lowest densities of SNVs and INDELs (especially frame-shift INDELs), the highest proportion of rare alleles with less enrichment of RPS SNVs. Notably, coding SNVs are also the least population differentiated.

Potentially functional SNVs are under natural selections

The putative functions of SNVs in the various genic and promoter regions are predicted using a variety of bioinformatics algorithms (see Additional file 1: Supplementary Methods). Approximately four hundred thousand (7%) pfSNVs in genic and promoter regions can potentially modulate gene expression and/or function. More than 93% of genes in the human genome contain at least one pfSNV (Fig. 2a). Each gene is predicted to contain an average of seven promoter SNVs capable of altering transcription factor binding sites (TFBS); eight intronic and two coding SNVs that may modulate splicing regulatory elements, i.e. intronic splicing regulatory element (ISRE) and exon splicing enhancers or silencers (ESE/ESS); and one coding SNV that is potentially deleterious to protein function and one SNV in 3′UTR that may alter miRNA binding site(s) (miRBS) (Fig. 2b).

To evaluate if pfSNVs are selectively constrained, the proportions of rare alleles (DAF < 0.05) of pfSNVs and non-functional SNVs (nfSNVs) in a specified region are compared. As evident in Fig. 2c, except for pfSNVs predicted to alter TFBS, most pfSNVs are enriched with rare alleles. Notably, pfSNVs predicted to be deleterious to protein function are more than 1.5-fold more enriched with rare alleles compared to nfSNVs in coding regions indicating that these pfSNVs may be under the strongest negative selection pressure.

Conversely, pfSNVs predicted to be deleterious to protein function are found to be the least significantly enriched with RPS SNVs (Fig. 2d, brown) (Bonferroni corrected p value < 0.001 by Fisher's exact test) consistent with the earlier observation indicating that these pfSNVs are under the strongest negative selection. The

other pfSNVs are not under any significant RPS except for pfSNVs predicted to alter ISRE (Bonferroni corrected p value = 0.019 by Fisher's exact test) (Fig. 2d). In addition, more than half of the RPS pfSNVs are predicted to affect ISRE (~ 53%) followed by TFBS (~ 30%) while least RPS pfSNVs (3%) are predicted to be deleterious to protein function (Fig. 2e).

Highly polymorphic vs conserved genes in the human genome

Amongst > 20,000 genes in the human genome, beta haemoglobin (*HBB*) gene is the most polymorphic gene, containing approximately 176 SNVs per kilobase (kb) with the highest density of SNVs within its coding region (Fig. 3a, red) (570 SNVs/kb). Several other haemoglobin genes (in green boxes) are also amongst the most polymorphic genes in the human genome with the majority of their SNVs residing within coding exons (red). Other highly polymorphic genes include the *MHC* family of genes (blue box) with most of their SNVs residing within introns (Fig. 3a, green) as well as the olfactory receptor (*OR*) gene family (orange box) where all the SNVs are also found within the coding region (Fig 3a, red).

The density of SNVs within each gene, normalized against their length, is determined for all > 20,000 protein-coding genes in the human genome. Most genes have approximately four SNVs per kilobase. Although ~ 97% of genes carry at least one SNV, 149 genes do not contain any polymorphism as SNV or INDEL. More than half of these 149 non-polymorphic genes were yet to be annotated. Nonetheless, the annotated non-polymorphic genes are significantly over-represented in histone 2A and 2B families and involved in nucleosome assembly (Fig. 3b, grey shaded; Additional file 1: Table S1).

We then focus on polymorphisms within the coding region since this region encodes the functional protein. Nearly 20% (4389) of genes in the human genome are found to be functionally conserved with no nsSNVs nor coding INDELs. These genes are enriched in various categories including GTPases and translational elongation (Fig. 3b, unshaded; Additional file 1: Table S2). Seventy genes are found to carry ultra-conserved elements (UCEs) [35] in their coding regions; hence, these genes are evolutionarily conserved. These ultra-conserved genes include the homeobox proteins and are primarily involved in the transcription factor activity, RNA splicing and pattern specification (Fig. 3b, shaded black; Additional file 1: Table S3).

Taken together, genes involved in the basic fundamental biological process, for example, gene regulation, are highly conserved during evolution, being least polymorphic within the human species as well as between species.

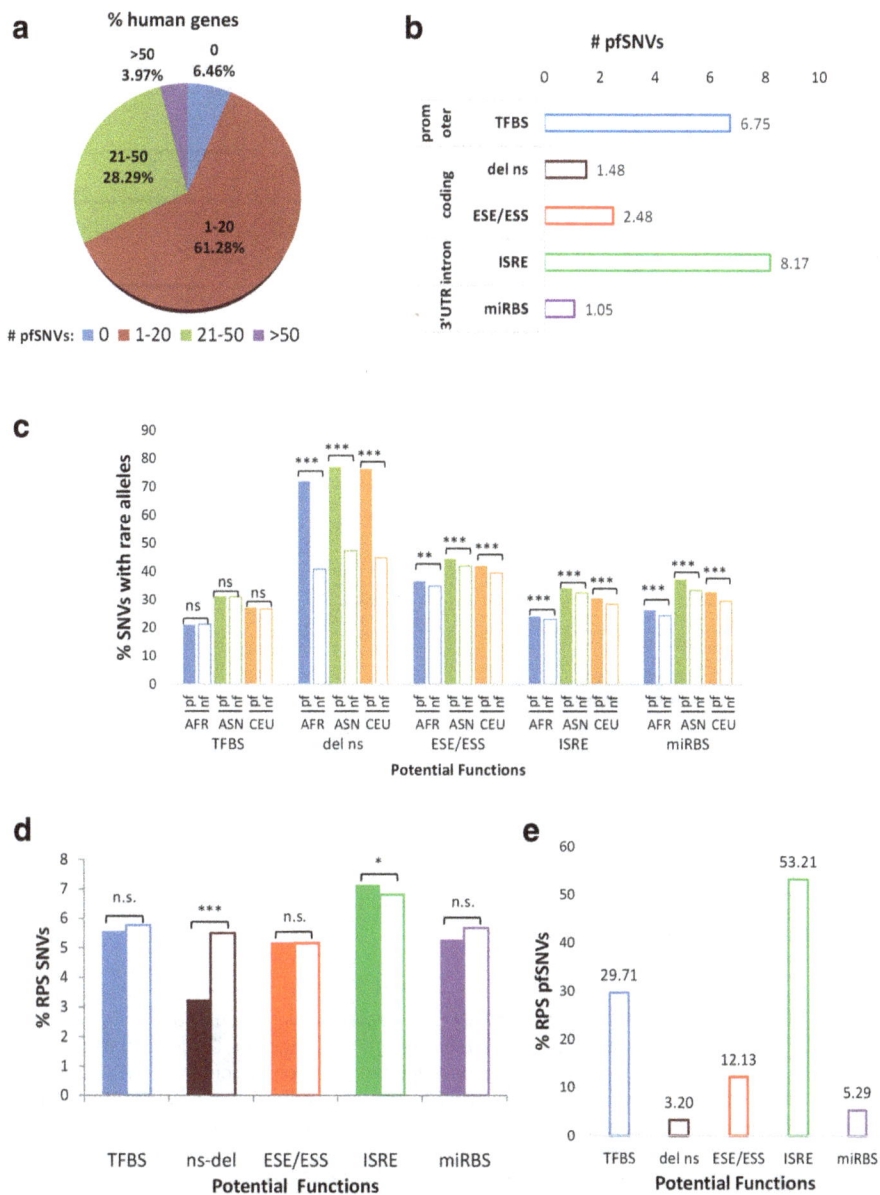

Fig. 2 Potentially functional SNVs. **a** Percentage of human genes containing a different number of pfSNVs. **b** Average numbers of pfSNVs with different potential functions in each transcript. **c** Percentage of SNVs with rare alleles (DAF < 0.05) amongst the pfSNVs and nfSNVs in the same genic regions, as determined in HapMap individuals from African (AFR), Asian (ASN) and European (EUR) population groups. **d** Percentage of pfSNVs or nfSNVs under RPS over all pf- or nfSNVs in that specific group. **e** Percentage of SNVs with different functions amongst all the RPS pfSNVs. pf, potentially functional; nf, non-functional; TFBS, SNVs that alter transcription factor binding sites; del ns, potentially deleterious nsSNV; ESE/ESS, SNVs that alter exon splice enhancers/silencers; ISRE, SNVs that alter intronic splicing regulatory elements; miRBS, SNVs that alter miRNA binding sites

Highly polymorphic genes are mainly involved in immune responses

A total of 512 highly polymorphic genes (see Additional file 1: Supplementary Methods) are identified. These genes are the most significantly over-represented in immune response pathways, as well as in the pathogenesis of a number of autoimmune diseases, including Graft-versus-host disease and type I diabetes mellitus (Fig. 3c, shaded grey; Additional file 1: Table S4). They are primarily in the *MHC* class I-related and class II-related protein families, involved in antigen presentation and processing. Majority of the *MHC* class I and class II genes are located on chromosome 6q21.3 which is the most polymorphic region in the human genome and facilitate the generation of diverse antigens to confer a selective advantage to fight infection [36]. The number of pfSNVs in the *MHC* genes (2234) is higher than the average number of pfSNVs in the

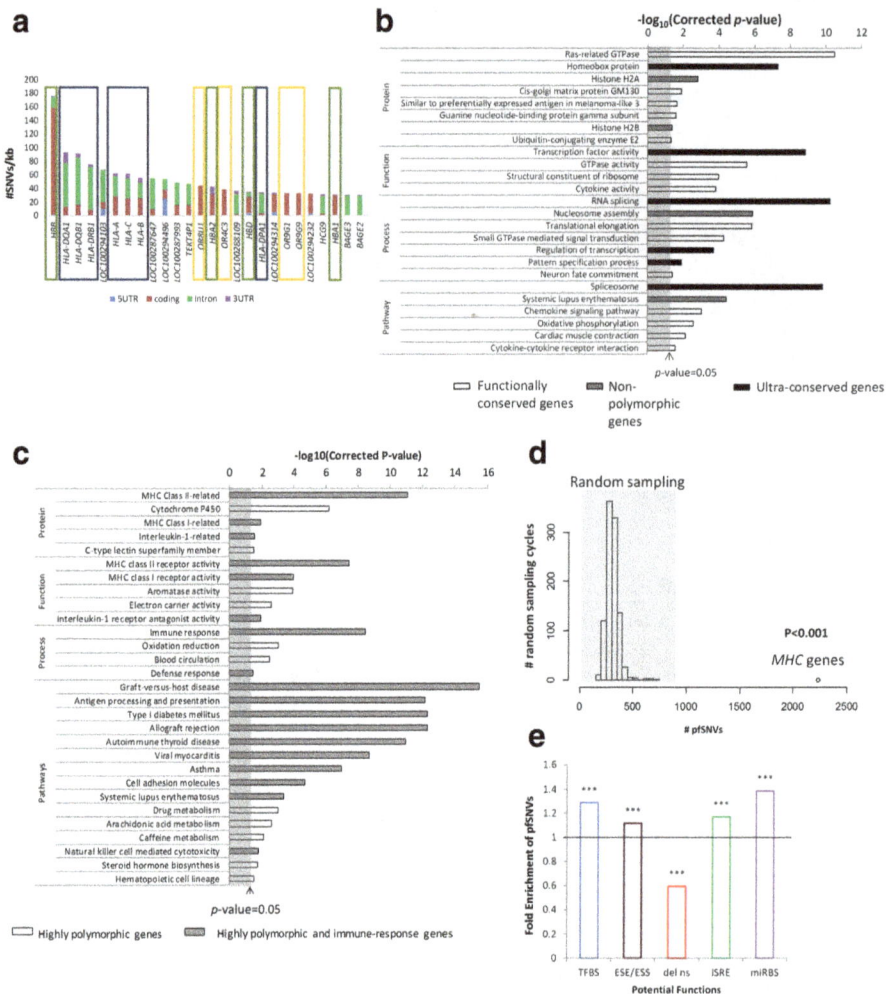

Fig. 3 Most polymorphic vs conserved genes in the human genome. **a** Genes having the highest number of SNVs normalized against gene length (> 30 SNVs/kb per gene). Green box: haemoglobin genes; blue box: *MHC* genes; orange box: *OR* genes. **b** Benjamini-corrected *p* values for the significantly enriched functional terms for non-polymorphic genes (grey bars), functionally conserved genes (white bars) which are genes without nsSNVs and coding INDELs as well as ultra-conserved genes (black bars) which are the genes with UCEs within their coding regions. **c** Benjamini-corrected *p* values for the functional terms that are significantly enriched by the highly polymorphic genes. Grey bars: the functional terms related to immune responses. **d** Empirical distribution of the numbers of pfSNVs obtained from 23 genes that are randomly sampled from all the human genes with lengths 3–15 kb for 1000 times. The number of pfSNVs in the 23 *MHC* genes is significantly higher than that in the randomly sampled gene sets (empirical *p* value < 0.001). **e** Enrichment of SNVs with different potential functions in *MHC* class I and class II genes. Fold enrichment is calculated as the percentage of pfSNVs in specific genic regions (e.g. coding SNVs that may alter ESE/ESS) in the *MHC* class I and class II genes against that for all the human genes. Deviation from one indicates that the pfSNVs are over- or under-represented in the *MHC* genes

other human genes (306). To determine if pfSNVs are significantly over-represented in this *MHC* gene family of 23 genes, sampling of 23 random human genes with similar gene length is performed 1000 times. The number of pfSNVs in the 23 random genes for each cycle is plotted to obtain an empirical distribution. The *MHC* family of genes, with 2234 pfSNVs, is found to carry significantly more pfSNVs than 1000 different sampling of 23 random human genes (empirical *p* value < 0.001 by random sampling test) (Fig. 3d). Interestingly, despite the enrichment of pfSNVs in the *MHC* gene family, there are significantly fewer

nsSNVs predicted to be deleterious in the *MHC* family of genes, compared to other human genes (Bonferroni corrected *p* value < 0.001 by Fisher's exact test) (Fig. 3e). Hence, this family of proteins can nimbly respond to different infection, through a diversity of different regulatory mechanism, including differential transcription factor/miRNA binding/splicing.

Lastly, highly polymorphic genes are also significantly enriched in drug metabolism, cytochrome P450 (*CYP450*), arachidonic acid and caffeine metabolism pathways (Fig. 3c, Additional file 1: Table S4).

Drug response genes are most affected by potentially deleterious polymorphisms in coding regions

Although > 90% (20,890/22,333) of the genes in the human genome have pfSNVs, ~ 54% contain at least one potentially deleterious coding polymorphism (Fig. 4a, shaded blue, dark blue and grey) while ~ 19% are functionally conserved with no nsSNVs nor coding INDELS (Fig. 4a, shaded orange). Potentially deleterious coding polymorphisms are under the strongest negative selection as suggested earlier (Fig. 2c) since they can potentially have a drastic effect on protein function. Approximately 5% (1104/22,333) of all genes in the human

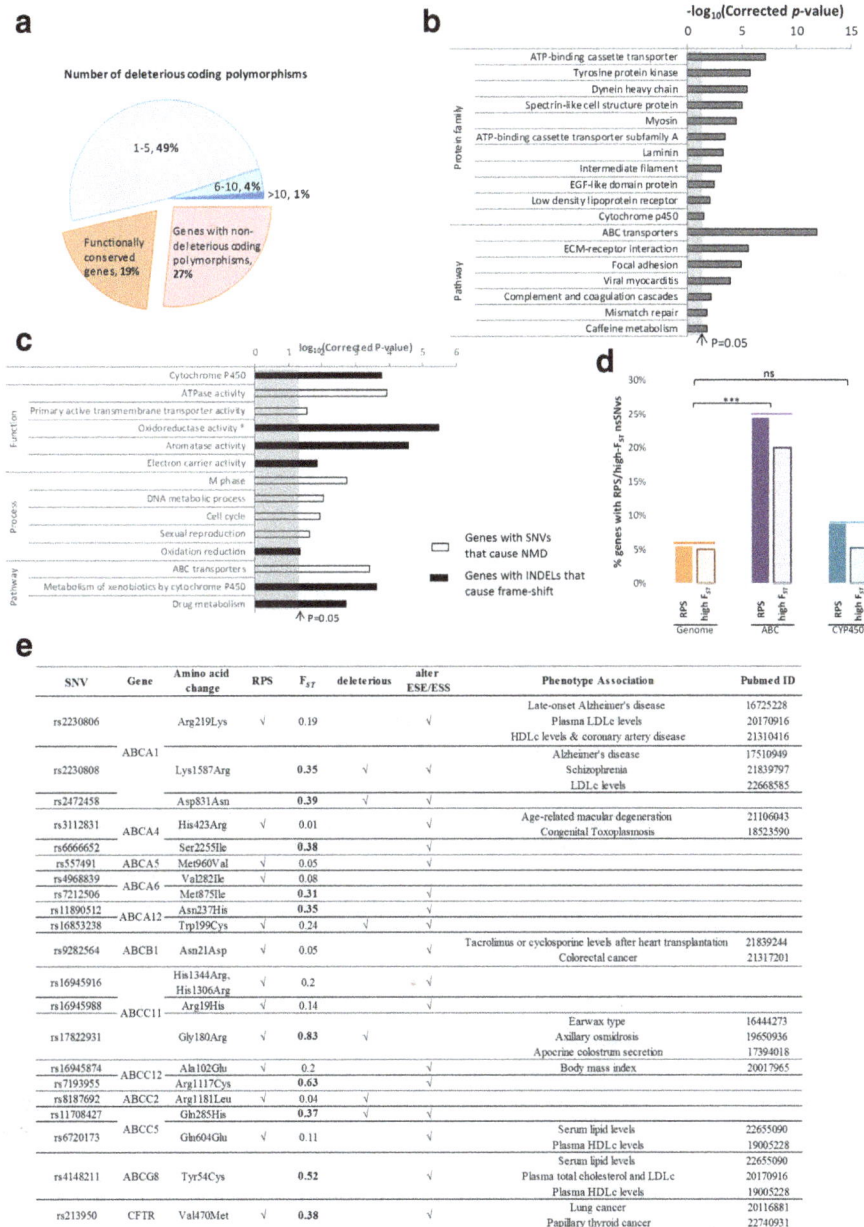

Fig. 4 Distribution of potentially deleterious coding polymorphisms in human genes. **a** Percentage of genes with different numbers of potentially deleterious coding polymorphisms in their coding regions. Genes without any potentially deleterious coding polymorphisms are divided into two groups: (1) functionally conserved genes, i.e. genes with no nsSNV nor INDELs in coding regions; (2) genes carrying non-deleterious SNVs in their coding regions. **b** Benjamini-corrected p values for the functional terms that show enrichment of the genes with more than five potentially deleterious nsSNVs. **c** Benjamini-corrected p values for the functional terms that show enrichment of the genes with SNVs that cause NMD (non-shaded) and the genes with coding INDELs that cause frame-shift (black). **d** Percentage of genes with RPS nsSNVs and genes carrying nsSNVs with high F_{ST} (> 0.3) in the whole genome, *ABC* transporter and *CYP450* family. **e** Recently positively selected and/or population-differentiated nsSNVs in the *ABC* transporters. F_{ST} scores in bold indicate F_{ST} > 0.3. *Oxidoreductase activity: oxidoreductase activity, acting on paired donors, with incorporation or reduction of molecular oxygen, reduced flavin or flavoprotein as one donor, and incorporation of one atom of oxygen

genome are highly enriched with more than five potentially deleterious nsSNVs in their coding regions (Fig. 4a, shaded blue and dark blue). These genes are significantly enriched in the ATP-binding cassette (*ABC*) transporter and the *CYP450* families, which play important roles in drug transport and metabolism (Fig. 4b, Additional file 1: Table S5). Notably, most of the common drug metabolizers including *CYP3A4* [37], *CYP1A1* [38] and *CYP2D6* [39] contain more than five potentially deleterious nsSNVs, while the *CYP* genes that metabolize endogenous substance, e.g. *CYP51A1* [40], are not affected by any potentially deleterious nsSNVs. Similarly, important xenobiotic transporters including *ABCB1*, *ABCC1* and *ABCG2* [41] have 8, 11 and 7 potentially deleterious nsSNVs, respectively, while genes in *ABCD* subfamily, which are peroxisomal transporters for very long chain fatty acids [42], contain fewer (1–3) potentially deleterious nsSNVs. In addition to drug metabolizer and transporter, other protein families enriched in genes with more than five predicted deleterious nsSNVs include tyrosine protein kinases, dynein heavy chains, spectrins and myosins.

Notably, not only are the *ABC* transporters significantly enriched in predicted deleterious coding SNVs (Fig. 4b), they are also enriched with nsSNVs that are predicted to cause nonsense-mediated decay (NMD) resulting in the degradation of the mRNA transcripts with premature stop codon (Fig. 4c, clear bars; Additional file 1: Table S6). Genes containing nsSNVs predicted to cause NMD are also significantly enriched in cell cycle processes including mitosis (Fig. 4c, clear bars; Additional file 1: Table S6).

While the *ABC* transporters are significantly enriched with nsSNVs predicted to cause NMD, the other family of genes involved in drug response, the *CYP450*, is significantly enriched with genes having another form of deleterious polymorphism, namely, INDELs that cause frame-shift, which have deleterious effect on protein function (Fig. 4c, black bars; Additional file 1: Table S7). Taken together, genes involved in the xenobiotic response, including drug transport and metabolism, are significantly enriched with potentially deleterious coding polymorphisms.

Signatures of natural selection on the nsSNVs in drug-response genes are investigated. Interestingly, unlike the *CYP450* family (5/57 genes), not only are the *ABC* transporters enriched with potentially deleterious coding polymorphisms, they are also significantly enriched (p value < 0.001 by Fisher's exact test) with genes carrying nsSNVs under RPS (11/45 genes) (p value = 0.24 by Fisher's exact test) compared to the other genes in the human genome (Fig. 4d). As genes under positive selection also show significant population differentiation [34], we evaluate if the drug response genes are

also enriched with nsSNVs that show significant population differentiation ($F_{ST} > 0.3$). Similar to the above observations, Fisher's exact test revealed that the *ABC* transporters (9/45) (p value < 0.001 by Fisher's exact test) but not the *CYP450* genes (3/57) (p value = 0.76 by Fisher's exact test) are significantly enriched with nsSNVs that show significant population differentiation (Fig. 4d). Hence, the nsSNVs in the *ABC* transporter family are under strong positive selection pressure.

As evident from the table in Fig. 4e, all, except one (rs4968839), of the nsSNVs at the *ABC* transporter family, which showed evidence of RPS or significant population differentiation, are predicted to either have a potentially deleterious effect on protein function or alter ESE/ESS modulating the proportion of the different splice forms. Notably, > 40% of these nsSNVs have been reported to be significantly associated with various phenotypes including clinically relevant ones [43–59] (Fig. 4e), highlighting the functional importance of the nsSNVs under natural selection at the *ABC* transporter gene family.

Discussion

In this study, we comprehensively investigate the architecture of genetic polymorphisms in the human genome and demonstrate that polymorphisms in coding regions, especially those affecting protein sequences and/or functions, are the most constrained in the human genome, consistent with previous observations [30]. In particular, frame-shift INDELs in coding regions are under strong purifying selection, consistent with the previous observation of the strongest depletion of frame-shift INDELs in coding regions, which are enriched with gene expression association possibly contributed by NMD [60].

Through the interrogation of nine global populations, we demonstrate that the median F_{ST} of SNVs at the coding regions is lower than that of the other regions of the genome (Fig. 1f). Moreover, the observations that coding regions have the lowest SNV density (Fig. 1b), excess of rare alleles (Fig. 1d) and enrichment of SNVs with no population differentiation (Fig. 1g) all indicate that coding SNVs are constrained by purifying selections. This is further strengthened by the observation that potentially deleterious nsSNVs show enrichment of rare alleles, compared to non-deleterious nsSNVs (Fig. 2c). Furthermore, coding regions contain significantly fewer INDELs that cause frame-shift (Fig. 1c). Hence, polymorphisms predicted to be deleterious to protein functions are under the strongest purifying selection.

In addition to the potentially deleterious nsSNVs, the potential functions and signatures of natural selections in the other polymorphisms are also investigated. Through computational prediction of the potential functions of SNVs, we observe that significantly more SNVs are predicted to alter TFBS than to code for a potentially deleterious nsSNV (Fig. 2b). Additionally, except for SNVs affecting TFBS, the other pfSNVs show more significant enrichment of rare alleles than nfSNVs in the same regions (Fig. 2c). Hence, pfSNVs are more constrained than the other SNVs in the same region, perhaps because they affect the functionally important regulatory sites. This observation is congruent with previous studies that reported stronger negative selection on conserved miRBS than other conserved 3′UTR sequences [31], though different prediction algorithms and SNV data were used. Levenstien and Klein also reported similar observation that SNVs in a few functional classes, e.g. non-synonymous, methylation sites and miRBS, are under negative selection compared to genome, and suggested that they are promising candidates for functional characterization [61]. While these previous studies examined SNVs residing within regulatory consensus sites of the promoter, this study focuses on SNVs that are predicted to either disrupt or create regulatory sites. Hence, the negative selective pressure on several different classes of pfSNVs suggests that pfSNVs are likely to influence gene functions and contribute to phenotypic changes.

This study also highlights that 'master regulators' of gene expression tend to be functionally conserved and maintained during evolution, while regulation of specific target genes is less constrained and flexible. This is evident from the observation that 'master regulator' involved in general gene regulation including epigenetics (e.g. histones), transcription/translation and splicing is significantly enriched with non-polymorphic, functionally conserved and ultra-conserved genes (Fig. 3b). In contrast, an average gene contains more SNVs affecting its own regulation than altering its function as evident from the observed enrichment of SNVs predicted to alter TFBS at promoters, ESE/ESS in coding regions and ISRE within introns compared to SNVs predicted to result in deleterious non-synonymous amino acid changes (Fig. 2b). This is consistent with previous observation that genetic variants occur more frequently in the miRNA target regions, compared to the functional regions within miRNAs [62]. In addition, Hsiao et al. demonstrated that alternative splicing events regulated by intronic genetic variants tend to be under positive selection [63], which is consistent with our results that intronic SNVs that potentially affect splicing mechanisms show enrichment of RPS SNVs, compared to the other functional classes (Fig. 2d). On the other hand, splicing factors are more conserved during evolution [63], and our study demonstrates that UCEs were enriched in the genes involved in RNA splicing (Fig. 3b).

On the other hand, genes that modulate response to environmental changes are the most polymorphic. The immune response *MHC* class I and class II genes, implicated in the pathogenesis of several autoimmune diseases, reside in the most polymorphic region of the human genome [36, 64] and carry the highest density of SNVs (Fig. 3c). Notably, this family of genes is significantly enriched in SNVs predicted to alter various regulatory elements including TFBS, ESE/ESS, ISRE and miRBS rather than protein function (Fig. 3e). In fact, while none of the RPS pfSNVs in the *MHC* family is predicted to cause a deleterious effect on protein function, 87 pfSNVs are found to display signature of RPS (Additional file 1: Table S8). Hence, the regulatory regions of the *MHC* family of genes are likely to be under strong positive selection, as previously suggested [65], and are functionally significant, regulating gene expression to modulate phenotypes. For example, a very well-studied polymorphism, rs9378249 upstream of the *HLA-B* gene, has previously been associated with bipolar disorder [66, 67] and hypertension [67]. This polymorphism is predicted to alter TFBS and exhibits the signature of RPS; hence, it may be a causal variant for the various diseases although the underlying molecular mechanism requires further validation.

Another class of genes that modulates response to the environment is the drug/xenobiotic response families of genes including the *ABC* transporter and the *CYP450* metabolism families of genes. Unlike the *MHC* immune genes, which are significantly enriched in regulatory SNVs predicted to modulate gene expression, these drug response gene families are enriched in SNVs that affect the functions of the proteins, namely nsSNVs predicted to be deleterious (Fig. 4b). Previous reports also highlighted the high SNV density and excess of rare nsSNVs of the *CYP450* pathway [68, 69] with 90–95% individuals carrying at least one actionable variant in *CYP450* genes [70]. Another report predicted that ~ 32% (1949/6165) of SNVs at the *CYP450* loci are putatively functional with *CYP4F12* carrying amongst the most novel putatively functional variants [71] which is consistent with our observations that *CYP4F12* is enriched with the highest number of pfSNVs (Additional file 1: Table S9). In addition to the pfSNVs, RPS and highly population differentiated ($F_{ST} > 0.3$) SNVs are significantly represented in the *ABC* transporter genes but not in the *CYP450* genes (Fig. 4d) suggesting that the *ABC* transporter genes may be under stronger positive selection than the *CYP450* genes. For example, rs17822931, a coding variant at the *ABCC11* earwax determinant gene [59], is found to be highly differentiated amongst populations, and the *A* allele is positively associated with adaptation to cold climate [72]. Greater than 40% of these RPS and/or population differentiated SNVs in the *ABC* transporter genes have been associated with phenotype modulation and even diseases, e.g. Alzheimer's and

Schizophrenia (table in Fig. 4e). Nearly all the coding SNVs at the *ABC* transporter gene family that display the signature of RPS or are significantly population differentiated are also predicted to alter ESE/ESS suggesting that differential splicing may also play an important role in the *ABC* genes to generate diverse splicing forms to respond to different environment.

Hence, adaptive genes that respond to environmental changes are likely to be highly polymorphic and subjected to strong positive selection pressures consistent with previous reports that variants associated with inflammatory diseases show evidence of RPS [73], and genes associated with pharmacogenomics show higher level of population differentiation, as a signature of positive selection [74]. Regulation of gene expression through variants that alter TFBS in the *MHC* gene family as well as modulation of protein function and/or splicing pattern in the *ABC* gene families highlight the different ways by different families of genes to adapt to the environment.

Conclusions

In conclusion, this study elucidates the overall architecture of the genetic polymorphisms, namely SNVs and INDELs, in the human genome. The coding region is found to be under strong negative selection, as being the least population differentiated, showing lowest densities of SNVs and INDELs (especially frame-shift INDELs), the highest proportion of rare alleles with less enrichment of RPS SNVs. SNVs predicted to be functional are found to be under negative selection with enrichment of rare alleles. Families of genes which are 'master regulators' of gene expression including those involved in epigenetics, transcription, translation or splicing are found to be least polymorphic, functionally conserved and/or enriched with ultra-conserved elements. Finally, genes that modulate response to the environment are the most polymorphic with the *MHC* gene family, which is involved in immune response, being the most polymorphic while genes involved in drug/xenobiotic response, including *ABC* transporter and *CYP450* genes, are the most enriched with functional nsSNVs.

Methods
Polymorphisms in the human genome

Polymorphisms from the dbSNP database (Build 131) were mapped to different genic regions (5'UTRs, coding regions, introns and 3'UTRs) of the human genes (NCBI Genome Build 37.1) with those residing outside genes classified as intergenic variants.

To minimize false-positive SNPs originating from highly paralogous sequences, which were estimated to be ~ 8% of biallelic coding SNVs in dbSNP129 [75], only polymorphisms, which mapped to a single location in the genome

and have been validated using a non-computational method or have allele frequency information (e.g. from 1000 Genomes project), were included in this study. In the 1000 genomes project, the variant assignment was restricted to 'accessible genome', whereby ambiguously placed reads or unexpectedly high or low numbers of aligned reads were excluded (~ 15% genome) to minimize the detection of false-positive variants [76]. To evaluate if our data is valid, SNV density data of this study was compared and found to be comparable to the SNV density data calculated from whole-genome sequencing of 179 HapMap individuals [76] of 1000 genomes project. For example, similar to our observations using dbSNP data, the MHC gene loci from the 1000 genomes sequencing data were also found to be significantly more polymorphic than other human genes ($p < 0.001$ by Mann-Whitney test). Hence, results from sequencing data from the 1000 genomes project were consistent with the findings in this study using dbSNP data, suggesting that, in spite of the potential ascertainment biases and sequencing artefacts inherent in the dbSNP database, our findings about the enrichment of SNPs in MHC genes are valid.

Two major forms of genetic polymorphisms, SNVs and INDELs, were investigated. SNV/INDEL density within a particular genic region, e.g. 5'UTR, was calculated as the number of SNVs/INDELs divided by the length of that region. For genes with multiple transcripts, the mean densities were taken. Genes lacking polymorphism in all genic regions (promoter, 5'UTR, coding, intron, 3'UTR) were regarded as non-polymorphic genes. Highly polymorphic genes were identified based on a binomial model as described in Additional file 1.

Allele frequency of SNVs in the human genome was determined in the three population (East Asian, African and European) groups (HapMap release 28) as described in Additional file 1.

F_{ST} statistics [28] using the pooled allele frequencies in the three population groups was then calculated for each of the genotyped and polymorphic loci. Two groups of SNVs, namely, (1) zero-F_{ST} SNVs ($F_{ST} = 0$) and (2) high-F_{ST} SNVs ($F_{ST} > 0.3$), were further analysed. The fold enrichment of zero-F_{ST} or high-F_{ST} SNVs in a specific genic region (e.g. coding region) was determined by calculating the percentage of these SNVs in the coding region divided by the percentage of all the genotyped SNVs in the same region, and the significance of enrichment is determined using the Fisher's exact test. Fold enrichment, which significantly deviates from one, indicates that these SNVs are under- or over-represented in these regions.

Natural selections

Genic regions that display signatures of negative selection were previously reported to have excess rare derived alleles [31]. Hence, to identify the regions of genes subjected to negative selection, we determined if there is a statistical enrichment of rare SNVs (DAF < 0.05) in each genic region using the Fisher's exact test.

SNVs displaying signatures of RPS were identified using LD- or haplotype-based methods as described in [26, 77]. To identify the regions enriched in RPS SNVs, the percentage of RPS SNVs within the region was compared with the percentage of RPS SNVs in the whole genome, and significance of difference was determined using the Fisher's exact test.

UCEs are sequences within the genome that are 100% identical to the sequences with the mouse and the rat genomes [35], hence displaying evolutionary conservation and signatures of strong negative selection. A total of 481 UCEs have been identified [35], of which 70 are evolutionarily conserved coding sequences, overlapping with coding regions.

Potential functions of SNVs

The pfSNP database (http://pfs.nus.edu.sg/) [26], which integrates a variety of bioinformatics prediction algorithms, was used to evaluate potential functions of all the SNVs in the human genome that alter TFBS, protein functions, splicing events and miRBS. The prediction algorithms employed in this study are described in [26] and Additional file 1.

Functional annotation

The Database for Annotation, Visualization and Integrated Discovery [78, 79] was utilized for functional annotation of the genes of interest. The enrichment of the genes in PANTHER protein family, GO-molecular function, GO-biological process and KEGG pathway was investigated. Benjamini-Hochberg-corrected p value < 0.05 signifies statistical significance.

Abbreviations

3'UTR: 3' Untranslated region; 5'UTR: 5' Untranslated region; *ABC*: ATP-binding cassette; *CYP450*: Cytochrome P450; DAF: Derived allele frequency; ESE/ESS: Exon splicing enhancer or silencer; *HBB*: Beta haemoglobin; INDEL: Insertion/deletion; ISRE: Intronic splicing regulatory element; *MHC*: Major histocompatibility complex; miRBS: MicroRNA binding site; miRNA: MicroRNA; nfSNV: Non-functional SNV; NMD: Nonsense-mediated decay; nsSNV: Non-synonymous SNV; *OR*: Olfactory receptor; pfSNV: Potentially functional SNV; RPS: Recent positive selection; SNV: Single-nucleotide variant; sSNV: Synonymous SNV; TFBS: Transcription factor binding site; UCE: Ultra-conserved element

Funding

This work was supported by a grant from the National Medical Research Council (NMRC) 1131/2007 and BioMedical Research Council - Science and Engineering Research Council (BMRC-SERC) 112 148 0008, as well as Block Funding from the National Cancer Centre Singapore and Duke-NUS Graduate Medical School Singapore to A/P GL Lee.

Authors' contributions

CGLL and SSC conceived and designed the study. YJ conducted the statistical analyses. JW predicted the potential functions of the SNVs and identified the SNVs under RPS. MB analysed the population differentiation levels of the SNVs. YJ and CGLL wrote the manuscript, and SSC helped edit the manuscript. All authors read and approved the final manuscript.

Consent for publication

Not applicable.

Competing interests

The authors declare that they have no competing interests.

Author details

[1]NUS Graduate School for Integrative Sciences and Engineering, National University of Singapore, Singapore 117456, Singapore. [2]Department of Biochemistry, National University of Singapore, Singapore 119077, Singapore. [3]Division of Medical Sciences, National Cancer Centre, Singapore 169610, Singapore. [4]Department of Paediatrics, Yong Loo Lin School of Medicine, National University of Singapore, Singapore 119228, Singapore. [5]Duke-NUS Graduate Medical School, Singapore 169547, Singapore.

References

1. Collins FS, Brooks LD, Chakravarti A. A DNA polymorphism discovery resource for research on human genetic variation. Genome Res. 1998;8: 1229–31.
2. Dimas AS, Deutsch S, Stranger BE, Montgomery SB, Borel C, Attar-Cohen H, Ingle C, Beazley C, Gutierrez Arcelus M, Sekowska M, et al. Common regulatory variation impacts gene expression in a cell type-dependent manner. Science. 2009;325:1246–50.
3. Montgomery SB, Sammeth M, Gutierrez-Arcelus M, Lach RP, Ingle C, Nisbett J, Guigo R, Dermitzakis ET. Transcriptome genetics using second generation sequencing in a Caucasian population. Nature. 2010;464:773–7.
4. Myers AJ, Gibbs JR, Webster JA, Rohrer K, Zhao A, Marlowe L, Kaleem M, Leung D, Bryden L, Nath P, et al. A survey of genetic human cortical gene expression. Nat Genet. 2007;39:1494–9.
5. Pickrell JK, Marioni JC, Pai AA, Degner JF, Engelhardt BE, Nkadori E, Veyrieras JB, Stephens M, Gilad Y, Pritchard JK. Understanding mechanisms underlying human gene expression variation with RNA sequencing. Nature. 2010;464:768–72.
6. Schadt EE, Molony C, Chudin E, Hao K, Yang X, Lum PY, Kasarskis A, Zhang B, Wang S, Suver C, et al. Mapping the genetic architecture of gene expression in human liver. PLoS Biol. 2008;6:e107.
7. Stranger BE, Forrest MS, Clark AG, Minichiello MJ, Deutsch S, Lyle R, Hunt S, Kahl B, Antonarakis SE, Tavare S, et al. Genome-wide associations of gene expression variation in humans. PLoS Genet. 2005;1:e78.
8. Stranger BE, Nica AC, Forrest MS, Dimas A, Bird CP, Beazley C, Ingle CE, Dunning M, Flicek P, Koller D, et al. Population genomics of human gene expression. Nat Genet. 2007;39:1217–24.
9. Veyrieras JB, Kudaravalli S, Kim SY, Dermitzakis ET, Gilad Y, Stephens M, Pritchard JK. High-resolution mapping of expression-QTLs yields insight into human gene regulation. PLoS Genet. 2008;4:e1000214.
10. Zeller T, Wild P, Szymczak S, Rotival M, Schillert A, Castagne R, Maouche S, Germain M, Lackner K, Rossmann H, et al. Genetics and beyond--the transcriptome of human monocytes and disease susceptibility. PLoS One. 2010;5:e10693.
11. Adzhubei IA, Schmidt S, Peshkin L, Ramensky VE, Gerasimova A, Bork P, Kondrashov AS, Sunyaev SR. A method and server for predicting damaging missense mutations. Nat Methods. 2010;7:248–9.
12. Karchin R, Diekhans M, Kelly L, Thomas DJ, Pieper U, Eswar N, Haussler D, Sali A. LS-SNP: large-scale annotation of coding non-synonymous SNPs based on multiple information sources. Bioinformatics. 2005;21:2814–20.
13. Yue P, Melamud E, Moult J. SNPs3D: candidate gene and SNP selection for association studies. BMC Bioinformatics. 2006;7:166.
14. Ng PC, Henikoff S. Accounting for human polymorphisms predicted to affect protein function. Genome Res. 2002;12:436–46.

15. Li B, Krishnan VG, Mort ME, Xin F, Kamati KK, Cooper DN, Mooney SD, Radivojac P. Automated inference of molecular mechanisms of disease from amino acid substitutions. Bioinformatics. 2009;25:2744–50.

16. Thomas PD, Kejariwal A, Guo N, Mi H, Campbell MJ, Muruganujan A, Lazareva-Ulitsky B. Applications for protein sequence-function evolution data: mRNA/protein expression analysis and coding SNP scoring tools. Nucleic Acids Res. 2006;34:W645–50.

17. Capriotti E, Calabrese R, Casadio R. Predicting the insurgence of human genetic diseases associated to single point protein mutations with support vector machines and evolutionary information. Bioinformatics. 2006;22: 2729–34.

18. Masica DL, Karchin R. Towards increasing the clinical relevance of in silico methods to predict pathogenic missense variants. PLoS Comput Biol. 2016; 12:e1004725.

19. Peterson TA, Doughty E, Kann MG. Towards precision medicine: advances in computational approaches for the analysis of human variants. J Mol Biol. 2013;425:4047–63.

20. Faber K, Glatting KH, Mueller PJ, Risch A, Hotz-Wagenblatt A. Genome-wide prediction of splice-modifying SNPs in human genes using a new analysis pipeline called AASsites. BMC Bioinformatics. 2011;12(Suppl 4):S2.

21. Yang JO, Kim WY, Bhak J. ssSNPTarget: genome-wide splice-site single nucleotide polymorphism database. Hum Mutat. 2009;30:E1010–20.

22. Kim BC, Kim WY, Park D, Chung WH, Shin KS, Bhak J. SNP@Promoter: a database of human SNPs (single nucleotide polymorphisms) within the putative promoter regions. BMC Bioinformatics. 2008;9(Suppl 1):S2.

23. Bao L, Zhou M, Wu L, Lu L, Goldowitz D, Williams RW, Cui Y. PolymiRTS Database: linking polymorphisms in microRNA target sites with complex traits. Nucleic Acids Res. 2007;35:D51–4.

24. Hiard S, Charlier C, Coppieters W, Georges M, Baurain D. Patrocles: a database of polymorphic miRNA-mediated gene regulation in vertebrates. Nucleic Acids Res. 2010;38:D640–51.

25. Gong J, Tong Y, Zhang HM, Wang K, Hu T, Shan G, Sun J, Guo AY. Genome-wide identification of SNPs in microRNA genes and the SNP effects on microRNA target binding and biogenesis. Hum Mutat. 2012;33:254–63.

26. Wang J, Ronaghi M, Chong SS, Lee CG. pfSNP: an integrated potentially functional SNP resource that facilitates hypotheses generation through knowledge syntheses. Hum Mutat. 2011;32:19–24.

27. Conde L, Vaquerizas JM, Santoyo J, Al-Shahrour F, Ruiz-Llorente S, Robledo M, Dopazo J. PupaSNP Finder: a web tool for finding SNPs with putative effect at transcriptional level. Nucleic Acids Res. 2004;32:W242–8.

28. Akey JM, Zhang G, Zhang K, Jin L, Shriver MD. Interrogating a high-density SNP map for signatures of natural selection. Genome Res. 2002;12:1805–14.

29. Sabeti PC, Varilly P, Fry B, Lohmueller J, Hostetter E, Cotsapas C, Xie X, Byrne EH, McCarroll SA, Gaudet R, et al. Genome-wide detection and characterization of positive selection in human populations. Nature. 2007; 449:913–8.

30. Barreiro LB, Laval G, Quach H, Patin E, Quintana-Murci L. Natural selection has driven population differentiation in modern humans. Nat Genet. 2008; 40:340–5.

31. Chen K, Rajewsky N. Natural selection on human microRNA binding sites inferred from SNP data. Nat Genet. 2006;38:1452–6.

32. Jha P, Lu D, Xu S. Natural selection and functional potentials of human noncoding elements revealed by analysis of next generation sequencing data. PLoS One. 2015;10:e0129023.

33. Enard D, Messer PW, Petrov DA. Genome-wide signals of positive selection in human evolution. Genome Res. 2014;24:885–95.

34. Nielsen R. Molecular signatures of natural selection. Annu Rev Genet. 2005; 39:197–218.

35. Bejerano G, Pheasant M, Makunin I, Stephen S, Kent WJ, Mattick JS, Haussler D. Ultraconserved elements in the human genome. Science. 2004;304:1321–5.

36. Vandiedonck C, Knight JC. The human major histocompatibility complex as a paradigm in genomics research. Brief Funct Genomic Proteomic. 2009;8: 379–94.

37. Zanger UM, Schwab M. Cytochrome P450 enzymes in drug metabolism: regulation of gene expression, enzyme activities, and impact of genetic variation. Pharmacol Ther. 2013;138:103–41.

38. Beresford AP. CYP1A1: friend or foe? Drug Metab Rev. 1993;25:503–17.

39. Wang B, Yang LP, Zhang XZ, Huang SQ, Bartlam M, Zhou SF. New insights into the structural characteristics and functional relevance of the human cytochrome P450 2D6 enzyme. Drug Metab Rev. 2009;41:573–643.

40. Lepesheva GI, Waterman MR. Sterol 14alpha-demethylase cytochrome P450 (CYP51), a P450 in all biological kingdoms. Biochim Biophys Acta. 2007;1770: 467–77.

41. Dean M, Rzhetsky A, Allikmets R. The human ATP-binding cassette (ABC) transporter superfamily. Genome Res. 2001;11:1156–66.

42. Baker A, Carrier DJ, Schaedler T, Waterham HR, van Roermund CW, Theodoulou FL. Peroxisomal ABC transporters: functions and mechanism. Biochem Soc Trans. 2015;43:959–65.

43. Abellan R, Mansego ML, Martinez-Hervas S, Martin-Escudero JC, Carmena R, Real JT, Redon J, Castrodeza-Sanz JJ, Chaves FJ. Association of selected ABC gene family single nucleotide polymorphisms with postprandial lipoproteins: results from the population-based Hortega study. Atherosclerosis. 2010;211:203–9.

44. Akey JM. Constructing genomic maps of positive selection in humans: where do we go from here? Genome Res. 2009;19:711–22.

45. Brion M, Sanchez-Salorio M, Corton M, de la Fuente M, Pazos B, Othman M, Swaroop A, Abecasis G, Sobrino B, Carracedo A, Spanish Multi-centre Group of AMD. Genetic association study of age-related macular degeneration in the Spanish population. Acta Ophthalmol. 2011;89:e12–22.

46. Chu LW, Li Y, Li Z, Tang AY, Cheung BM, Leung RY, Yik PY, Jin DY, Song YQ. A novel intronic polymorphism of ABCA1 gene reveals risk for sporadic Alzheimer's disease in Chinese. Am J Med Genet B Neuropsychiatr Genet. 2007;144B:1007–13.

47. Jamieson SE, de Roubaix LA, Cortina-Borja M, Tan HK, Mui EJ, Cordell HJ, Kirisits MJ, Miller EN, Peacock CS, Hargrave AC, et al. Genetic and epigenetic factors at COL2A1 and ABCA4 influence clinical outcome in congenital toxoplasmosis. PLoS One. 2008;3:e2285.

48. Jordan de Luna C, Herrero Cervera MJ, Sanchez Lazaro I, Almenar Bonet L, Poveda Andres JL, Alino Pellicer SF. Pharmacogenetic study of ABCB1 and CYP3A5 genes during the first year following heart transplantation regarding tacrolimus or cyclosporine levels. Transplant Proc. 2011;43:2241–3.

49. Junyent M, Tucker KL, Smith CE, Garcia-Rios A, Mattei J, Lai CQ, Parnell LD, Ordovas JM. The effects of ABCG5/G8 polymorphisms on plasma HDL cholesterol concentrations depend on smoking habit in the Boston Puerto Rican Health Study. J Lipid Res. 2009;50:565–73.

50. Kolovou V, Marvaki A, Karakosta A, Vasilopoulos G, Kalogiani A, Mavrogeni S, Degiannis D, Marvaki C, Kolovou G. Association of gender, ABCA1 gene polymorphisms and lipid profile in Greek young nurses. Lipids Health Dis. 2012;11:62.

51. Li Q, Yin RX, Wei XL, Yan TT, Aung LH, Wu DF, Wu JZ, Lin WX, Liu CW, Pan SL. ATP-binding cassette transporter G5 and G8 polymorphisms and several environmental factors with serum lipid levels. PLoS One. 2012;7:e37972.

52. Ma XY, Liu JP, Song ZY. Associations of the ATP-binding cassette transporter A1 R219K polymorphism with HDL-C level and coronary artery disease risk: a meta-analysis. Atherosclerosis. 2011;215:428–34.

53. Miura K, Yoshiura K, Miura S, Shimada T, Yamasaki K, Yoshida A, Nakayama D, Shibata Y, Niikawa N, Masuzaki H. A strong association between human earwax-type and apocrine colostrum secretion from the mammary gland. Hum Genet. 2007;121:631–3.

54. Nakano M, Miwa N, Hirano A, Yoshiura K, Niikawa N. A strong association of axillary osmidrosis with the wet earwax type determined by genotyping of the ABCC11 gene. BMC Genet. 2009;10:42.

55. Oh IH, Oh C, Yoon TY, Choi JM, Kim SK, Park HJ, Eun YG, Chung DH, Kwon KH, Choe BK. Association of CFTR gene polymorphisms with papillary thyroid cancer. Oncol Lett. 2012;3:455–61.

56. Ota M, Fujii T, Nemoto K, Tatsumi M, Moriguchi Y, Hashimoto R, Sato N, Iwata N, Kunugi H. A polymorphism of the ABCA1 gene confers susceptibility to schizophrenia and related brain changes. Prog Neuro-Psychopharmacol Biol Psychiatry. 2011;35:1877–83.

57. Sainz J, Rudolph A, Hein R, Hoffmeister M, Buch S, von Schonfels W, Hampe J, Schafmayer C, Volzke H, Frank B, et al. Association of genetic polymorphisms in ESR2, HSD17B1, ABCB1, and SHBG genes with colorectal cancer risk. Endocr Relat Cancer. 2011;18:265–76.

58. Sundar PD, Feingold E, Minster RL, DeKosky ST, Kamboh MI. Gender-specific association of ATP-binding cassette transporter 1 (ABCA1) polymorphisms with the risk of late-onset Alzheimer's disease. Neurobiol Aging. 2007;28: 856–62.

59. Yoshiura K, Kinoshita A, Ishida T, Ninokata A, Ishikawa T, Kaname T, Bannai

M, Tokunaga K, Sonoda S, Komaki R, et al. A SNP in the ABCC11 gene is the determinant of human earwax type. Nat Genet. 2006;38:324–30.

60. Montgomery SB, Goode DL, Kvikstad E, Albers CA, Zhang ZD, Mu XJ, Ananda G, Howie B, Karczewski KJ, Smith KS, et al. The origin, evolution, and functional impact of short insertion-deletion variants identified in 179 human genomes. Genome Res. 2013;23:749–61.

61. Levenstien MA, Klein RJ. Predicting functionally important SNP classes based on negative selection. BMC Bioinformatics. 2011;12:26.

62. Saunders MA, Liang H, Li WH. Human polymorphism at microRNAs and microRNA target sites. Proc Natl Acad Sci U S A. 2007;104:3300–5.

63. Hsiao YH, Bahn JH, Lin X, Chan TM, Wang R, Xiao X. Alternative splicing modulated by genetic variants demonstrates accelerated evolution regulated by highly conserved proteins. Genome Res. 2016;26:440–50.

64. Trowsdale J. The MHC, disease and selection. Immunol Lett. 2011;137:1–8.

65. Suo C, Xu H, Khor CC, Ong RT, Sim X, Chen J, Tay WT, Sim KS, Zeng YX, Zhang X, et al. Natural positive selection and north-south genetic diversity in East Asia. Eur J Hum Genet. 2012;20:102–10.

66. Jiang Y, Zhang H. Propensity score-based nonparametric test revealing genetic variants underlying bipolar disorder. Genet Epidemiol. 2011;35: 125–32.

67. Ross KA. Evidence for somatic gene conversion and deletion in bipolar disorder, Crohn's disease, coronary artery disease, hypertension, rheumatoid arthritis, type-1 diabetes, and type-2 diabetes. BMC Med. 2011;9:12.

68. Genomes Project C, Abecasis GR, Auton A, Brooks LD, DePristo MA, Durbin RM, Handsaker RE, Kang HM, Marth GT, McVean GA. An integrated map of genetic variation from 1,092 human genomes. Nature. 2012;491:56–65.

69. Zhou Y, Ingelman-Sundberg M, Lauschke VM. Worldwide distribution of cytochrome P450 alleles: a meta-analysis of population-scale sequencing projects. Clin Pharmacol Ther. 2017;102:688–700.

70. Bank PCD, Swen JJ, Guchelaar HJ. Implementation of pharmacogenomics in everyday clinical settings. Adv Pharmacol. 2018;83:219–46.

71. Fujikura K, Ingelman-Sundberg M, Lauschke VM. Genetic variation in the human cytochrome P450 supergene family. Pharmacogenet Genomics. 2015;25:584–94.

72. Ohashi J, Naka I, Tsuchiya N. The impact of natural selection on an ABCC11 SNP determining earwax type. Mol Biol Evol. 2011;28:849–57.

73. Raj T, Kuchroo M, Replogle JM, Raychaudhuri S, Stranger BE, De Jager PL. Common risk alleles for inflammatory diseases are targets of recent positive selection. Am J Hum Genet. 2013;92:517–29.

74. Amato R, Pinelli M, Monticelli A, Marino D, Miele G, Cocozza S. Genome-wide scan for signatures of human population differentiation and their relationship with natural selection, functional pathways and diseases. PLoS One. 2009;4:e7927.

75. Musumeci L, Arthur JW, Cheung FS, Hoque A, Lippman S, Reichardt JK. Single nucleotide differences (SNDs) in the dbSNP database may lead to errors in genotyping and haplotyping studies. Hum Mutat. 2010;31:67–73.

76. Genomes Project C, Abecasis GR, Altshuler D, Auton A, Brooks LD, Durbin RM, Gibbs RA, Hurles ME, McVean GA. A map of human genome variation from population-scale sequencing. Nature. 2010;467:1061–73.

77. Voight BF, Kudaravalli S, Wen X, Pritchard JK. A map of recent positive selection in the human genome. PLoS Biol. 2006;4:e72.

78. Huang d W, Sherman BT, Lempicki RA. Systematic and integrative analysis of large gene lists using DAVID bioinformatics resources. Nat Protoc. 2009;4: 44–57.

79. Huang d W, Sherman BT, Lempicki RA. Bioinformatics enrichment tools: paths toward the comprehensive functional analysis of large gene lists. Nucleic Acids Res. 2009;37:1–13.

19

Predicting the combined effect of multiple genetic variants

Mingming Liu[1], Layne T. Watson[1,2,3] and Liqing Zhang[1*]

Abstract

Background: Many genetic variants have been identified in the human genome. The functional effects of a single variant have been intensively studied. However, the joint effects of multiple variants in the same genes have been largely ignored due to their complexity or lack of data. This paper uses HMMvar, a hidden Markov model based approach, to investigate the combined effect of multiple variants from the 1000 Genomes Project. Two tumor suppressor genes, TP53 and phosphatase and tensin homolog (PTEN), are also studied for the joint effect of compensatory indel variants.

Results: Results show that there are cases where the joint effect of having multiple variants in the same genes is significantly different from that of a single variant. The deleterious effect of a single indel variant can be alleviated by their compensatory indels in TP53 and PTEN. Compound mutations in two genes, β-MHC and MyBP-C, leading to severer cardiovascular disease compared to single mutations, are also validated.

Conclusions: This paper extends the functionality of HMMvar, a tool for assigning a quantitative score to a variant, to measure not only the deleterious effect of a single variant but also the joint effect of multiple variants. HMMvar is the first tool that can predict the functional effects of both single and general multiple variations on proteins. The precomputed scores for multiple variants from the 1000 Genomes Project and the HMMvar package are available at https://bioinformatics.cs.vt.edu/zhanglab/HMMvar/

Introduction

Identifying the deleterious effects of a variant is significant for disease studies. Different types of variation data have been identified with advances in sequencing technologies. Single nucleotide polymorphism (SNP) is the largest group of mutations in the variants identified so far in humans, and numerous methods have been developed for predicting the functional effects of SNPs. The second most common type of mutations is indel, referring to insertion or deletion of nucleotide bases. More and more indels have been discovered to be associated with diseases or cancers. Frameshift indels are expected to have large effects on protein functions (loss of function), since they change the reading frame of a gene thus change amino acids and probably the functions of proteins. Compared to SNPs, less work has been done on predicting the functional effect of indels.

Methods for predicting the functional effects of different types of variants are typically grouped into two classes [1], conservation-based predictor and trained classifier. Previous studies mainly concern SNPs, and a few dozen computer programs and web servers are devoted to predicting the effects of SNP variants. For example, SIFT SNP [2] is a conservation-based predictor and PolyPhen [3] is a trained classifier. Recent indel prediction studies include an evolutionary conservation-based approach for both coding and noncoding regions [4], a trained classifier method for frameshift variants [5], and another evolutionary conservation-based method for multiple types of variation [6]. A limitation of all these methods is that they only predict the effect of a single variant and cannot measure the functional effect of a set of variants in their entirety. Complex diseases are likely to be caused by multiple genes and/or multiple mutations on individual genes [7], so quantitatively measuring the effect of multiple variants together should be helpful for detecting causal genes/mutations for diseases. For example, it has been shown that the correlation between breast

*Correspondence: lqzhang@cs.vt.edu
[1] Department of Computer Science, Virginia Polytechnic Institute and State University, Blacksburg, VA, USA
Full list of author information is available at the end of the article

cancer and multiple SNPs of the ORAI1 gene is more significant than that with single SNPs [8]. The authors use a genetic algorithm to find combinations of SNPs along with their genotypes that are significantly different between the case group and the control group. The results reveal that new insights in cancer studies are possible by considering the joint effect of multiple variants or the associations among genetic variants. Other work [9, 10] concerns the variants C677T (alanine to valine) in the catalytic domain and A1298C (glutamate to alanine) in the regulatory domain of the methylenetetrahydrofolate reductase (MTHFR) gene, known to decrease the activity of the MTHFR gene and that patients could be inappropriately counseled for being at high risk for thrombotic episodes due to the difficulty of distinguishing between cis compound heterozygotes and trans compound heterozygotes. Therefore, it is important to study the joint effect of multiple variants.

This paper focuses on predicting the joint effect of variants from a single gene using a previously proposed hidden Markov model, HMMvar [11]. As the hidden Markov model is computed from the multiple protein sequence alignment for homologous proteins from different species, it reflects extent of evolutionary conservation naturally by its probabilistic profile. The probabilistic profile can be used to compute and compare the likelihood of generating mutant baring sequences given the HMM with the likelihood of generating mutant free sequences, i.e., wild type sequences, given the HMM. The lower the former compared with the latter, the more deleterious the mutants are likely. Therefore, HMMvar is able to predict the functional effect of a single mutation, as well as the joint effect of multiple mutations in coding regions.

To demonstrate the effectiveness of HMMvar, data from the 1000 Genomes project is used to identify genes that have multiple mutations, and HMMvar is used to predict the effect of multiple mutations on the genes identified. In addition, indels from two tumor suppressor genes, TP53 and phosphatase and tensin homolog (PTEN), are also used to investigate the effect of multiple indels from a single gene. If a frameshift indel occurs, it is possible that a nearby second indel rescues the gene by restoring the reading frame. There is a very limited knowledge about this kind of compensatory indels, but these are important because the deleterious effect of frameshift indels could be minimized by nearby compensatory indels. Hu and Pauline [5] claim that frameshift indels near each other are more likely to restore the translation frame. The present work found compensatory indel sets for TP53 and PTEN and measured the functional effects of individual indels and compensatory indel sets using HMMvar.

Materials and methods

Determine haplotypes by genotypes for the 1000 Genomes Project data

All the variants from the 1000 Genomes project Phase I along with their genotypes and ancestry alleles are collected to find the determined haplotype of an individual from a single gene. In order to quantitatively measure the effect of multiple variants on the same gene, the variant sets are formed in terms of their genotypes and the corresponding ancestral alleles. Given a certain gene and an individual sample, the variants are grouped into four classes based on the location and genotype. Figure 1 illustrates the classification of variants from a gene with three transcripts.

- Class 1: variants that are in the coding regions and the genotypes are homozygous and different from the ancestry allele, as the red variants shown in Fig. 1.
- Class 2: variants that are in the coding regions and the genotypes are homozygous and the same as the ancestry allele, as the green variants shown in Fig. 1.
- Class 3: variants that are in the coding regions and the genotypes are heterozygous, as the blue variants shown in Fig. 1.
- Class 4: variants that are not in the coding regions, such as 3'-utr, 5'-utr, or intron regions, as the orange variants shown in Fig. 1.

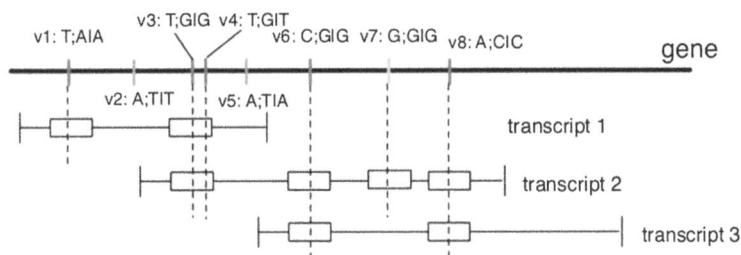

Fig. 1 An example of variant classification in terms of genotypes. The *colored sticks* on the gene represent variants at different locations. *Colors* represent different classes of variants. The format v1 : T; A|A means variant v1's ancestral allele is T and the genotype is A|A, the same as other variants. The *boxes* on the transcripts represent exon regions. The gene and the transcripts share the same coordinate system

Only the variants in class 1 are kept as a set to be scored, because all the variants in Class 1 are homozygous and are mutants compared to the ancestral alleles. They can form a determined haplotype for a sample individual. As shown in the Fig. 1 example, a variant set is formed for each of the transcripts: Transcript 1 contains variant set $\{v_1, v_3\}$; Transcript 2 contains variant set $\{v_3, v_6, v_8\}$; Transcript 3 contains variant set $\{v_6, v_8\}$. Finally, these sets will be scored against the corresponding transcripts by HMMvar. The homozygotes detected in individual samples along with the set score are available in the database for further analysis.

For each gene (transcript by considering alternative splicing), all homologous genotype variants that are different from the ancestry allele are identified based on an individual sample. As a result, a transcript related to a certain gene might be associated with multiple variant sets due to the difference of genotypes among samples, and a variant set can also be associated with multiple transcripts due to alternative splicing. Table 1 shows an example illustrating the relationship between individual, gene, and variant set. Only the records related to two individuals are shown here as an example (there are actually 2566 records related to gene ABCB5). As shown, gene ABCB5 is associated with multiple variant sets and even the same transcript (NM_178559.5) is associated with multiple variant sets due to the difference of genotypes of different individuals. The same variant set corresponds to multiple transcripts and multiple individuals. Finally, processing all genes that contain at least one variant set with size greater than one yielded 67,109 variant sets from 8021 genes (14,917 transcripts) involving 1092 individual samples.

Compensatory indels in TP53 and PTEN

The indels from two tumor suppressor genes, TP53 and PTEN, are collected from two databases, International Agency for Research on Cancer (IARC) [12] and Catalogue of Somatic Mutations in Cancer (COSMIC) [13].

The 4736 variations (3565 for TP53 and 1171 for PTEN) include frameshift or in-frame insertions, deletions, and complexes (both insertion and deletion take place simultaneously in one location) in coding regions (Table 2).

The effect of a deleterious mutation at the sequence level could be compensated for or alleviated by another mutation. For example, frameshift caused by a one base pair deletion could be recovered by a one base pair insertion nearby. A compensatory indel set is two or more indels that combine to preserve the open reading frame [14]. To simplify the search for compensatory indels, we restrict the consideration of compensatory indel sets, preserving the open reading frame, to those satisfying four conditions: (i) the number of nucleotides inserted or deleted per indel is less than five; (ii) the length of each indel is not divisible by three; (iii) the combined length of all indels is divisible by three; and (iv) all indels in the set occur within 20 base pairs. A single variant in a compensatory indel set is corrected (preserves the reading frame) by combining all other variants in the set. This paper considers the compensatory indel sets that satisfy the above four conditions for each of the TP53 variants and PTEN variants. Dynamic programming was used to find compensatory indel sets for single variants, which is similar to a subset sum problem [15], but with three different sums (-3, 0, and 3). To bound the computational effort, the maximum size of a compensatory indel set is bounded at 10, and the maximum number of compensatory indel sets for each valid length (sums -3, 0, and 3) is bounded at 20. The effects of compensatory indels are evaluated by comparing the HMMvar score of a single variant (as the mutant type) with the HMMvar score of a compensatory indel set (as the mutant type).

HMMvar

According to the theory of natural selection, different regions of a functional sequence are subject to different selective pressures. Multiple sequence alignment reveals this by residual conservation in certain positions. Some

Table 1 Variants sets related to gene ABCB5

Individual ID	Transcript ID	Set ID
NA20805	NM_001163941.1	7619
NA20805	NM_001163942.1	3062
NA20805	NM_001163993.2	3062
NA20805	NM_178559.5	7619
NA20806	NM_178559.5	2807
NA20806	NM_001163993.2	3062
NA20806	NM_001163942.1	3062
NA20806	NM_001163941.1	2807
...	...	...

Table 2 Data description

Type	Database	
	IARC (TP53)	COSMIC (PTEN)
Insertion (in-frame)	90	7
Insertion (frameshift)	419	116
Deletion (in-frame)	364	43
Deletion (frameshift)	1016	283
Complex (in-frame)	94	8
Complex (frameshift)	53	19
Total	2036	476

positions are more conserved than others and less tolerant to mutations. HMMvar [11] embodied this theory by using profile hidden Markov models [16] to predict the effect of mutations. A profile HMM captures the characteristics of a multiple sequence alignment, from which quantitative conservation information (a probability) is obtained. Thus, a high probability of generation from the profile HMM for the wild type sequence and a low probability for the mutant sequence suggest that the mutation might be deleterious. HMMvar measures the fitness of a sequence against the profile HMM that represents a set of homologous proteins (ideally only orthologous proteins from different species). So, it is natural for HMMvar to score a mutant sequence with one or multiple variants. This property also enables scoring the joint effect of compensatory indels (defined later in the section). The odds ratio

$$O = \frac{P_w/(1-P_w)}{P_m/(1-P_m)} \quad (1)$$

is used to score the effect of indel mutations, where P_w (P_m) is the probability that the wild type (mutated type) protein sequence could have been generated by the profile HMM trained on a homologous protein sequence set, usually calculated by the Viterbi algorithm. The higher the O is, the more deleterious the mutation or mutations. In general, when $O > 1$, the mutation or mutations might be deleterious, when $O \approx 1$, the mutation or mutations might be neutral, and when $O < 1$, the mutation or mutations might be beneficial. For scoring the effects of SNP mutation or mutations [11], the odds ratio score is not as reliable as the bit score (computed by the HMMER3 package) difference $\log_2(P_w/P_m)$ used here for SNPs.

Results

Scoring multiple SNPs

Processing the variants from the 1000 Genomes project resulted in scoring 67,109 SNP sets. A SNP set may be formed from different transcripts, which results in multiple scores for a set (there are 91,970 set scores in total). For a SNP set and transcript pair, HMMvar measures the deleterious effect of the SNP set using the original transcript as the wild type sequence. 291,662 single variants from those SNP sets were gathered and scored. The mean set score distribution is significantly different from the single variant score distribution (one-tailed Wilcoxon rank-sum test, $p < 2.2 \times 10^{-16}$). One thousand SNP set scores and 1000 single SNP scores are repeatedly sampled from 91,970 set scores and 275,840 single SNP scores. The cumulative distribution functions of the means of the set scores and single scores are shown in Fig. 2. The SNP sets are more likely to be scored higher than those of single SNPs. The density of SNP set scores tend to be higher than the density for individual SNP scores on both ends.

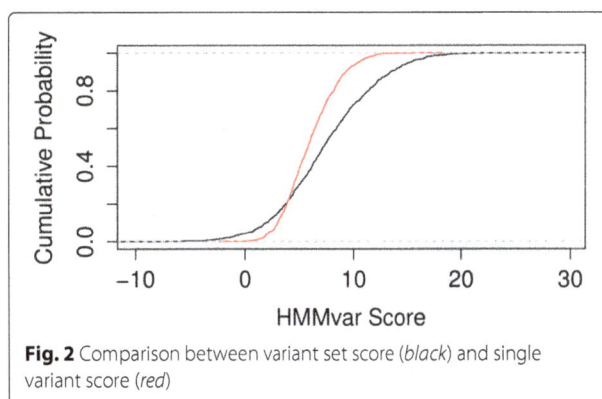

Fig. 2 Comparison between variant set score (*black*) and single variant score (*red*)

Let $V = \{v_1, v_2, \ldots, v_n\}$ be a set of variants v_i ($1 \leq i \leq n$), S denote the HMMvar score of the set V, and s_1, s_2, ..., s_n be the corresponding single variant scores of v_1, v_2, ..., v_n, respectively. Define V as a compensatory mutation (CM) set if $S \leq \min\{s_1, s_2, \ldots, s_n\} - 1.5$ ($\max\{s_1, s_2, \ldots, s_n\} - \min\{s_1, s_2, \ldots, s_n\}$). One hundred eighteen CM sets were obtained from the data set. The CM sets indicate that the deleterious effect of a single variant is compensated by combining it with other variants.

Define V as a noncompensatory mutation (nonCM) set if $S \geq \max\{s_1, s_2, \ldots, s_n\} + 1.5$ ($\max\{s_1, s_2, \ldots, s_n\} - \min\{s_1, s_2, \ldots, s_n\}$). Two thousand three hundred ninety-two nonCM sets were obtained from the data set. The nonCM sets indicate the joint effect of multiple neutral variants could possibly result in deleterious effect.

To investigate the single variants in the CM and nonCM sets, all the single variants from all the CM sets and all the nonCM sets are gathered, respectively. The allele frequency distributions from these two groups are compared in Fig. 3. When the allele frequency is less than 0.1, the proportion of the nonCM variants is greater than that

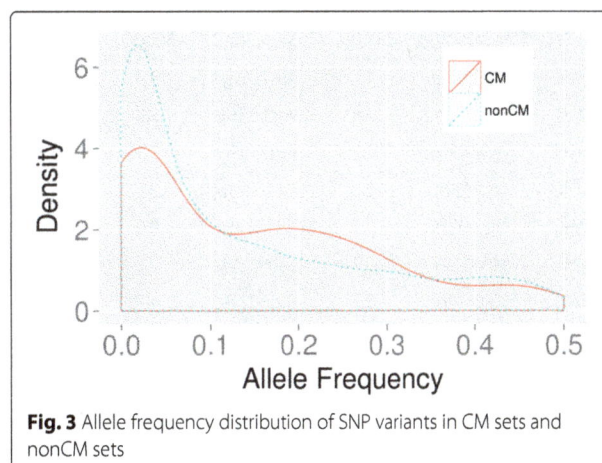

Fig. 3 Allele frequency distribution of SNP variants in CM sets and nonCM sets

of the CM variants. This is probably because the single variants are so deleterious that in most of cases, the joint effect of these deleterious variants is still deleterious. However, when the allele frequency is in the range of 0.1 to 0.3, the signal of the compensatory mutation effect is boosted.

As a test case for HMMvar's capability in predicting the effect of multiple variants compared to the effect of single variants, the multiple mutations that have been shown to increase the severity of cardiovascular disease from single mutations are scored, in β-myosin heavy chain (MHC) and myosin-binding protein C (MyBP-C) genes. Studies have shown that single mutations in these two genes can lead to genetic cardiovascular disease, and multiple mutations on these same genes can lead to more severe cardiovascular disorders and even death [17]. As shown in Table 3 for both genes, compound mutations all have higher HMMvar scores than single mutations, consistent with the notion that compound mutations in these genes cause more severe cardiovascular disease than single mutations. The set score effectively reflects the cumulative effects of the single mutations. The maximum score for compound missense mutations in the βMHC gene is the combination of Arg719Trp and Met349Thr, which has been reported causing sudden death [17].

Scoring compensatory indels
From TP53, 850 variants were found that met the criterion for belonging to a compensatory indel set, out of 3565 variants. The deleterious functional effects caused by these variants can be greatly weakened by compensatory indels as measured by HMMvar scores. There may be different compensatory indel sets for a given single variant due to different combinations. Figure 4a shows the HMMvar score of a single variant versus the median of the HMMvar scores of the corresponding compensatory indel sets. It is obvious that most of the deleterious

variants (high HMMvar scores) are neutralized by the compensatory indel sets (low HMMvar scores $\approx$ 1).

PTEN is also an intensively studied tumor suppressor gene. Figure 4b shows the HMMvar score of 246 variants versus the median HMMvar score of the corresponding compensatory indel sets, which shows the same trend as the TP53 variants. This scoring procedure provides candidate compensatory indel sets, which when substituted for the indel, ameliorate the deleterious effect of that single mutation. For instance, the deleterious variant c.142delA (COSMIC428080) associated with skin cancer [18] has HMMvar score 1.75; however, with compensatory indels, the deleterious effect can be lessened to a HMMvar score of 1.07. At the same time, the results here demonstrate the importance of scoring multiple variants together, instead of individually, to understand their joint effect.

Discussion
A single mutation, if not detrimental, can still exist in populations with low frequency. Over time, other mutations can also occur and thus multiple mutations can accumulate on the same gene. Compared to individual mutations, multiple mutations can be either more deleterious or less deleterious, the latter being known as compensatory mutation. Although it is not known which scenario is more prevalent in evolution, both scenarios have ample literature. Multiple mutations on the same gene, also called compound mutations, have been found to contribute or be linked to various genetic diseases (cf. [19–21]). In fact, a recent survey [17] of genetic cardiovascular disease led the authors to propose that multiple mutations, as opposed to single mutations, can be used as the genetic marker for the severity of cardiovascular disease, illustrating the importance of taking into account multiple mutations in disease outcome predictions. However, the current algorithms for predicting variant effect are limited to a single variant, a SNP or an indel. To fill this research gap, the present work proposes extending HMMvar, a hidden Markov model-based scoring method [11], to predict the effect of any number of mutations in any combinations (i.e., SNPs and/or indels).

Results show that multiple mutations do tend to have different effects on genes compared to single mutations, as reflected by the significant difference in the distributions of the HMMvar scores (one-tailed Wilcoxon rank-sum test, $p < 2.2 \times 10^{-16}$). The HMMvar scores of multiple SNPs tend to be larger than those of single SNPs, suggesting that many of these multiple mutations exacerbate the deleterious effect of single mutations. Note that while scoring the SNPs discovered in the 1000 Genomes project, the scored variants are identified by the next-generation sequencing (NGS) data where short sequences are generated and compared to the human reference genome to identify variants. Therefore, the genotype of

Table 3 Scoring multiple mutations in β MHC and MyBP-C genes

Gene	Mutation1	Score1	Mutation2	Score2	Set score
βMHC	Val39Met	1.7	Arg723Cys	3.4	5.0
βMHC	Pro211Leu	2.4	Arg663His	2.2	4.5
βMHC	Met349Thr	2.2	Arg719Trp	3.1	5.3
βMHC	Arg663His	2.2	Val763Met	2.3	4.4
βMHC	Arg719Gln	1.7	Thr1513Ser	0.0	1.6
βMHC	Asp906Gly	2.7	Leu908Val	2.0	4.6
MyBP-C	Gly5Arg	1.7	Arg502Trp	4.9	5.7
MyBP-C	Arg502Trp	3.9	Ser858Asn	2.4	6.4
MyBP-C	Alu542Gln	2.2	Ala851Val	2.2	4.4
MyBP-C	Asp745Aly	3.9	Pro873His	4.0	7.9
MyBP-C	Arg810His	2.9	Arg820Gln	2.6	5.5

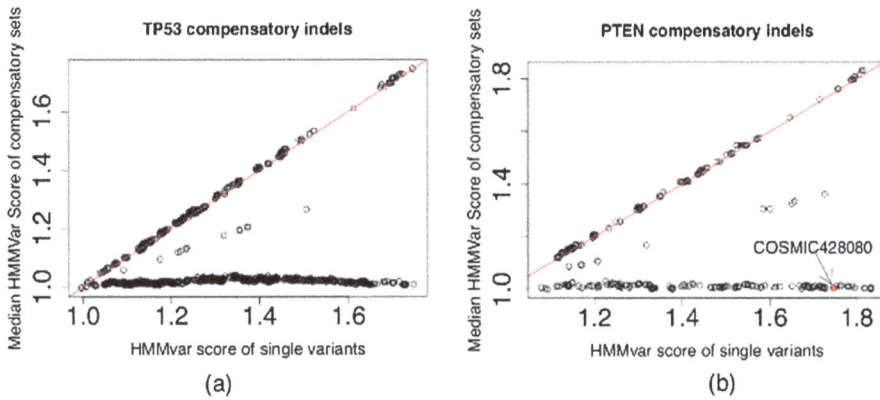

Fig. 4 Scatter plot of HMMvar score of a single variant versus the median HMMvar score of the corresponding compensatory indel sets for the TP53 gene and the PTEN gene. The *red line* is $y = x$. **a** TP53 compensatory indels. **b** PTEN compensatory indels. The *red solid circle* marks the COSMIC variant with ID 428080

the mutant individual is unknown, as is whether multiple mutations exist on the same allele or different alleles. To circumvent this problem, only those variants that are in homozygous state are scored. Figure 5 shows the zygosity of disease-causing mutations or any mutations in general. Single variants could be in a heterozygous (a) or homozygous (b) state. For multiple variants on the same gene (Fig. 5c–e shows two mutations as an example), there are three possible scenarios: trans compound heterozygous (c), cis compound heterozygous (d), or compound homozygous (e). This study scored compound mutations as scenario (e), so the two mutations are linked on the same allele. Homozygous mutations, single or multiple, cannot be detrimental as individuals with them in homozygous state will not be able to survive.

When multiple mutations occur and accumulate on the same gene, it is possible that though deleterious by themselves, they come together and become less deleterious or even beneficial to the carrier due to either recovery of the original gene function or gain of new function. This type of mutation, known as compensatory mutation, has been documented in the literature with many of the cases found in bacteria and viruses [22–24]. Potential compensatory indels were identified in two tumor suppressor genes, TP53 and PTEN, where compensatory indels are composed of frameshift indels that can recover the original reading frame. Results show that the HMMvar scores for the effect of compensatory indels are indeed much lower than the scores of the frameshift indels, with many of them close to one (Fig. 4), suggesting that compensatory indels can rescue the deleterious effect of frameshift indels. Similarly, Fig. 3 shows that SNPs with putative compensatory effect (CM) tend to have higher frequencies in the 1000 Genomes data than those SNPs predicted to be noncompensatory (nonCM, Fig. 3).

HMMvar can predict the effect of a set of multiple variants in its entirety. This is especially useful when multiple variants occur in a protein, each of which may have deleterious effects on the protein function, but the combination of them may be less deleterious due to a compensatory

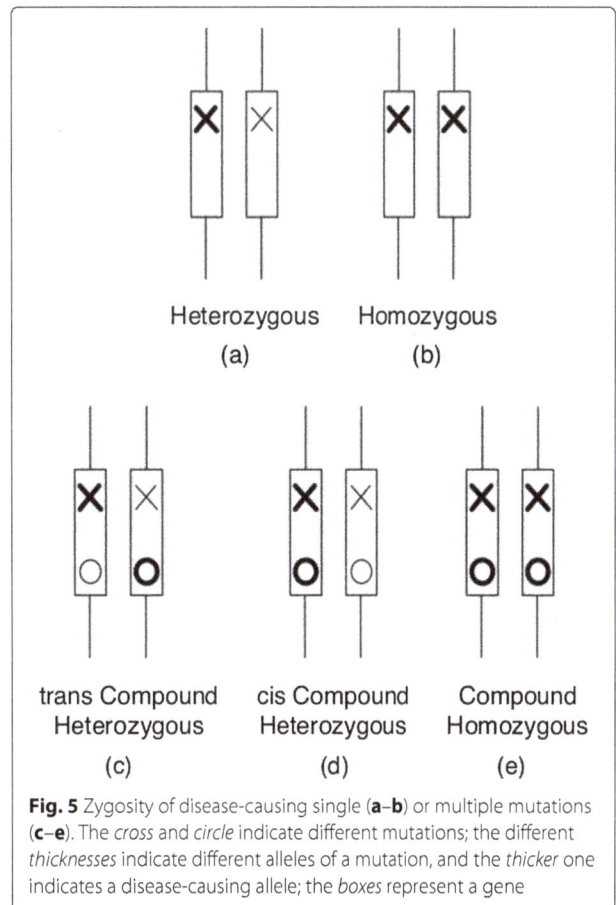

Fig. 5 Zygosity of disease-causing single (**a–b**) or multiple mutations (**c–e**). The *cross* and *circle* indicate different mutations; the different *thicknesses* indicate different alleles of a mutation, and the *thicker* one indicates a disease-causing allele; the *boxes* represent a gene

effect. Profile HMMs, used as proposed, have the capability to predict the joint effect of multiple mutations along the gene given a specific haplotype. Due to current technological limitations, inferring genotypes of a gene is still a challenge and little data exists that can be used for understanding the effect of multiple variations on the same gene. With future sequencing technology, long sequences may be generated and genotypes of a gene may be determined with certainty, in which case the HMMvar method will be of great use in understanding the joint impact of multiple mutations, in addition to single mutations, and better identification of disease contributing/causing variations.

Competing interests

The authors declare that they have no competing interests.

Authors' contributions

ML, LTW, and LZ wrote the paper. ML performed the computational experiments. LTW proposed the HMMvar D and O scores. LZ proposed the use of HMMs for variant effect prediction. All authors read and approved the final manuscript.

Acknowledgements

The work was partially supported by NIH grant AI085091 to Zhang. We thank Tony Reiter for the technical support for constructing the website. This article is partially supported by Virginia Tech's Open Access Subvention Fund.

Author details

[1] Department of Computer Science, Virginia Polytechnic Institute and State University, Blacksburg, VA, USA. [2] Department of Mathematics, Virginia Polytechnic Institute and State University, Blacksburg, VA, USA. [3] Department of Aerospace and Ocean Engineering, Virginia Polytechnic Institute and State University, Blacksburg, VA, USA.

References

1. Cooper G, Shendure J. Needles in stacks of needles: finding disease-causal variants in a wealth of genomic data. Nat Rev Genet. 2011;12:628–40.
2. Pauline C, Henikoff S. Predicting deleterious amino acid substitutions. Genome Res. 2011;11:863–74.
3. Ramensky V, Bork P, Sunyaev S. Human non-synonymous snps: server and survey. Nucleic Acids Res. 2002;30:3894–900.
4. Zia A, Moses A. Ranking insertion, deletion and nonsense mutations based on their effect on genetic information. BMC Bioinformatics. 2011;12:299.
5. Hu J, Pauline C. Predicting the effects of frame shifting indels. Genome Biol. 2012;13:2.
6. Choi Y, Sims G, Murphy S. Predicting the functional effect of amino acid substitutions and indels. PLoS ONE. 2012;7:10.
7. Loeb KR, Loeb LA. Significance of multiple mutations in cancer. Carcinogenesis. 2000;21(3):379–85.
8. Chang WC, Fang YY, Chang HW1, Chuang LY, Lin YD, Hou MF, et al. Identifying association model for single-nucleotide polymorphisms of orai1 gene for breast cancer. Cancer Cell Int. 2014;14:29.
9. Zetterberg H, Regland B, Palmer M, Ricksten A, Palmqvist L, Rymo, L, et al. Increased frequency of combined methylenetetrahydrofolate reductase c677t and a1298c mutated alleles in spontaneously aborted embryos. Eur J Hum Genet. 2002;10:113–8.
10. Brown NM, Pratt VM, Buller A, Pike-Buchanan L, Redman JB, Sun W, et al. Detection of 677ct/1298ac "double variant" chromosomes: implications for interpretation of mthfr genotyping results. Genet Med. 2005;7:278–82.
11. Liu M, Watson LT, Zhang L. Quantitative prediction of the effect of genetic variation using hidden Markov models. BMC Bioinformatics. 2014;15:5.
12. Petitjean A, Mathe E, Kato S, Ishioka C, Tavtigian SV, Hainaut P, et al. Impact of mutant p53 functional properties on tp53 mutation patterns and tumor phenotype: lessons from recent developments in the iarc tp53 database. Hum Mutat. 2007;28:622–9.
13. Forbes SA, Bindal N, Bamford S, Cole C, Kok CY, Beare D, et al. Cosmic: mining complete cancer genomes in the catalogue of somatic mutations in cancer. Nucleic Acids Res. 2010;39:945–50.
14. Williams LE, Wernegreen JJ. Sequence context of indelmutations and their effect on protein evolution in a bacterial endosymbiont. Bioinformatics. 2013;5:599–605.
15. Martello S, Toth P. 4 subset-sum problem. In: Knapsack problems: algorithms and computer interpretations. Chichester: Wiley Interscience; 1990. p. 105–36.
16. Eddy S. Profile hidden markov models. Bioinformatics. 1998;14:755–63.
17. Kelly M, Seminarian C. Multiple mutations in genetic cardiovascular disease a marker of disease severity? Circ Cardiovasc Genet. 2009;2: 182–90.
18. Konopka B, Paszko Z, Janiec-Jankowska A, Goluda M. Assessment of the quality and frequency of mutations occurrence in pten gene in endometrial carcinomas and hyperplasias. Cancer Lett. 2002;178:43–51.
19. Kamphans T, Sabri P, Zhu N, Heinrich V, Mundlos S, Robinson PN, et al. Filtering for compound heterozygous sequence variants in non-consanguineous pedigrees. PLOS ONE. 2013;8:70151.
20. Rafi MA, Coppola S, Liu SL, Rao HZ, Wenger DA. Disease-causing mutations in cis with the common arylsulfatase a pseudodeficiency allele compound the difficulties in accurately identifying patients and carriers of metachromatic leukodystrophy. Mol Genet Metab. 2013;79(2):83–90.
21. Carlsen BC, Meldgaard M, Johansen JD, Thyssen JP, Menne T, Szecsi PB, et al. Filaggrin compound heterozygous patients carry mutations in trans position. Exp Dermatol. 2013;22(9):572–5.
22. Wu NC, Young AP, Dandekar S. Systematic identification of h274y compensatory mutations in influenza a virus neuraminidase by high-throughput screening. J Virol. 2013;87(2):1193–1199.
23. Poon A, Chaom L. The rate of compensatory mutation in the dna bacteriophage ϕx174. Genetics. 2005;170(3):989–99.
24. Gonzalez-Ortega E, Ballana E, Badia R, Clotet B, Este JA. Compensatory mutations rescue the virus replicative capacity of virip-resistant hiv-1. Antivir Res. 2011;92(3):479–83.

20

Ranking non-synonymous single nucleotide polymorphisms based on disease concepts

Hashem A Shihab[1], Julian Gough[2], Matthew Mort[3], David N Cooper[3], Ian NM Day[1] and Tom R Gaunt[1*]

Abstract

As the number of non-synonymous single nucleotide polymorphisms (nsSNPs) identified through whole-exome/whole-genome sequencing programs increases, researchers and clinicians are becoming increasingly reliant upon computational prediction algorithms designed to prioritize potential functional variants for further study. A large proportion of existing prediction algorithms are 'disease agnostic' but are nevertheless quite capable of predicting when a mutation is likely to be deleterious. However, most clinical and research applications of these algorithms relate to specific diseases and would therefore benefit from an approach that discriminates between functional variants specifically related to that disease from those which are not. In a whole-exome/whole-genome sequencing context, such an approach could substantially reduce the number of false positive candidate mutations. Here, we test this postulate by incorporating a disease-specific weighting scheme into the Functional Analysis through Hidden Markov Models (FATHMM) algorithm. When compared to traditional prediction algorithms, we observed an overall reduction in the number of false positives identified using a disease-specific approach to functional prediction across 17 distinct disease concepts/categories. Our results illustrate the potential benefits of making disease-specific predictions when prioritizing candidate variants in relation to specific diseases. A web-based implementation of our algorithm is available at http://fathmm.biocompute.org.uk.

Keywords: SNV, nsSNPs, Disease-causing, Disease-specific, FATHMM, HMMs, SIFT, PolyPhen, Bioinformatics

Background

The average human exome harbours around 20,000 single nucleotide variants (SNVs), of which approximately half are annotated as non-synonymous single nucleotide polymorphisms (nsSNPs) [1]. However, characterizing the functional consequences of nsSNPs by direct laboratory experimentation is both time consuming and expensive. Therefore, computational prediction algorithms capable of predicting and/or prioritizing putatively functional variants for further experimentation are becoming increasingly important.

There is a plethora of computational prediction algorithms capable of analysing the functional consequences of nsSNPs [2]. One of these methods is the Functional Analysis through Hidden Markov Models (FATHMM) algorithm [3]: a sequence-based method which combines evolutionary conservation in homologous (both

orthologous and/or paralogous) sequences with 'pathogenicity weights', representing the overall tolerance of proteins (and their component domains) to mutations. Using our original weighting scheme (adjusted for inherited disease mutations), we observed an improved predictive performance over alternative computational prediction algorithms using a 'gold-standard' benchmark [4]. Nonetheless, these algorithms, including our own, were not designed to discriminate between nsSNPs influencing a specific disease (disease-specific) and other putative disease-causing/functional mutations (non-specific). For example, when tasked with discriminating between cancer-associated and other germline polymorphisms, these algorithms are capable of identifying a high proportion of cancer-promoting mutations. However, a large proportion of putative disease-causing (non-neoplastic) mutations are misclassified as having a role in carcinogenesis [5]. In both a clinical and a research context, these tools are commonly used to investigate the aetiology of specific diseases. We therefore believe that there is a significant need for disease-specific functional variant predictions.

---

* Correspondence: Tom.Gaunt@bristol.ac.uk
[1]Bristol Centre for Systems Biomedicine and MRC Integrative Epidemiology Unit, School of Social and Community Medicine, University of Bristol, Oakfield House, Oakfield Grove, Bristol BS8 2BN, UK

To the best of our knowledge, computational prediction algorithms have been explored exclusively in a *gene-specific* manner, e.g. predicting the effects of nsSNPs in mismatch repair proteins [6,7]. The sole context in which *disease-specific* predictions have been developed is in the prediction of cancer-associated mutations [8-11]. In our previous work, we adapted our original algorithm by means of a cancer-specific weighting scheme and observed improved predictive performances over alternative (cancer-specific) computational prediction algorithms when predicting the functional consequences of cancer-associated nsSNPs [12]. We have now extended this concept to a novel and more comprehensive 'disease-specific' weighting scheme to investigate whether such an approach is capable of prioritizing nsSNPs based on 17 disease concepts/categories.

Results

In order to assess the potential benefits of making disease-specific predictions, we compared the performance of our disease-specific weighting scheme with the performance of our original algorithm (weighted for inherited disease mutations) and two (generic) computational prediction algorithms: SIFT [13] and PolyPhen-2 [14]. In our analysis, all generic prediction algorithms were found to be capable of discriminating between disease-causing mutations (i.e. both disease-specific and non-specific disease-causing mutations) and putative neutral polymorphisms (see Additional file 1: Supp. Info 1). However, our analysis showed that no distinction could be made between disease-specific and other non-specific disease-causing mutations when using these algorithms. For example, generic algorithms are incapable of discriminating between musculoskeletal-related variants and other disease-associated variants, thereby leading to high false positive rates (i.e. other disease-causing variants being incorrectly identified as being pathogenic with respect to musculoskeletal-related disease). On the other hand, it appears that a disease-specific approach to functional prediction is capable of distinguishing between disease-specific and other disease-causing mutations, thereby reducing the number of false positives identified and improving the overall performance of the algorithm. While our disease-specific approach is more

Table 1 Performance of computational prediction algorithms when discriminating between disease-specific variants and other disease-causing/neutral variants

Algorithm	tp	fp	tn	fn	Accuracy	Precision	Specificity	Sensitivity	NPV	MCC	AUC
Musculoskeletal											
SIFT	4,730	37,701	23,323	944	0.61	0.57	0.38	0.83	0.70	0.24	0.64
PolyPhen-2	5,278	44,047	34,859	714	0.66	0.61	0.44	0.88	0.79	0.36	0.71
FATHMM	5,902	51,596	29,202	201	0.66	0.60	0.36	*0.97*	*0.92*	0.41	0.73
Disease-Specific	4,120	3,123	77,675	1,983	*0.82*	*0.95*	*0.96*	0.68	0.75	*0.66*	*0.93*
Disease-Specific (20-fold)	-	-	-	-	0.80	0.92	0.94	0.66	0.74	0.63	-
Developmental											
SIFT	845	41,586	23,983	284	0.56	0.54	0.37	0.75	0.59	0.12	0.56
PolyPhen-2	920	48,405	35,337	236	0.61	0.58	0.42	0.80	0.67	0.23	0.63
FATHMM	1,006	52,429	33,278	188	0.62	0.58	0.39	*0.84*	*0.71*	0.26	0.59
Disease-Specific	621	710	84,997	573	*0.76*	*0.98*	*0.99*	0.52	0.67	*0.58*	*0.90*
Disease-Specific (20-fold)	-	-	-	-	0.74	0.97	0.99	0.49	0.66	0.55	-
Endocrine											
SIFT	3,084	39,347	23,443	824	0.58	0.56	0.37	0.79	0.64	0.18	0.60
PolyPhen-2	2,890	46,435	35,031	542	0.64	0.60	0.43	0.84	0.73	0.30	0.67
FATHMM	3,597	49,466	33,522	316	0.66	0.61	0.40	*0.92*	*0.83*	0.38	0.71
Disease-Specific	2,392	1,015	81,973	1,521	*0.80*	*0.98*	*0.99*	0.61	0.72	*0.65*	*0.94*
Disease-Specific (20-fold)	-	-	-	-	0.79	0.97	0.98	0.60	0.71	0.63	-
Metabolic											
SIFT	10,731	31,700	21,913	2,354	0.61	0.58	0.41	0.82	0.69	0.25	0.64
PolyPhen-2	11,337	37,988	33,788	1,785	0.67	0.62	0.47	0.86	0.78	0.36	0.72
FATHMM	13,068	39,914	33,271	648	0.70	0.64	0.45	*0.95*	*0.91*	0.47	0.80
Disease-Specific	10,767	3,209	69,976	2,949	*0.87*	*0.95*	*0.96*	0.78	0.82	*0.75*	*0.95*
Disease-Specific (20-fold)	-	-	-	-	0.86	0.94	0.95	0.77	0.81	0.74	-

specific than generic computational prediction algorithms, it would appear that this approach is also less sensitive. This general trend of greater specificity/less sensitivity was observed throughout the 17 disease concepts we tested (see Table 1 and Figure 1—data shown for musculoskeletal, developmental, endocrine and metabolic disorders; see Additional file 1: Supp. Info 2–18 for additional performance comparisons pertinent to the remaining disease concepts). These results illustrate the potential benefit of using a disease-specific approach to functional prediction when assessing nsSNPs in relation to specific diseases (by reducing the number of false positives identified); however, further work is needed to reduce the number of false negatives identified and improve sensitivity.

In the above, tp, fp, tn and fn refer to the number of true positives, false positives, true negatives and false negatives observed, respectively. Accuracy, precision, specificity, sensitivity, negative predictive value (NPV) and Matthew's correlation coefficient (MCC) were calculated using normalized numbers. Italic font corresponds to the best performing method for a given statistic.

As our weighting scheme was derived using the same mutation data used to assess our method (albeit using a leave-one-out analysis), we recognize the potential for bias. Therefore, we also performed a 20-fold cross-validation analysis (see Table 1 and Additional file 1: Supp. Info 2–18). We observed no significant deviations in the performance measures reported and therefore concluded that

the performance of our disease-specific approach is not an artefact of over-fitting. We also recognize that most of our algorithm's predictive power comes from our weighting scheme, i.e. it is the weighting scheme that allows us to differentiate between disease-associated variants and other disease-causing mutations. Therefore, we also compared our approach to a naive weighting scheme. Here, we used our weighting scheme (omitting sequence conservation) to derive a prediction score. Proteins, and their constituent domains, with a higher proportion of disease-associated mutations would predict all variants falling within them as *disease*, and those with a higher proportion of other disease-causing mutations/neutral polymorphisms would predict all variants as *neutral*. Overall, we observed a similar performance to that of our algorithm (see Additional file 1: Supp. Info 19–36). However, it should be noted that a naive approach is incapable of reliably discriminating between disease-associated mutations and other disease-causing variants as the weighting scheme becomes more balanced, whereas our disease-specific approach (which incorporates sequence conservation for prediction) appears to be less susceptible to balanced weights.

In order to facilitate the replication of our work, we have annotated SwissProt/TrEMBL disease variants (Release 2014_06) with the disease concepts used in our analysis and make this resource publically available at our website (http://fathmm.biocompute.org.uk). Using this dataset to train and test our algorithm, we observed similar

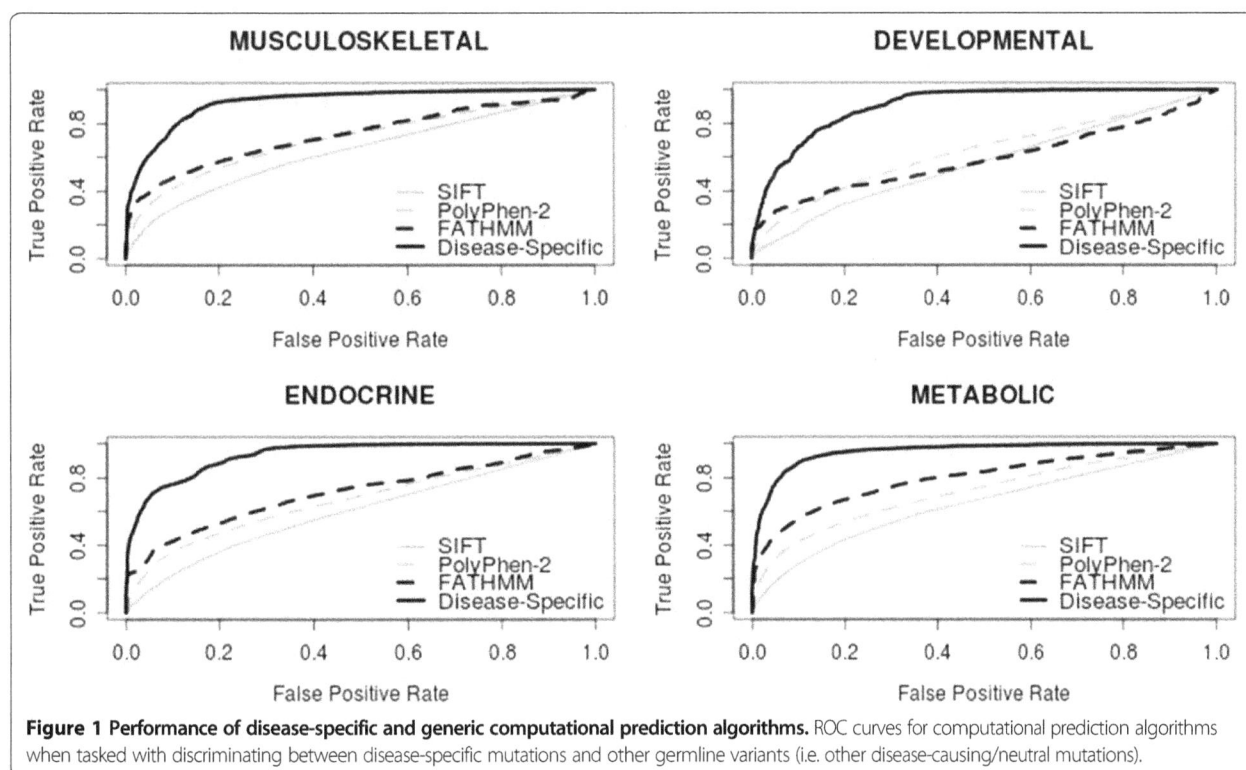

Figure 1 Performance of disease-specific and generic computational prediction algorithms. ROC curves for computational prediction algorithms when tasked with discriminating between disease-specific mutations and other germline variants (i.e. other disease-causing/neutral mutations).

performances to those reported above (see Additional file 2).

Discussion

There is a plethora of computational prediction algorithms available to predict the functional consequences of nsSNPs [2]. However, these algorithms are not designed to distinguish between mutations related to a specific disease, or a group of related diseases (disease-specific), and other putative disease-causing (non-specific) mutations. As the cost of whole-exome/whole-genome sequencing falls, making these methods more amenable to use in a research or clinical context, the challenge of filtering true disease-causing candidate variants from other putative functional variants is likely to become increasingly important. In this work, we assessed the potential benefits of making disease-specific predictions (relevant to 17 disease categories) using the Functional Analysis through Hidden Markov Models (FATHMM) framework and observed an overall reduction in the number of false positives identified, thereby leading to improved specificity over traditional algorithms. However, we also observed an increase in the number of false negatives identified and conclude that additional work is needed to improve sensitivity and enhance the utility of our disease-specific approach. Nevertheless, there is potential to extend this approach to more specific categories for the purposes of enhancing clinical prediction.

An important consideration when evaluating the performance of computational prediction algorithms is the cross-validation dataset. Here, the performance of such algorithms should be trained and tested using different datasets (cross-validation). In order to alleviate the potential for bias in our results, we performed a 20-fold cross-validation procedure across our 17 disease concepts. From this analysis, we observed no significant deviations in the reported performance measures and therefore conclude that the performances observed were not an artefact of our disease-specific weighting scheme.

One of the major limitations of our disease-specific approach is that, in extreme cases, there is potential for dominating pathogenicity weights which could bias or exaggerate the effects of variants, e.g. when prioritizing variants in proteins and/or domains which have very strong associations with the disease concept under investigation. Here, the pathogenicity weights used could dominate the underlying amino acid probabilities (used to measure sequence conservation) and therefore bias the prediction. For example, when these weights are biased towards the disease concept, neutral polymorphisms falling within diverse regions of a protein/domain would be classified as 'damaging' as opposed to being classified as 'benign'. As a consequence, our disease-specific models are best suited as a whole-genome/whole-exome prioritization method (hypothesis-free) and should be used with caution

when prioritizing variants in a gene-specific manner. In an attempt to alleviate the potential effects of dominating pathogenicity weights, measures of sequence conservation are presented alongside our rankings so that spurious predictions can be assessed and ignored.

An alternative approach to our disease-specific weighting scheme is to filter putative disease-causing nsSNPs using the Gene Ontology [15]. However, this approach is dependent upon protein annotations being made available whereas our algorithm does not require prior information on protein function. Furthermore, users adopting this approach are required to select from a range of technical phrases, e.g. 'negative regulation of cellular macromolecule biosynthetic process' (GO: 2000113). In contrast, our disease-specific models do not require any formal knowledge on GO terms and biological processes, just an understanding of which model/concept best represents the disease under investigation. Our disease-specific models, including a high-throughput web-based implementation of our algorithm and a standalone software package, are available at http://fathmm.biocompute.org.uk.

Methods

Predicting the functional consequences of nsSNPs

The procedure for predicting the functional consequences of nsSNPs has been described in Shihab et al. [3]. In brief, an *ab initio* hidden Markov model (HMM), representing the multiple sequence alignment of homologous (both orthologous and/or paralogous) sequences within the SwissProt/TrEMBL [16] database, is constructed using the HMMER3 [17] software suite. In conjunction, protein domains from the SUPERFAMILY [18] and Pfam (Pfam-A and Pfam-B) [19] databases are annotated onto the full-length protein sequence. If the mutation falls within an annotated region, then the corresponding model is extracted and used alongside our *ab initio* model. Next, our algorithm combines sequence conservation, within the most informative model (as measured by the Kullback-Leibler divergence [20] from the SwissProt/TrEMBL amino acid composition), with pathogenicity weights, representing the overall tolerance of the corresponding model to mutations (Equation 1).

$$\ln \frac{(1.0-P_\text{w})(W_\text{n}+1.0)}{(1.0-P_\text{m})(W_\text{d}+1.0)} \quad (1)$$

In Equation 1, P_w and P_m represent the probabilities for the wild-type and mutant amino acid residues, respectively, whereas W_d and W_n represent the relative frequencies of disease-associated and functionally neutral nsSNPs mapping onto the corresponding model, respectively. Here, we use inherited disease-causing nsSNPs annotated as DMs (damaging mutations) in the Human Gene Mutation Database (HGMD Pro 12.4 [21]) and putative neutral polymorphisms from the SwissProt/TrEMBL database [16] (Release 2013_04) to derive W_d and W_n, respectively. The effect of our weighting scheme is as follows: when using pure conservation-based prediction methods, nsSNPs falling within diverse regions of the protein (or domain) are typically considered 'neutral/benign'. However, our weighting scheme assesses the tolerance of the corresponding model (representing a protein or domain) to mutation and then adjusts a conservation-based prediction accordingly. For example, nsSNPs falling within *P53* (a well-established cancer gene) are penalized according to the gene's intolerance to mutation whereas nsSNPs falling within *MHC* (known to contain hypervariable regions) are not penalized given the gene's apparent tolerance to mutation.

Incorporating a disease-specific weighting scheme

In order to derive a disease-specific weighting scheme, the phenotypes reported for inherited disease-causing nsSNPs listed as DMs (damaging mutations) in the Human Gene Mutation Database (HGMD Pro 12.4 [21]) were annotated using natural language processing against the Unified Medical Language System (UMLS [22]). These mutations were then grouped into 1 (or more) of 17 different root disease concepts, e.g. digestive disorders ([23] —see Table 2

Table 2 Summary of nsSNPs used in our disease-specific mutation datasets

Dataset	Number of proteins	Number of amino acid substitutions
Human Gene Mutation Database (HGMD)		
Blood	99	1,474
Blood coagulation	45	3,508
Developmental	188	1,199
Digestive	116	1,850
Ear, nose and throat	113	943
Endocrine	192	3,913
Eye	227	3,031
Genitourinary	166	3,031
Heart	247	3,743
Immune	75	1,293
Metabolic	485	13,797
Musculoskeletal	309	6,110
Nervous system	473	8,553
Psychiatric	163	747
Reproductive	88	883
Respiratory	44	775
Skin	164	3,183
SwissProt/TrEMBL		
Putative neutral polymorphisms	11,601	37,488

for the complete list). For disease-specific predictions, our original weighting scheme (see Equation 1) is replaced with the relative frequencies of disease-specific mutations and other non-specific disease-causing mutations/neutral polymorphisms mapping onto the model, i.e. our pathogenic training set consists of disease-causing mutations related to the disease concept whereas our neutral training set comprises all other disease-causing mutations (not related to the corresponding disease concept) and putative neutral mutations. This disease-specific weighting scheme has the same effect as our original weighting scheme (i.e. to penalize specific variants); however, this approach penalizes just those variants falling within disease-specific susceptible proteins or domains and treats other disease-causing mutations as neutral polymorphisms (with respect to the disease concept under investigation).

Performance statistics

In accordance with published guidelines [24], the following six parameters are used to assess the performance of our disease-specific models:

$$\text{Accuracy} = \frac{\text{tp} + \text{tn}}{\text{tp} + \text{tn} + \text{fp} + \text{fn}}$$

$$\text{Precision} = \frac{\text{tp}}{\text{tp} + \text{fp}}$$

$$\text{Sensitivity} = \frac{\text{tp}}{\text{tp} + \text{fn}}$$

$$\text{Specificity} = \frac{\text{tn}}{\text{fp} + \text{tn}}$$

$$\text{NPV} = \frac{\text{tn}}{\text{tn} + \text{fn}}$$

$$\text{MCC} = \frac{(\text{tp} \cdot \text{tn}) - (\text{fn} \cdot \text{fp})}{\sqrt{(\text{tp} + \text{fn})(\text{tp} + \text{fp})(\text{tn} + \text{fn})(\text{tn} + \text{fp})}}$$

In the aforementioned data, tp and fp refer to the number of true positives and false positives reported, respectively, whereas tn and fn refer to the number of true negatives and false negatives reported, respectively. Receiver operating characteristic (ROC) and area under the curve (AUC) analyses were performed using the ROCR software suite [25].

Additional files

Additional file 1: Performance of computational prediction algorithms. This file reports the performance of computational prediction algorithms when tasked with discriminating between inherited disease-causing mutations, disease-specific mutations and neutral polymorphisms.

Additional file 2: Performance of computational prediction algorithms using SwissProt/TrEMBL. This file reports the performance of our disease-specific algorithm and two generic computational prediction algorithms: SIFT and PolyPhen-2, when tasked with discriminating between disease-specific mutations and other disease-causing mutations/neutral polymorphisms in SwissProt/TrEMBL.

Competing interests

The authors declare that they have no competing interests.

Authors' contributions

HAS participated in the design of the study and performed the analysis. TRG, JG and INMD participated in the design and coordination of the study. DNC and MM provided the training data and corresponding disease annotations. All authors read and approved the final manuscript.

Acknowledgements

This work was supported by the UK Medical Research Council (MRC) [MC_UU_12013/8 and G1000427/1] and was carried out in the Bristol Centre for Systems Biomedicine (BCSBmed) Doctoral Training Centre (director INMD) using the computational facilities of the Advanced Computing Research Centre, University of Bristol—http://www.bris.ac.uk/acrc. JG's contribution was supported by the Biotechnology and Biological Sciences Research Council (BBSRC) [BB/G022771]. MM and DNC gratefully acknowledge the financial support of BIOBASE GmbH.

Author details

[1]Bristol Centre for Systems Biomedicine and MRC Integrative Epidemiology Unit, School of Social and Community Medicine, University of Bristol, Oakfield House, Oakfield Grove, Bristol BS8 2BN, UK. [2]Department of Computer Science, University of Bristol, The Merchant Venturers Building, Bristol BS8 1UB, UK. [3]Institute of Medical Genetics, School of Medicine, Cardiff University, Cardiff CF14 4XN, UK.

References

1. Bamshad MJ, Ng SB, Bigham AW, Tabor HK, Emond MJ, Nickerson DA, Shendure J: Exome sequencing as a tool for Mendelian disease gene discovery. Nat Rev Genet 2011, 12:745–755.
2. Thusberg J, Olatubosun A, Vihinen M: Performance of mutation pathogenicity prediction methods on missense variants. Hum Mutat 2011, 32:358–368.
3. Shihab HA, Gough J, Cooper DN, Stenson PD, Barker GLA, Edwards KJ, Day INM, Gaunt TR: Predicting the functional, molecular, and phenotypic consequences of amino acid substitutions using hidden Markov models. Hum Mutat 2013, 34:57–65.
4. Sasidharan Nair P, Vihinen M: VariBench: a benchmark database for variations. Hum Mutat 2013, 34:42–49.
5. Kaminker JS, Zhang Y, Waugh A, Haverty PM, Peters B, Sebisanovic D, Stinson J, Forrest WF, Bazan JF, Seshagiri S, Zhang Z: Distinguishing cancer-associated missense mutations from common polymorphisms. Cancer Res 2007, 67:465–473.
6. Ali H, Olatubosun A, Vihinen M: Classification of mismatch repair gene missense variants with PON-MMR. Hum Mutat 2012, 33:642–650.
7. Thompson BA, Greenblatt MS, Vallee MP, Herkert JC, Tessereau C, Young EL, Adzhubey IA, Li B, Bell R, Feng B, Mooney SD, Radivojac P, Sunyaev SR, Frebourg T, Hofstra RM, Sijmons RH, Boucher K, Thomas A, Goldgar DE, Spurdle AB, Tavtigian SV: Calibration of multiple in silico tools for predicting pathogenicity of mismatch repair gene missense substitutions. Hum Mutat 2013, 34:255–265.
8. Kaminker JS, Zhang Y, Watanabe C, Zhang Z: CanPredict: a computational tool for predicting cancer-associated missense mutations. Nucleic Acids Res 2007, 35:W595–W598.
9. Carter H, Chen S, Isik L, Tyekucheva S, Velculescu VE, Kinzler KW, Vogelstein B, Karchin R: Cancer-specific high-throughput annotation of somatic mutations: computational prediction of driver missense mutations. Cancer Res 2009, 69:6660–6667.
10. Reva B, Antipin Y, Sander C: Predicting the functional impact of protein mutations: application to cancer genomics. Nucleic Acids Res 2011, 39:e118.
11. Gonzalez-Perez A, Deu-Pons J, Lopez-Bigas N: Improving the prediction of the functional impact of cancer mutations by baseline tolerance transformation. Genome Med 2012, 4:89.
12. Shihab HA, Gough J, Cooper DN, Day INM, Gaunt TR: Predicting the functional consequences of cancer-associated amino acid substitutions. Bioinformatics 2013, 29:1504–1510.
13. Ng PC, Henikoff S: Predicting deleterious amino acid substitutions. Genome Res 2001, 11:863–874.
14. Adzhubei IA, Schmidt S, Peshkin L, Ramensky VE, Gerasimova A, Bork P, Kondrashov AS, Sunyaev SR: A method and server for predicting damaging missense mutations. Nat Methods 2010, 7:248–249.
15. Ashburner M, Ball CA, Blake JA, Botstein D, Butler H, Cherry JM, Davis AP, Dolinski K, Dwight SS, Eppig JT, Harris MA, Hill DP, Issel-Tarver L, Kasarskis A, Lewis S, Matese JC, Richardson JE, Ringwald M, Rubin GM, Sherlock G: Gene ontology: tool for the unification of biology. The Gene Ontology Consortium. Nat Genet 2000, 25:25–29.
16. Apweiler R, Bairoch A, Wu CH, Barker WC, Boeckmann B, Ferro S, Gasteiger E, Huang H, Lopez R, Magrane M, Martin MJ, Natale DA, O'Donovan C, Redaschi N, Yeh LS: UniProt: the Universal Protein knowledgebase. Nucleic Acids Res 2004, 32:D115–D119.
17. Eddy SR: A new generation of homology search tools based on probabilistic inference. Genome Inform 2009, 23:205–211.
18. Gough J, Karplus K, Hughey R, Chothia C: Assignment of homology to genome sequences using a library of hidden Markov models that represent all proteins of known structure. J Mol Biol 2001, 313:903–919.
19. Sonnhammer EL, Eddy SR, Durbin R: Pfam: a comprehensive database of protein domain families based on seed alignments. Proteins 1997, 28:405–420.
20. Kullback S, Leibler RA: On information and sufficiency. Ann Math Stat 1951, 22:79–86.
21. Stenson PD, Mort M, Ball EV, Howells K, Phillips AD, Thomas NS, Cooper DN: The Human Gene Mutation Database: 2008 update. Genome Med 2009, 1:13.
22. Lindberg DA, Humphreys BL, McCray AT: The Unified Medical Language System. Methods Inf Med 1993, 32:281–291.
23. Mort M, Evani US, Krishnan VG, Kamati KK, Baenziger PH, Bagchi A, Peters BJ, Sathyesh R, Li B, Sun Y, Xue B, Shah NH, Kann MG, Cooper DN, Radivojac P, Mooney SD: In silico functional profiling of human disease-associated and polymorphic amino acid substitutions. Hum Mutat 2010, 31:335–346.
24. Vihinen M: Guidelines for reporting and using prediction tools for genetic variation analysis. Hum Mutat 2013, 34:275–282.
25. Sing T, Sander O, Beerenwinkel N, Lengauer T: ROCR: visualizing classifier performance in R. Bioinformatics 2005, 21:3940–3941.

A study of the role of GATA4 polymorphism in cardiovascular metabolic disorders

Nzioka P Muiya, Salma M Wakil, Asma I Tahir, Samya Hagos, Mohammed Najai, Daisy Gueco, Nada Al-Tassan, Editha Andres, Nejat Mazher, Brian F Meyer and Nduna Dzimiri*

Abstract

Background: The study was designed to evaluate the association of *GATA4* gene polymorphism with coronary artery disease (CAD) and its metabolic risk factors, including dyslipidaemic disorders, obesity, type 2 diabetes and hypertension, following a preliminary study linking early onset of CAD in heterozygous familial hypercholesterolaemia to chromosome 8, which harbours the *GATA4* gene.

Results: We first sequenced the whole *GATA4* gene in 250 individuals to identify variants of interest and then investigated the association of 12 single-nucleotide polymorphisms (SNPs) with the disease traits using Taqman chemistry in 4,278 angiographed Saudi individuals. Of the studied SNPs, rs804280 (1.14 (1.03 to 1.27); $p = 0.009$) was associated with CAD (2,274 cases vs 2,004 controls), hypercholesterolaemia (1,590 vs 2,487) (1.61 (1.03–2.52); $p = 0.037$) and elevated low-density lipoprotein-cholesterol (hLDLC) (575 vs 3,404) (1.87 (1.10–3.15); $p = 0.020$). Additionally, rs3729855_T (1.52 (1.09–2.11); $p = 0.013$)) and rs17153743 (AG + GG) (2.30 (1.30–4.26); $p = 0.005$) were implicated in hypertension (3,312 vs 966), following adjustments for confounders. Furthermore, haplotypes CCCGTGCC ($x^2 = 4.71$; $p = 0.041$) and GACCCGTG ($x^2 = 3.84$; $p = 0.050$) constructed from the SNPs were associated with CAD and ACCCACGC ($x^2 = 6.58$; $p = 0.010$) with myocardial infarction, while hypercholesterolaemia ($x^2 = 3.86$; $p = 0.050$) and hLDLC ($x^2 = 4.94$; $p = 0.026$) shared the AACCCATGT, and AACCCATGTC was associated with hLDLC ($x^2 = 4.83$; $p = 0.028$). A 10-mer GACCCGCGCC ($x^2 = 7.59$; $p = 0.006$) was associated with obesity (1,631 vs 2,362), and the GACACACCC ($x^2 = 4.05$; $p = 0.044$) was implicated in type 2 diabetes mellitus 2,378 vs 1,900).

Conclusion: Our study implicates GATA4 in CAD and its metabolic risk traits. The finding also points to the possible involvement of yet undefined entities related to GATA4 transcription activity or gene regulatory pathways in events leading to these cardiovascular disorders.

Keywords: GATA4 gene polymorphism, Haplotypes, Coronary artery disease, Dyslipidaemia, High-density lipoprotein-cholesterol, Hypercholesterolaemia, Hypertriglyceridaemia, Low-density lipoprotein-cholesterol

Background

The GATA binding proteins constitute a family of cell-restricted zinc-finger transcription factors (TFs), which recognize the GATA motif present in the promoters of many genes. This family, comprising six developmental/cell-type specific transcription factors, GATA1-6, is critical to the development of diverse tissues [1-4] and acts in cooperation with more widely expressed factors to direct lineage-specific gene expression [5-11]. Their transcriptional activity is modulated through interactions with nuclear proteins, including the zinc finger proteins of the Kruppel and FOG/U-shaped families, general co-activators of the p300 and cAMP-response element-binding (CREB) protein (CBP), the myocardial-expressed protein Nkx2.5, and NF-AT3 [12-15]. In particular, in the cardiac tissue where its expression is more than 20-fold greater than in other tissues [16], it is a critical regulator of cardiac angiogenesis and gene expression, and is involved in modulating cardiomyocyte differentiation and adaptive responses of the adult heart [5,17,18]. Additionally, GATA4 is abundantly expressed in the endocardium and endothelial cells, pointing to an important regulatory control in their development and function [11,19]. The

* Correspondence: dzimiri@kfshrc.edu.sa
Genetics Department, King Faisal Specialist Hospital and Research Centre, Riyadh 11211, Saudi Arabia

abundance and localization of the GATA4 to the myocardial tissue suggests a pivotal cardiovascular functional input, which if perturbed, may lead to various cardiac disorders and coronary artery disease (CAD)-related events. However, while the importance of this gene is now well appreciated in congenital heart diseases [20-26] and some other cardiac malformations [27-29], little is known about its potential role in vascular activity-related diseases, such as dyslipidaemia and diabetes mellitus. Dyslipidaemia can be triggered as a result of a genetic predisposition, secondary causes, or a combination of both. However, the genetic causes of this disorder remain to be identified. In a preliminary study investigating the genomic linkage to early onset of CAD in two Saudi families with heterozygous familial hypercholesterolaemia (HFH), we identified a locus on chromosome 8, which harbours the GATA4 gene, as a plausible candidate for CAD, HFH and harbouring of low high-density lipoprotein levels. This led to the notion that GATA4 presents a potential candidate for CAD onset, especially in dyslipidaemic conditions. Given the lack of information on the role of this gene in the etiology of atherosclerosis, our study sought to comprehensively investigate the likelihood of GATA4 polymorphism predisposing individuals to acquiring the cardiovascular metabolic risk traits, particularly dyslipidaemia, obesity, type 2 diabetes mellitus and hypertension, as a potential trigger for the disease onset related to these disorders. To this effect, we first sequenced the gene in the two HFH families and an additional 250 individuals from the Saudi general population to identify potentially informative variants, and then performed a population-based association study for selected variants with CAD and its risk traits in a larger cohort of angiographed Saudi individuals.

Results

Linkage analysis for dyslipidaemia and early onset coronary artery disease

The initial scanning of the HFH yielded several peaks in different genomic regions including Chromosome (Chr) 8, among others, giving a logarithm of the odds (LOD) score of 1.8 that isolated at least three of the affected siblings with the early onset of CAD in both families. One of the potential culprits at this locus is GATA4, which we elected to pursue further for its role in dyslipidaemia-related onset of CAD. Subsequent sequencing of the genes in all members of the two HFH families and an additional 250 individuals led to the selection of 12 single-nucleotide polymorphisms (SNPs), rs2740434_G > A (minor allele frequency = 0.26), rs17153743_A > G (0.02), rs13264774_C > T (0.14), rs56298569_G > C (0.44), rs804280 A > C (0.04), rs3729855 C > T (0.18), rs3729856_A > G (0.45), rs1062219 C > T (0.45), rs12825_C > G (0.09), rs804291_C > T (0.32), rs11785481 C > T (0.12), rs3203358 C > G (0.20) for further case-control studies. Selection of the SNPs was based partly on the prevalence in our general population and partly on currently available information of their role in the disease. Furthermore, these SNPs reside in the later portion of the gene (also encompassing its three prime untranslated region, 3'-UTR), which encodes the C-terminal of the protein and harbours gene regulatory motifs, both of which are thought to be important in the transcriptional activity of the GATA4 (Figure 1). We were also curious to understand the potential role of changes in the 3'-UTR in the disease. Figure 2 displays the linkage disequilibrium (LD) structure of the ten SNPs included in the haplotyping.

GATA4 genotyping and disease

Since GATA4 constitutes an established risk gene for congenital heart disease, it was necessary to rule out the

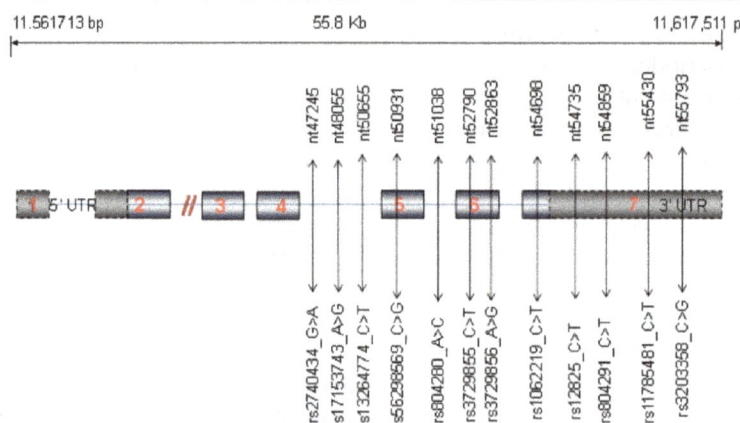

Figure 1 Schematic of GATA4 gene sequence on chromosome 8p23.1. Diagram shows relative loci of studied variants and their positions (drawn not to scale). Numbers in rectangles represent the exons.

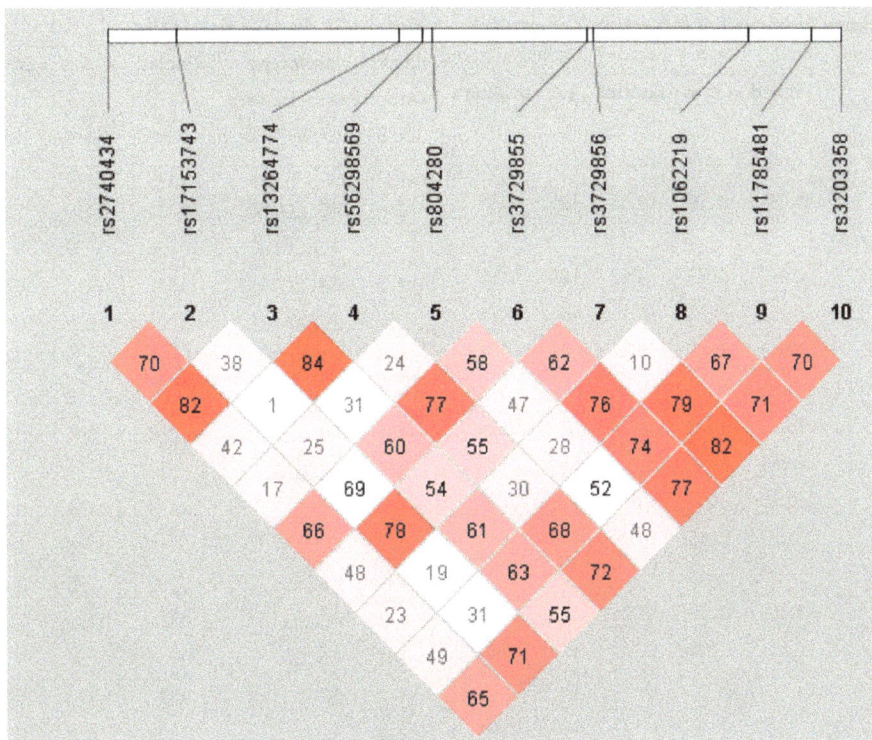

D'=0.013-0.83; r²=0.002-0.089

Figure 2 Linkage disequilibrium (LD) structure of ten SNPs included in haplotyping. D', coefficient of linkage disequilibrium; r, regression coefficient.

likelihood of such potential relationships masking those pertaining to disease trait-gene interactions under investigation. We therefore first tested for associations of the gene variants with congenital heart disease as an independent patient group ($n = 113$) in our study population, using a homogenous set of individuals with dilated cardiomyopathy ($n = 272$) as controls. The results pointed to a significant causative association for the rs3729856 A > G (p.S377G) (Odds ratio (95% confidence interval) = 1.76 (1.19–2.61); $p = 0.005$), the rs3729856_GG (5.09 (2.14–12.06); $p < 0.000001$), and rs11785481_TT (6.23 (1.93–20.17); $p < 0.002$) with congenital heart disease. Interestingly, in the multivariate analyses for congenital heart disease versus the rest of the studied population (113 cases vs 3,969 non-disease individuals), only the rs11785481_T (1.78 (1.06–3.00); $p = 0.030$) retained its significant association (Table 1), pointing to a strong link that could potentially mask the relationships for the other diseases in our current study setting. Furthermore, multivariate analyses showed an association of rs804280_C (1.14 (1.03–1.27); $p = 0.009$) with CAD (Table 1), while the rs1062219C > T ($p = 0.036$) lost this property, following the adjustment for confounders. Interestingly, the rs804291_TT was associated with both hypercholesterolaemia (hChol) (1.61 (1.03–2.52)

and elevated low-density lipoprotein-cholesterol (hLDLC) (1.87 (1.10–3.15); $p = 0.020$). In contrast, the rs3203358_G was protective against acquiring both CAD (0.88 (0.78–0.99); $p = 0.042$) and hChol (0.86 (0.75–0.98); $p = 0.026$) (Table 1). Besides, the rs3729855_T (p.N352N) (1.52 (1.09–2.11) and rs17153743 (AG + GG) (2.30 (1.30–4.26); $p = 0.005$) were implicated in hypertension, while the former appeared to be protective against type 2 diabetes mellitus (0.68 (0.53–0.88); $p = 0.003$) (Table 2), also following the Bonferroni corrections for age, sex, and other confounders (Additional file 1, GATA4 Suppl data).

GATA4 haplotyping and cardiovascular disease traits
Since several SNPs were associated with the different cardiovascular disease traits, we were interested in testing the likelihood of these relationships being delineable at the haplotype level. We employed the most frequent 10-mer GACACACCCG (frequency = 0.145) as the baseline for the analyses. While the 9-mer haplotype, ACCCGTGCC ($x^2 = 3.80$; $p = 0.051$) was weakly associated with CAD, its 8-mer derivative CCCGTGCC ($x^2 = 4.71$; $p = 0.041$) was associated with the disease. This relationship became stronger with the shortening of these haplotype sequences, culminating in the 4-mer ACCC ($x^2 = 7.17$; $p = 0.007$) displaying the most significant

Table 1 Association of GATA4 gene variants with coronary artery disease, myocardial infarction congenital heart disease

Block	Haplotype	Pooled	Cases	Control	χ^2	p value
Hypercholesterolaemia						
1–9	AACCCATGT	0.063	0.071	0.060	3.86	0.050
5–10	CATGTC	0.088	0.096	0.082	4.25	0.039
High low-density lipoprotein						
1–10	AACCCATGTC	0.062	0.079	0.062	4.83	0.028
	GACCCATGCC	0.015	0.024	0.015	4.94	0.026
1–9	AACCCATGT	0.063	0.078	0.062	4.31	0.038
Hypertriglyceridaemia						
2–7	ACACAT	0.083	0.072	0.088	4.79	0.029
2–6	ACCTA	0.013	0.018	0.011	4.65	0.031
3–6	CCTA	0.016	0.021	0.014	4.34	0.037
1–5	AACCT	0.01	0.015	0.009	4.49	0.034
Hypertension						
1–8	GACCCACG	0.02	0.018	0.026	4.26	0.039
1–6	GACCCA	0.103	0.098	0.119	6.78	0.009
Type 2 diabetes mellitus						
1–9	GACACACCC	0.18	0.192	0.175	4.05	0.044
Obesity						
1–10	GACCCGCGCC	0.025	0.032	0.022	7.59	0.006**
	GACCCGCGC	0.026	0.033	0.022	7.42	0.006**
2–10	ACCCGCGCC	0.026	0.031	0.022	6.55	0.011
3–10	CCCGCGCC	0.026	0.031	0.022	6.55	0.011
1–8	GACCCGCG	0.028	0.033	0.024	5.27	0.022
1–7	GACCCGC	0.028	0.033	0.024	6.14	0.013
4–10	CCGCGCC	0.026	0.031	0.022	5.97	0.015
3–9	CCCGCGC	0.027	0.033	0.023	6.88	0.009
2–8	ACCCGCG	0.028	0.033	0.025	4.53	0.033
3–8	CCCGCG	0.027	0.032	0.024	5.23	0.022
3–7	CCCGC	0.029	0.034	0.025	5.33	0.021
5–9	CGCGC	0.030	0.035	0.026	5.08	0.024

The table shows selected haplotypes associated with the disease. The most frequent 10-mer haplotype (0.14) was employed as the baseline to determine the relative effects of the other haplotypes. The studied SNPs are rs2740434 (also denoted as 1), rs17153743 (2), rs13264774 (3), rs56298569 (4), rs804280 (5), rs3729855 (6), rs3729856 (7), rs1062219 (8), rs11785481 (9) and rs3203358 (10) arranged sequentially by their chromosomal positions, whereby blocks represent the range of variants constituting the respective haplotypes. $*p < 0.01$; $**p < 0.005$ by χ^2 test.

Table 2 Association of GATA4 gene variants with metabolic disease risk traits

Block	Haplotype	Pooled	Cases	Control	χ^2	p value
Coronary artery disease						
1–8	GACCCGTG	0.108	0.114	0.101	3.84	0.050
3–10	CCCGTGCC	0.110	0.116	0.102	4.17	0.041
2–9	ACCCGTGC	0.112	0.119	0.105	3.87	0.049
	ACCCATGC	0.026	0.029	0.022	4.08	0.044
1–7	GACCCGT	0.109	0.116	0.102	3.93	0.048
	CCCATGC	0.026	0.029	0.022	3.81	0.050
2–8	ACCCGTG	0.113	0.120	0.105	4.51	0.034
4–10	CCGTGCC	0.111	0.118	0.104	3.91	0.048
4–9	ACATCC	0.069	0.064	0.075	4.60	0.032
3–8	CCCGTG	0.114	0.121	0.106	4.63	0.031
2–7	ACCCGT	0.119	0.126	0.111	4.66	0.031
3–7	CCCGT	0.119	0.126	0.110	4.96	0.026
1–5	GACCC	0.239	0.250	0.228	5.48	0.019
4–8	CCGTG	0.116	0.122	0.108	4.21	0.040
1–4	GACC	0.244	0.254	0.232	5.69	0.017
4–7	CCGT	0.121	0.129	0.113	4.73	0.030
2–5	ACCC	0.403	0.416	0.387	7.17	0.007*
Myocardial infarction						
1–10	AACACATCCC	0.04	0.038	0.049	6.01	0.014
1–9	AACACATCC	0.041	0.038	0.050	6.33	0.012
1–8	AACACATC	0.042	0.038	0.050	6.54	0.011**
2–9	ACCCACGC	0.029	0.033	0.023	6.58	0.010**
3–9	CCCACGC	0.029	0.032	0.023	5.97	0.015
2–8	ACCCACG	0.033	0.037	0.026	6.76	0.009**
6–10	GTCCC	0.014	0.016	0.011	3.84	0.050
4–8	CCACG	0.034	0.038	0.028	5.60	0.018
4–8	ATAC	0.013	0.010	0.018	7.98	0.005**
2–5	ACCC	0.403	0.410	0.387	4.07	0.044

The table shows selected haplotypes associated with disease. The most frequent 10-mer haplotype (0.14) was employed as the baseline to determine the relative effects of the other haplotypes. The studied SNPs are rs2740434 (also denoted as 1), rs17153743 (2), rs13264774 (3), rs56298569 (4), rs804280 (5), rs3729855 (6), rs3729856 (7), rs1062219 (8), rs11785481 (9) and rs3203358 (10) arranged sequentially by their chromosomal positions, and blocks represent the range of variants constituting the respective haplotypes. $*p < 0.01$; $**p < 0.005$ by χ^2 test.

relationship with disease (Table 3; Additional file 2, GATA4 Haplo Suppl data). A similar trend was observed for the 8-mer GACCCGTG ($\chi^2 = 3.84$; $p = 0.050$) which also culminated in the 4-mer GACC ($\chi^2 = 5.69$; $p = 0.017$) showing the most significant association with the disease. Put together, these trends indicate that the SNPs rs17153743 (2), rs13264774 (3), rs56298569 (4) and rs804280 (5) constituted the primary core for

this association with CAD. Myocardial infarction was equipotently linked to the 8-mer ACCCACGC ($\chi^2 = 6.58$; $p = 0.010$) and its 7-mer derivative ACCCACG ($\chi^2 = 6.76$; $p = 0.009$), thereby sharing with CAD, the same three causative nucleotides of the four SNPs comprising the core of the relationships. Notable relationships also included the 10-mer AACACATCCC ($\chi^2 = 6.01$; $p = 0.014$) and the 4-mer ATAC ($\chi^2 = 7.98$; $p = 0.005$) showing the most significant protective property against acquiring MI.

Since the outgoing notion was that GATA4 may mediate dyslipidaemia-mediated onset of CAD, it was interesting to investigate whether haplotyping might reveal any link with lipidaemic disorders. Interestingly, hChol (x^2 = 3.86; p = 0.050) and elevated hLDLC levels (x^2 = 4.94; p = 0.026) shared the common 9-mer AACCCATGT, while the 10-mer AACCCATGTC was associated with hLDLC (x^2 = 4.83; p = 0.028), but only weakly so with hChol (x^2 = 3.24; p = 0.072) (Table 4; Additional file 2, GATA4 Haplo Suppl data). Hypertriglyceridaemia (hTG) was linked to the 5-mer ACCTA (x^2 = 4.65; p = 0.031), while low high-density lipoprotein was associated with the CCCAC (x^2 = 6.54; p = 0.011).

We then proceeded to investigate further potential relationships of these haplotypes with other metabolic risk traits, particularly hypertension, type 2 diabetes and obesity. The results showed that a 10-mer GACCCGCGCC (x^2 = 7.59; p = 0.006) was associated with obesity, while a 9-mer GACACACCC (x^2 = 4.05; p = 0.044) was implicated

in type 2 diabetes mellitus (Table 4). We also observed several protective haplotypes for all disease traits, which were primarily complementary to the core of the causative sequences (See also Additional file 2, GATA4 Haplo Suppl data). However, we could not establish any significant causative link with hypertension.

Discussion

In the present study, we first screened two families with HFH in which the primary probands presented with severe phenotypes of early onset CAD. We identified several potential loci of which Chr 8 appealed as the most attractive choice for detailed investigation on the genetic basis for dyslipidaemia-related onset of the disease. This locus harbours the *GATA4* gene at Chr 8p23.1-22, which we elected to study first as the most likely candidate. Screening the complete coding and non-coding areas of the gene revealed several mutations, among which we selected 12 for the case-control study. The study has

Table 3 Association of GATA4 haplotypes with coronary artery disease/myocardial infarction

Variant	Genotype/allele	Controls	Cases	Univariate analysis		Multivariate analysis	
				p value	Exp (B)(95% CI)	Corrected *p* value	Exp (B″)(95% CI)
Hypercholesterolaemia							
rs804291CT	TT	0.016	0.027	0.023*	1.66(1.07–2.57)	0.037*	1.61(1.03–2.52)
rs11785481CT	CT + TT	0.196	0.220	0.065	1.16(0.99–1.34)	0.087	1.15(0.98–1.35)
rs3203358CG	CG + GG	0.358	0.323	0.020*	0.85(0.75–0.98)	0.026*	0.86(0.75–0.98)
Hypertriglyceridaemia							
rs3729855CT	T	0.030	0.041	0.023	1.35(1.01–1.78)	0.004**	1.49(1.14–1.94)
	CT + TT	0.053	0.070	0.005**	1.34(1.09–1.64)	0.010*	1.47(1.10–1.96)
rs1062219CT	TT	0.220	0.182	0.011*	0.79(0.67–0.98)	0.016*	0.80(0.67–0.99)
Elevated low-density lipoprotein-cholesterol							
rs804291CT	TT	0.018	0.033	0.016*	1.90(1.13–3.21)	0.020*	1.87(1.10–3.15)
rs11785481CT	CT + TT	0.201	0.229	0.032*	1.18(1.01–1.37)	0.136	1.18(0.95–1.45)
Low high-density lipoprotein-cholesterol							
rs2740434CT	CT + TT	0.084	0.069	0.013*	0.81(0.69–0.96)	0.744	0.93(0.62–1.41)
Hypertension							
rs3729855CT	T	0.027	0.035	0.089	1.30(0.95–1.77)	0.013*	1.52(1.09–2.11)
rs17153743AG	AG + GG	0.017	0.031	0.022*	1.87(1.10–3.18)	0.005*	2.30(1.30–4.26)
rs13264774CT	T	0.041	0.024	0.008*	0.59(0.40–0.87)	0.085	0.68(0.44–1.05)
Type 2 diabetes mellitus							
rs3729855CT	T	0.040	0.028	0.003*	0.70(0.55–0.89)	0.003*	0.68(0.53–0.88)
rs13264774CT	CT + TT	0.035	0.023	0.025*	0.66(0.045–0.95)	0.199	0.76(0.51–1.16)
Obesity							
rs17153743AG	GG	0.138	0.070	0.031	0.73(0.55–0.98)	0.357	0.94(0.82–1.07)
rs11785481CT	TT	0.029	0.022	0.033*	0.73(0.55–0.98)	0.903	0.84(0.73–0.95)
rs2740434GA	GA + AA	0.086	0.065	0.013*	0.73(0.50–0.94)	0.011*	0.72(0.56–0.93)

The table summarizes the univariate and multivariate analyses for the relationships of the gene variants with various cardiovascular disease traits. Bonferroni tests have been performed to adjust for age and sex and adjustment made for the confounders in the multivariate tests. *p < 0.05; **p < 0.005.

unequivocally identified two SNPs, the rs1062219 and rs804280 as risk variants for CAD, both of which were also linked to congenital heart disease in the present study as well as other previous studies in different ethnic populations [25,26,30]. Therefore, our study does not only furnish strong support for an important role for these SNPs in congenital heart disease, but also points to a possible sharing of common disease pathways involved in the etiologies of CAD and congenital heart disease at the GATA4 signaling level. Notably, by far, the majority of GATA4 mutations studied to date seem to point to their influencing its transcriptional activity, and have been associated primarily with cardiac malformations, such as congenital heart disease. Thus, the difference in the nature of the congenital heart disease and CAD etiologies renders it very intriguing why these two disorders would share the GATA4, or any other signalling pathway for that matter, as a common disease pathway. A number of speculations have recently been advanced on how changes in GATA4 gene may influence disease pathways. One of these suggests an impediment of the GATA4 transcription activity in congenital heart disease as involving at least two of the variants, p.A348A and p.S377G [26], which were also included in the present study. Since GATA4 transcription activity is subject to regulation at the level of gene expression and through post-translational modifications of protein, the processes involved in these regulatory mechanisms are therefore worthy considering as possible culprits in our study. On the other hand, however, in our study, while the former was implicated in MI, the latter was actually protective against acquiring CAD. Accordingly, these observations seem to point to delineable differences between CAD and congenital heart disease in their interactions with the GATA4 variants. Besides, the ubiquity of GATA4 in the myocardium also raises fundamental questions with respect to possible diversity in its cardiac-related function(s) and its involvement in other cardiovascular disease pathways, such as those leading to atherosclerosis. Hence, the primary question raised in our study was whether a link existed between the impact of GATA4 on CAD manifestation and the presence of metabolic disorders, as our linkage study in HFH seemed to suggest.

To begin with, our results pointed to an association of the rs804291 and rs11785481 with both hChol and hLDLC, which appeared to follow primarily an autosomal recessive mode of inheritance. A closer analysis of the data also indicated that while the rs11785481 was not directly related to CAD per se, the probability of an individual acquiring CAD increased greatly in hypercholesterolaemic individuals. We also noted that two other SNPs that were not directly associated with hChol became so in the presence of MI, similarly indicative of a possible interaction of the dyslipidaemic disease traits with changes in GATA4 as a possible link to

CAD/MI manifestation in these individuals. Moreover, the analyses for the other metabolic risk factors revealed the association for two variants with hypertension, further linking risk traits for metabolic syndrome to GATA4 polymorphism. Altogether, these observations furnish support for the notion of a contribution of some interactions of metabolic risk traits with GATA4 to the disease pathways leading to atherosclerosis, an assertion which requires further investigation.

The various ways in which the GATA4 variants relate to alterations in lipid levels and CAD/MI in this study lead to important questions regarding the possible mechanisms or pathways linking them with one another. Based on our results, it appears that such mechanisms may involve events associated with the harbouring of lHDLC, for example, being linked to pathways directly influencing circulating lipid levels and possibly leading to acquiring CAD in dyslipidaemic individuals, as indicated by the inverse relationship of some of these traits with the different GATA4 variants. Accordingly, the processes appear to be separable from those leading to cardiac malformations. Hence, the novel finding implicating the GATA4 in dyslipidaemia seems to point to yet undefined entities, possibly involving the regulation of the GATA4 functional state, as playing an important role in the disease process. In this regard, perhaps the most revealing observation of the present study is the linking of noncoding variants of the gene to disease. Of particular interest was the finding of causative haplotypes for CAD/MI as well as the metabolic risk traits encompassing variants in the 3'-UTR of the gene. Thereby, isolating the two coding SNPs from those residing in the 3'-UTR did not alter the significance level for the latter, possibly pointing to the changes at this chromosomal locus rather than the individual variants as the underlying genetic basis for these manifestations. Notably, the significance levels for the haplotype associations were higher than those of individual constituent variants, pointing to the importance of these genomic sequences in revealing the impact of a gene on disease that would otherwise remain uncovered through associations with changes at individual loci. Implications are that these events may be due to changes other than those directly pertaining to a functional motif of the GATA4 protein. Furthermore, the close proximity of these variants may also suggest the presence of sequences encoding some other yet unidentified entities as the potential culprits, especially considering the fact that several 3-mer haplotypes at this locus were shared by some of the traits. Thus, it is plausible to postulate that alteration at this genomic locus, and not necessarily in the GATA4 gene function per se, may offer an explanation for the observed link to alterations in the metabolic risk traits and CAD manifestation.

A number of speculations have been raised with regard to the possible ways in which changes in the 3'-UTR of the GATA4 gene may influence disease pathways. For example, both germline and somatic mutations have recently been described in this region that were predicted to affect RNA folding as cause for congenital heart disease [25]. This is in line with the notion that the mechanisms by which *GATA4* contributes to hChol metabolism may be related to the regulation of the gene itself or its mRNA maturation rather than the transcriptional activity of its protein product. These mechanisms are likely to be the result of complex interactions with yet unidentified cofactors, as has been suggested recently by some studies [6]. While the study has produced interesting data that may contribute to our knowledge of GATA4 interaction with cardiovascular disease, there are some limitations in the extent of its applicability. Thus, one of the main limitations is that no such potential mechanisms were tested to verify the notion that the 3′-UTR of the *GATA4* gene plays an important role in the discussed cardiovascular disease pathways. Furthermore, like other association studies on complex diseases, the potential impact of the present findings may be limited to ethnic Arab populations due to inter-ethnic variations in prevalent epigenetic and environmental factors. Besides, it is not likely that our present findings per se can be exploited as predictive markers for the dyslipidaemic disorders.

Conclusion

In summary, our study identified the GATA4 transcription factor as an independent risk factor for congenital heart disease and CAD/MI and a metabolic risk trait for cardiovascular diseases. Thus, apart from the demonstrated association of *GATA4* polymorphism with dyslipidaemia, our results also point to interrelationships of the two disease components with CAD/MI, which may explain partly how these diseases lead to atherosclerosis. The finding of several causative haplotypes for MI/CAD embracing the 3′UTR of the *GATA4* gene points to important roles for this chromosomal locus in the etiology of CAD/MI and the possible involvement of yet undefined entities

related to GATA4 transcription activity or gene regulatory pathways in events leading to these cardiovascular disorders.

Materials and methods
Study population

The present study was performed in three stages involving three independent groups of Saudi individuals. The first group comprised two families of a total of 22 members in which HFH was prevalent (Figure 3). The index case of the first family was the third (S3) of seven sons and two daughters born to unrelated parents, who underwent triple coronary artery bypass surgery at the age of 14 years. On admission, this propositus had reported with a chest wound and xanthomas, as well as clinical features of bilateral carotid artery disease and a very severe form of familial hyperlipidaemia (FH). He presented with very high cholesterol (Chol) level of 10.1 mmol/L (desirable <5.2 mmol/l) and LDL-Chol level of 7.9 mmol/L (optimal <2.59 mmol/l). Furthermore, he harboured low HDL-Chol levels (0.51 mmol/l; normal 1.04–1.55). The father displayed the characteristics of borderline HFH (Chol 6.1 mmol/L, LDL 4.0 mmol/L), while the mother, two other sons (S4 and S6) and a daughter (D2) had clinical phenotypes of the disease (combination of high Chol (≥6.2 mmol/L) and high LDL-Chol (≥4.12 mmol/L)). Two other sons with otherwise normal Chol and LDL-Chol levels also displayed low HDL-Chol levels (<1.04 mmol/L), while none of the family members had isolated elevated LDL-Chol levels. In the second family, two index cases, two daughters of a family of seven daughters and two sons concurrently presented with early onset CAD at the age of 17 and 15 years. The two candidates had identical exceedingly high Chol levels of 22.5 mmol/l and LDL-Chol levels of 19.6 (D6) and 18.0 mmol/l (D7), in addition to harbouring low HDL levels of 0.93 and 0.65 mmol/l, respectively. The father (8.6 mmol/l), mother (7.7 mmol/l) and a third daughter (D3; 7.6 mmol/l) were

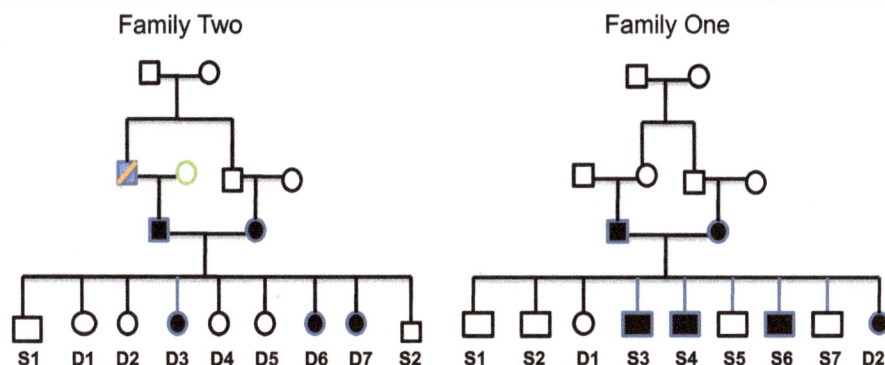

Figure 3 Pedigrees of two families with heterozygous hyperlipidaemia. Proband S3 in Family One underwent triple coronary artery bypass at the age of 14 years, as well as D6 and D7 in Family Two presented with early onset coronary artery disease at the age of 17 and 15 years.

also hypercholesterolaemic. The criteria for the diagnosis of familial hyperlipidaemia (FH) were adapted from our Institutional Guidelines, employing the reference values approved by the USA National Cholesterol Education Program (NCEP). Following the identification of Chr 8p23 region, which harbours the *GATA4* gene (8p23.1-22), as a potential culprit for the disease through linkage study of the above two families, we elected to sequence the gene in the two families and a group of 250 healthy blood donors visiting our Blood Bank Clinic to identify potentially informative SNPs of interest for the case-control association study.

The case-control study was performed in a total of 4,274 Saudi individuals consisting of 2,274 CAD patients (1,736 males and 538 females, mean age 60.8 ± 0.4 years) with angiographically determined narrowing of the coronary vessels by at least 50% and 2,004 angiographed controls (1,074 male and 930 female, mean age 50.2 ± 0.5 years) (Table 5). These controls (CON) were individuals undergoing surgery for valvular disease or reporting for follow-up on various other cardiac procedures, but were established to have no significant narrowing of the coronary vessels. Among the study population, 1,590 patients were hypercholesterolaemic (hChol) patients (1,051 male and 539 female, mean age 58.2 ± 0.2 years) undergoing anti-

hyperlipidaemic therapy and/or known to have high levels of total cholesterol (Chol >6.20 mmol/L), 1,776 patients harboured lHDLC levels (<1.04 mmol/L), 1,088 had high triglyceride (hTG) levels (>2.25 mmol/L), and 575 exhibited high low-density lipoprotein levels. These were denoted as dyslipidaemia study patients. Another subset of interest comprised 2,378 individuals having type 2 diabetes mellitus (T2DM; formerly known as non-insulin-dependent diabetes mellitus or adult-onset diabetes). The National Diabetes Data Group of the USA and the second World Health Organization Expert Committee on Diabetes Mellitus (1998) [31] define type 2 diabetes mellitus as a metabolic disorder that is characterized by high blood glucose (generally defined as fasting glucose level >126 mg/dL) in the context of insulin resistance and relative insulin deficiency. Furthermore, 3,312 individuals had primary (essential) hypertension (HTN) (Table 5), defined as ≥140 mm Hg systolic blood pressure or ≥90 mm Hg diastolic pressure based on The Sixth Report of the Joint National Committee on Prevention, Detection, Evaluation, and Treatment of High Blood Pressure (JNC VI) criteria [32]. Accordingly, essential, primary, or idiopathic hypertension is defined as high blood pressure in which secondary causes such as renovascular disease, renal failure, pheochromocytoma, aldosteronism, or other causes of secondary hypertension or

Table 4 Association of GATA4 haplotypes with metabolic disease risk traits

	Controls			Cases		
	All	Male	Female	All	Male	Female
Gender	2,004	1,074 (53.6)	930 (46.4)	2,274	1,736 (76.3)	538 (23.7)
Age	50.2 ± 0.5	50.81 ± 0.5	49.66 ± 0.5	60.8 ± 0.4	59.7 ± 0.3	61.8 ± 0.5
BMI	29.4 ± 0.2	28.01 ± 0.2	30.69 ± 0.2	29.7 ± 0.2	28.3 ± 0.1	31.1 ± 0.3
MI	1,388	684 (0.49)	704 (0.51)	2,890	2,126 (0.74)	764 (0.26)
T2DM	1,900	1,220 (0.64)	680 (0.36)	2,378	1,590 (0.67)	788 (0.33)
HTN	966	650(0.67)	314 (0.33)	3,312	2,158 (0.65)	1,154 (0.35)
lHDLC	2,209	1,262 (0.57)	947 (0.43)	1,776	1,371 (0.77)	405 (0.23)
hLDL	3,404	2,267 (0.67)	1,137 (0.33)	575	364 (0.63)	211 (0.37)
hTG	2,896	1,859 (0.64)	1,037 (0.36)	1,088	773 (0.71)	315 (0.29)
hChol	2,487	1,639 (0.66)	848 (0.34)	1,590	1,051 (0.66)	539 (0.34)
CHD	4,128	2,741 (0.66)	1,387 (0.34)	150	69 (0.46)	81 (0.54)
FH	3,421	2,255 (0.66)	1,166 (0.34)	857	555 (0.65)	302 (0.35)
OBS	2,362	1,729 (0.73)	633 (0.27)	1,631	895 (0.55)	736 (0.45)
Smokers	2,575	1,223 (0.47)	1,352 (0.53)	1,619	1,547 (0.96)	72 (0.04)
VD						
One	0	0	0	847	614 (0.72)	233 (0.28)
Two	0	0	0	456	355 (0.78)	101 (0.22)
More than two	0	0	0	971	767 (0.79)	204 (0.21)

The numbers in brackets give the percentages of the total (all) values of the group. BMI, body mass index; CHD, congenital heart disease; FH, family history of CAD; MI, myocardial infarction; lHDLC, low high-density lipoprotein-cholesterol level; hTG, hypertriglyceridaemia; hChol, hypercholesterolaemia; HTN, hypertension; T2DM, type 2 diabetes mellitus; VD, number of diseased vessels.

Table 5 Patient Demographics and clinical data

Variant	Genotype/allele	Controls	Cases	Univariate analysis		Multivariate analysis	
				p value	Exp (B)(95% CI)	Corrected p value	Exp (B″)(95% CI)
Coronary artery disease							
rs3729855CT	T	0.037	0.030	0.047*	0.78(0.62–1.00)	0.315	0.87(0.67–1.14)
rs3203358CG	G	0.206	0.191	0.074	0.90(0.82–1.00)	0.042*	0.88(0.78–0.99)
rs1062219CT	CT + TT	0.663	0.694	0.034*	1.15 (1.01–1.31)	0.091	1.13(0.98–1.31)
rs17153743AG	AG + GG	0.033	0.022	0.036*	0.67(0.47–0.98)	0.032*	0.67(0.43–0.96)
rs804280AC	C	0.420	0.446	0.018*	1.11(1.02–1.21)	0.009*	1.14(1.03–1.27)
	AC + CC	0.647	0.688	0.005**	1.20(1.06–1.36)	0.012*	1.20(1.03–1.39)
Myocardial infarction							
rs3729855CT	T	0.040	0.030	0.020*	0.75(0.59–0.96)	0.323	0.86(0.86–1.56)
rs3729856AG	GG	0.031	0.041	0.024*	1.34(1.04–1.72)	0.146	1.42(0.88–2.33)
rs13264774CT	CT + TT	0.269	0.247	0.029*	0.89(0.80–0.99)	0.103	0.84(0.69–1.04)
rs804280AC	AC + CC	0.645	0.681	0.020*	1.17(1.07–1.29)	0.511	1.07(0.88–1.28)
Congenital heart disease							
rs3729856AG	G	0.145	0.215	0.012*	1.61(1.11–2.30)	0.081	1.51(0.95–2.40)
	AG + GG	0.280	0.362	0.018*	1.45(1.07–1.99)	0.108	1.57(0.90–2.73)
rs12825CG	CG + GG	0.777	0.854	0.010*	1.68(1.13–2.50)	0.108	1.71(0.89–3.29)
rs11785481CT	T	0.115	0.154	0.037*	1.41(1.02–1.94)	0.030*	1.78(1.06–3.00)
	CT + TT	0.190	0.275	0.053*	1.61(1.15–2.27)	0.183	1.50(0.83–2.75)
rs2740434CT	CT + TT	0.073	0.123	0.020*	1.78(1.09–2.89)	0.403	0.70(0.30–1.63)
rs13264774CT	CT + TT	0.267	0.205	0.057	0.71(0.049–1.01)	0.238	0.69(0.38–1.27)

The table summarizes the univariate and multivariate analyses for the relationships of the gene variants with coronary artery disease, myocardial infarction and congenital heart disease in the studied 4,278 individuals. Bonferroni tests have been performed to adjust for age and sex, and adjustment made for the confounders in the multivariate tests.*$p < 0.05$; **$p < 0.005$.

Mendelian forms (monogenic) are absent [32]. The fourth group comprised 1,631 obese candidates with body-mass index (BMI) of ≥ 30.0 kg/m^2, and classified as the obesity subset. Among these subsets of patients, some patients harboured a combination of two or possibly three of the cardiovascular risk traits. All individuals with major cardiac rhythm disturbances, incapacitating or life-threatening illness, major psychiatric illness or substance abuse, history of cerebral vascular disease, neurological disorders, and administration of psychotropic medication or any other disorders likely to interfere with variables under investigation were excluded from the study. This study was performed in accordance with the regulations laid down by the King Faisal Specialist Hospital and Research Centre Ethics Committee in compliance with the Helsinki Declaration [33] and all participants signed an informed consent.

Five millilitres of peripheral blood was sampled in EDTA tubes after obtaining written consent, and genomic DNA extracted from leukocytes by the standard salt methods using PUREGENE DNA isolation kit (Qiagen, Germantown, MD, USA).

Linkage analysis and gene sequencing

Whole genome-wide scanning of two families with heterozygous familial hypercholesterolaemia was performed using the Affymetrix Gene Chip 250 Sty1 mapping array (Affymetrix, Inc., Santa Clara, CA, USA), and multipoint parametric linkage analysis for estimating the LOD scores performed using the GeneHunter Easy Linkage analysis software 4.0 (Affymetrix, Inc., Santa Clara, CA, USA) as described previously [34]. A recessive model of inheritance was used with a population-disease allele frequency of 0.0001, based on the Asian SNP allele frequencies, and the Copy Number Analyzer for GeneChip® (CNAG) Ver. 3.0 (Affymetrix, Inc., Santa Clara, CA, USA) software was employed in order to search for shared chromosomal regions of homozygosity. Following the identification of Chr 8 as a potential risk locus, screening for GATA4 mutations of interest was accomplished by sequencing on the MegaBACE DNA analysis system (Amersham Biosciences, Sunnyvale, CA, USA). Briefly, the DNA was subjected to PCR amplification by standard methods, following which the PCR products were sequenced and the data analyzed for SNPs by Lasergene software (DNASTAR, Inc. Madison, WI, USA) as described previously [35].

Association studies

Following the identification of possible informative variants based on the screening of our general population, genotyping was accomplished using Taqman chemistry on the Applied Biosystems Prism 7900HT Sequence Detection System (ABI Inc., Foster City, CA, USA). Primers and the TaqMan probes were procured from Applied Biosystems (ABI, Warrington, UK), and assays run by standard methods (Eurogentec, Seraing, Belgium) on the ABI2720 thermocycler (ABI Inc., Foster City, CA, USA). The plates were then scanned for FRET signal and data analyzed using Applied Biosystems SDS 2.4 software.

Statistical analysis

Comparison of genotypes and alleles between different groups for continuous dependent variables was achieved by Analysis of Variance (ANOVA) or Student's t test as appropriate. Categorical variables were analyzed by Chi-square test, and logistic regression analysis was used to compute odds ratios and their 95% confidence intervals. The Bonferroni test was employed to correct for the potential impact of the classical confounders, such as age, sex, smoking and disease history for all studied disease traits. Initially, univariate analysis was performed for individual SNPs to test for their possible association with the different traits. For the SNPs, that showed a significant value at $p < 0.05$, we then performed two tests, multivariate and multinomial regression analyses to test for possible confounding effects of the different disease traits in the model. The Haplo.stats package [36] in the R Statistical Computing software [37] was used to perform haplotype-based association analysis. Odd ratios for haplotypes were calculated using the baseline 10-mer haplotype GACACACCCG (global score statistics = 14.4) as reference, and the Haplotype Score statistic for the association of a haplotype with the binary trait was calculated as in [38] and [39]. Significance of association was determined between haplotypes and the case-control status - a binomial trait denoting whether or not a patient had the disease. All other statistical analyses were performed using the SPSS software Version 20 (SPSS Inc., Chicago, IL, USA), and data are expressed as mean ± SEM. Associations with a two-tailed p value < 0.05 was considered statistically significant.

Abbreviations

CAD: coronary artery disease; CHD: congenital heart disease; CBP: cAMP-response element-binding (CREB) protein; FH: family history of CAD; GATA4: GATA binding protein 4; hChol: hypercholesterolaemia; HFH: familial hypercholesterolaemia; hLDLC: high low-density lipoprotein-cholesterol; hTG: hypertriglyceridaemia; HTN: hypertension; lHDLC: low high-density lipoprotein-cholesterol level; MI: myocardial infarction; T2DM: type 2 diabetes mellitus; TFs: transcription factors.

Competing interests

The authors declare that they have no competing interests.

Authors' contributions

NPM was involved in running the Affymetrix assays, designing probes, screening for gene mutations as well as participating in the write-up of the manuscript. SMW performed Affymetrix genotyping analysis. AIT performed the statistical analysis. SH ran the Affymetrix assays. MN and DG were responsible for the overall running of the Taqman assays. NAT contributed to statistical analysis. EA was responsible for procuring patient material and data. NM was responsible for the clinical patient data and material acquisition. BFM contributed to the write-up of the manuscript. ND is the Principal Investigator, with the overall responsibility for the project and preparation of the manuscript. All authors read and approved the final manuscript.

Acknowledgements

This work was supported through a Cardiovascular Research Grant 2010020 of the King Faisal Specialist Hospital and Research Centre. The authors wish to express their gratitude for the support.

References

1. Narita N, Bielinska M, Wilson DB: Wild-type endoderm abrogates the ventral developmental defects associated with GATA-4 deficiency in the mouse. Dev Biol 1997, 189(2):270–274.
2. Fujikura J, Yamato E, Yonemura S, Hosoda K, Masui S, Nakao K, Miyazaki Ji J, Niwa H: Differentiation of embryonic stem cells is induced by GATA factors. Genes Dev 2002, 16(7):784–789.
3. Morrisey EE, Ip HS, Tang Z, Parmacek MS: GATA-4 activates transcription via two novel domains that are conserved within the GATA-4/5/6 subfamily. J Biol Chem 1997, 272(13):8515–8524.
4. Morrisey EE, Ip HS, Tang Z, Lu MM, Parmacek MS: GATA-5: a transcriptional activator expressed in a novel temporally and spatially-restricted pattern during embryonic development. Dev Biol 1997, 183(1):21–36.
5. Watt AJ, Battle MA, Li J, Duncan SA: GATA4 is essential for formation of the proepicardium and regulates cardiogenesis. Proc Natl Acad Sci USA 2004, 101(34):12573–12578.
6. Pikkarainen S, Tokola H, Kerkela R, Ruskoaho H: GATA transcription factors in the developing and adult heart. Cardiovasc Res 2004, 63(2):196–207.
7. Zeisberg EM, Ma Q, Juraszek AL, Moses K, Schwartz RJ, Izumo S, Pu WT: Morphogenesis of the right ventricle requires myocardial expression of Gata4. J Clin Invest 2005, 115(6):1522–1531.
8. Crispino JD, Lodish MB, Thurberg BL, Litovsky SH, Collins T, Molkentin JD, Orkin SH: Proper coronary vascular development and heart morphogenesis depend on interaction of GATA-4 with FOG cofactors. Genes Dev 2001, 15(7):839–844.
9. Reiter JF, Alexander J, Rodaway A, Yelon D, Patient R, Holder N, Stainier DY: Gata5 is required for the development of the heart and endoderm in zebrafish. Genes Dev 1999, 13(22):2983–2995.
10. Molkentin JD, Lin Q, Duncan SA, Olson EN: Requirement of the transcription factor GATA4 for heart tube formation and ventral morphogenesis. Genes Dev 1997, 11(8):1061–1072.
11. Charron F, Nemer M: GATA transcription factors and cardiac development. Semin Cell Dev Biol 1999, 10(1):85–91.
12. Durocher D, Grepin C, Nemer M: Regulation of gene expression in the endocrine heart. Recent Prog Horm Res 1998, 53:7–23. Discussion 22–23.
13. Shiojima I, Komuro I, Oka T, Hiroi Y, Mizuno T, Takimoto E, Monzen K, Aikawa R, Akazawa H, Yamazaki T, Kudoh S, Yazaki Y: Context-dependent transcriptional cooperation mediated by cardiac transcription factors Csx/Nkx-2.5 and GATA-4. J Biol Chem 1999, 274(12):8231–8239.
14. Fox AH, Kowalski K, King GF, Mackay JP, Crossley M: Key residues characteristic of GATA N-fingers are recognized by FOG. J Biol Chem 1998, 273(50):33595–33603.
15. Molkentin JD: The zinc finger-containing transcription factors GATA-4, -5, and −6. Ubiquitously expressed regulators of tissue-specific gene expression. J Biol Chem 2000, 275(50):38949–38952.

16. GNF Normal Tissue Atlas. GATA4 (GATA binding protein 4). 2013. http://biogps.org/#goto=genereport&id=2626.

17. Heineke J, Auger-Messier M, Xu J, Oka T, Sargent MA, York A, Klevitsky R, Vaikunth S, Duncan SA, Aronow BJ, Robbins J, Crombleholme TM, Molkentin JD: Cardiomyocyte GATA4 functions as a stress-responsive regulator of angiogenesis in the murine heart. J Clin Invest 2007, 117(11):3198–3210.

18. Oka T, Maillet M, Watt AJ, Schwartz RJ, Aronow BJ, Duncan SA, Molkentin JD: Cardiac-specific deletion of Gata4 reveals its requirement for hypertrophy, compensation, and myocyte viability. Circ Res 2006, 98(6):837–845.

19. Rivera-Feliciano J, Lee KH, Kong SW, Rajagopal S, Ma Q, Springer Z, Izumo S, Tabin CJ, Pu WT: Development of heart valves requires Gata4 expression in endothelial-derived cells. Development 2006, 133(18):3607–3618.

20. Hatcher CJ, Diman NY, McDermott DA, Basson CT: Transcription factor cascades in congenital heart malformation. Trends Mol Med 2003, 9(12):512–515.

21. Hirayama-Yamada K, Kamisago M, Akimoto K, Aotsuka H, Nakamura Y, Tomita H, Furutani M, Imamura S, Takao A, Nakazawa M, Matsuoka R: Phenotypes with GATA4 or NKX2.5 mutations in familial atrial septal defect. Am J Med Genet A 2005, 135(1):47–52.

22. Sarkozy A, Conti E, Neri C, D'Agostino R, Digilio MC, Esposito G, Toscano A, Marino B, Pizzuti A, Dallapiccola B: Spectrum of atrial septal defects associated with mutations of NKX2.5 and GATA4 transcription factors. J Med Genet 2005, 42(2):e16.

23. Okubo A, Miyoshi O, Baba K, Takagi M, Tsukamoto K, Kinoshita A, Yoshiura K, Kishino T, Ohta T, Niikawa N, Matsumoto N: A novel GATA4 mutation completely segregated with atrial septal defect in a large Japanese family. J Med Genet 2004, 41(7):e97.

24. Maslen CL: Molecular genetics of atrioventricular septal defects. Curr Opin Cardiol 2004, 19(3):205–210.

25. Reamon-Buettner SM, Cho SH, Borlak J: Mutations in the 3'-untranslated region of GATA4 as molecular hotspots for congenital heart disease (CHD). BMC Med Genet 2007, 8:38.

26. Tomita-Mitchell A, Maslen CL, Morris CD, Garg V, Goldmuntz E: GATA4 sequence variants in patients with congenital heart disease. J Med Genet 2007, 44(12):779–783.

27. Saffirio C, Marino B, Digilio MC: GATA4 as candidate gene for pericardial defects. Ann Thorac Surg 2007, 84(6):2137.

28. Rajagopal SK, Ma Q, Obler D, Shen J, Manichaikul A, Tomita-Mitchell A, Boardman K, Briggs C, Garg V, Srivastava D, Goldmuntz E, Broman KW, Benson DW, Smoot LB, Pu WT: Spectrum of heart disease associated with murine and human GATA4 mutation. J Mol Cell Cardiol 2007, 43(6):677–685.

29. Azakie A, Fineman JR, He Y: Myocardial transcription factors are modulated during pathologic cardiac hypertrophy in vivo. J Thorac Cardiovasc Surg 2006, 132(6):1262–1271.

30. Reamon-Buettner SM, Borlak J: GATA4 zinc finger mutations as a molecular rationale for septation defects of the human heart. J Med Genet 2005, 42(5):e32.

31. American Heart Association: Arteriosclerosis, Thrombosis, and Vascular Biology Annual Conference 2007. Thromb Vasc Biol 2007, 27(6):e35–e137.

32. Dzimiri N, Al-Najai M, Elhawari S, Gueco D, Vigilla MG, Andres E, Muiya P, Mazher N, Alshahid M, Meyer BF: The interaction of AGT gene polymorphism with cardiovascular risk traits in atherosclerosis. Arterioscler Thromb Vasc Biol 2012, 32(5_MeetingAbstracts):A524.

33. WMA Declaration of Helsinki - ethical principles for medical research involving human subjects. [http://www.wma.net/en/30publications/10policies/b3/index.html].

34. Alshahid M, Wakil SM, Al-Najai M, Muiya NP, Elhawari S, Gueco D, Andres E, Hagos S, Mazher N, Meyer BF, Dzimiri N: New susceptibility locus for obesity and dyslipidaemia on chromosome 3q22.3. Hum Genomics 2013, 7:15.

35. Elhawari S, Al-Boudari O, Muiya P, Khalak H, Andres E, Al-Shahid M, Al-Dosari M, Meyer BF, Al-Mohanna F, Dzimiri N: A study of the role of the myocyte-specific enhancer factor-2A gene in coronary artery disease. Atherosclerosis 2010, 209(1):152–154.

36. Schaid DJ: Statistical genetics and genetic epidemiology. [http://mayoresearch.mayo.edu/mayo/research/schaid_lab/software.cfm].

37. The R Project for Statistical Computing. [http://www.r-project.org/].

38. Schaid DJ, Rowland CM, Tines DE, Jacobson RM, Poland GA: Score tests for association between traits and haplotypes when linkage phase is ambiguous. Am J Hum Genet 2002, 70(2):425–434.

39. Lake SL, Lyon H, Tantisira K, Silverman EK, Weiss ST, Laird NM, Schaid DJ: Estimation and tests of haplotype-environment interaction when linkage phase is ambiguous. Hum Hered 2003, 55(1):56–65.

CER1 gene variations associated with bone mineral density, bone markers, and early menopause in postmenopausal women

Theodora Koromila[1], Panagiotis Georgoulias[2], Zoe Dailiana[3], Evangelia E Ntzani[4], Stavroula Samara[3], Chris Chassanidis[3], Vassiliki Aleporou-Marinou[1] and Panagoula Kollia[1*]

Abstract

Background: Osteoporosis has a multifactorial pathogenesis characterized by a combination of low bone mass and increased fragility. In our study, we focused on the effects of polymorphisms in *CER1* and *DKK1* genes, recently reported as important susceptibility genes for osteoporosis, on bone mineral density (BMD) and bone markers in osteoporotic women. Our objective was to evaluate the effect of *CER1* and *DKK1* variations in 607 postmenopausal women. The entire *DKK1* gene sequence and five selected *CER1* SNPs were amplified and resequenced to assess whether there is a correlation between these genes and BMD, early menopause, and bone turnover markers in osteoporotic patients.

Results: Osteoporotic women seem to suffer menopause 2 years earlier than the control group. The entire *DKK1* gene sequence analysis revealed six variations. There was no correlation between the six *DKK1* variations and osteoporosis, in contrast to the five common *CER1* variations that were significantly associated with BMD. Additionally, osteoporotic patients with rs3747532 and rs7022304 *CER1* variations had significantly higher serum levels of parathyroid hormone and calcitonin and lower serum levels of osteocalcin and IGF-1.

Conclusions: No significant association between the studied *DKK1* variations and osteoporosis was found, while *CER1* variations seem to play a significant role in the determination of osteoporosis and a potential predictive role, combined with bone markers, in postmenopausal osteoporotic women.

Keywords: *CER1*, *DKK1*, SNPs, Bone markers, Fracture, Menopause

Introduction

Osteoporosis is a complex multifactorial disease characterized by low bone mass with a consequent increase in bone fragility, especially in the hips, spine, and wrist [1]. According to evidence arising from large observational studies [2,3] that is already part of the World Health Organization (WHO) and European guidelines for the management of osteoporosis, the clinical significance of osteoporosis is its established association with fracture risk, which is also mediated by a number of other epidemiological and clinical factors [4]. Apart from these traditional risk factors and due to a knowledge gap

regarding fracture susceptibility, various bone-related biomarkers have also been proposed as potential fracture risk factors [5,6]. Several bone markers are measured in the serum in order to evaluate the bone turnover and to predict the fracture risk in elderly women [7,8].

The recently evolved novel concept of fat-bone interactions suggests that adipose tissue might profoundly affect bone formation and/or resorption [9]. Adipokines such as leptin have recently emerged as mediators of the protective effects of fat on bone tissue [10,11]. Moreover, serum osteocalcin (OC) has been considered as a specific marker of osteoblast function since OC levels correlate with bone formation rates. Insulin-like growth factor-1 (IGF-1) is also essential for the development and growth of the skeleton and maintenance of bone

* Correspondence: pankollia@biol.uoa.gr
[1]Laboratory of Human Genetics, Department of Genetics & Biotechnology, Faculty of Biology, National and Kapodistrian University of Athens, Athens 15701, Greece
Full list of author information is available at the end of the article

mass. IGF-1 promotes chondrogenesis and increases bone formation by regulating the functions of differentiated osteoblasts [12]. Furthermore, parathyroid hormone (PTH) is an important regulator of bone turnover because of the indirect stimulation of bone resorption through osteoclasts. While PTH increases the concentration of calcium in the blood, calcitonin (CT) reduces blood calcium and inhibits osteoclast activity in the bone.

In recent years, numerous gene polymorphisms (single nucleotide polymorphisms (SNPs)) have been associated with bone mineral density (BMD) and/or risk of fracture, identified either by a candidate gene approach or by genome-wide association studies (GWAS) [13,14]. The transforming growth factor beta (TGFbeta) and Wnt signaling pathways have a functional role in bone mass regulation, influencing both osteoblasts and osteoclasts.

The dickkopf Wnt signaling pathway inhibitor 1 (DKK1) gene in humans is located in 10q11.2 (NM_012242.2). The DKK1 gene belongs to a small gene family of four members (DKK1–4) that encodes secreted proteins that typically inhibit canonical Wnt signaling by binding to the receptors of two different families, namely LRP5-LRP6 [15] and Kremen 1-Kremen 2 [16]. The extracellular regions of LRP5-LRP6 interact with the Wnt antagonists DKK1 and sclerostin (SOST). In molecular network analyses, SOST shows a strong, positive correlation with DKK1 [17,18]. Mice overexpressing Dkk1 develop severe osteopenia, in part due to diminished bone formation [19]. Finally, overexpression of DKK1 in glucocorticoid-induced osteoporosis [18,20,21] as well as in osteosarcoma and osteolytic metastatic bone disease in multiple myeloma [22-24] led to the hypothesis that DKK1 is a strong candidate gene for the regulation of bone homeostasis.

Additionally, bone morphogenetic proteins (BMPs) are multifunctional growth factors that belong to the TGFbeta superfamily and have a significant role in bone remodeling. The activity of BMPs is controlled at different molecular levels [25]. A series of BMP antagonists bind BMP ligands and inhibit BMP functions. The human cerberus 1, DAN family BMP antagonist gene (CER1; NM_005454.2), a candidate gene for osteoporosis located in 9p23-p22, belongs to a distinct group of BMP antagonists (ligand traps) that can bind directly to BMPs and inhibit their activity [26-33].

In this case−control study, the whole DKK1 gene sequence was replicated for the first time, as a possible regulator of bone mass as previously reported on GWAS [14]. Furthermore, the five common genetic variations of the CER1 gene previously reported by Koromila et al. [32] were verified in a larger cohort. The correlation among the aforementioned SNPs with BMD, osteocalcin, and some bone turnover regulators as well as with menopause age of Greek postmenopausal women revealed significant conclusions.

Results
General characteristics of the assessed cohort

We analyzed 457 osteoporotic and 150 healthy postmenopausal women. As expected, the two groups revealed a statistically significantly difference ($p < 0.001$) in the mean T-score and the fracture record. The two groups were found to be similar in their other general characteristics, with the exception of mean years since menopause ($p < 0.05$). The majority of the osteoporotic group (78.9%) suffered from at least one fracture (vertebral, hip, or other fractures). Further details of both the osteoporotic and control groups are presented in Table 1.

DKK1 and CER1 gene variants

The analysis of the whole DKK1 gene sequence revealed six SNPs. Among the DKK1 SNPs, rs11001560, rs11815201, rs112910014, and rs1569198 are intron-located; rs74711339 is located in the 3′ untranslated region (UTR); and the synonymous variation rs2241529 is located in exon 2. No significant association for the identified DKK1 variants and BMD was found. Moreover, we found no significant association between DKK1 and age, body mass index (BMI), smoking, early menopause, or bone markers.

Genotype distributions of all CER1 alleles were in Hardy-Weinberg equilibrium ($p < 0.05$). Although, among the five CER1 SNPs, rs3747532 and rs1494360 are not independent ones ($r^2 > 0.8$) while the other three SNPs are not on any array according to SNAP analysis, we observed a statistically significant association for all five CER1 SNPs (Table 2). Specifically, the rs1494360 SNP was independently associated with hip fractures ($p = 0.043$) or the

Table 1 Characteristics of the osteoporotic ($N = 457$) and control ($N = 150$) groups

	Control	Osteoporotic
Age (years), mean [SD]	68.3 [11.2]	70.1 [11.3]
BMI (kg/m^2), mean [SD]	28.1 [5.3]	26.9 [5]
Smoking (%)		
No	80.3	82.7
Yes	19.7	16.3
Years since menopause, mean [SD]	18.1 [11.7]	21.0 [12]
T-score, mean [SD]	−0.6 [0.3]	−2.8 [0.6]
Vertebral fracture (%)		
No	100	88.3
Yes	0	11.7
Hip fracture (%)		
No	100	56.2
Yes	0	43.8
Other fractures (%)		
No	100	76.6
Yes	0	23.4

Table 2 Association of *CER1* genotypes with *T*-score and multiple logistic regression analysis for fracture prediction

CER1 genotypes	Total cohort			Osteoporotic patients						
	T-score			Vertebral fracture		Hip fracture		Any fracture	
	Mean	SD	*P*	OR (95% CI)	*P*	OR (95% CI)	*P*	OR (95% CI)	*P*
rs3747532: (C/C)	−1.1	1.6	*<0.05*	1.71 (0.2–3.13)	NS	1.63 (0.76–3.48)	NS	1.79 (0.79–4.09)	NS
(C/G), (G/G)	−2.0	1.0							
rs1494360: (G/G)	−1.1	1.4	*<0.05*	2.12 (0.23–13.54)	NS	*1.98 (0.3–23.22)*	*<0.05*	*1.38 (0.49–15.5)*	*<0.01*
G/T), (T/T)	−2.4	1.1							
rs7022304: (A/A)	−1.1	1.5	*<0.05*	0.90 (0.23–3.49)	NS	1.22 (0.49–3.01)	NS	1.47 (0.49–8.1)	NS
(A/G), (G/G)	−2.2	1.2							
rs17289263: (A/A)	−1.0	1.4	*<0.05*	1.55 (0.42–5.04)	NS	1.15 (0.48–2.75)	NS	2.18 (0.61–5.2)	NS
(A/G), (G/G)	−2.2	1.2							
rs74434454: (T/T)	−1.0	1.5	*<0.05*	1.13 (0.46–3.72)	NS	1.90 (0.34–10.56)	NS	1.98 (0.3–13.22)	NS
(T/C), (C/C)	−2.2	1.1							

Significant values are shown in italics; *NS* not significant.

presence of any fracture ($p < 0.01$) when multiple logistic regression analysis was performed for the prediction of fractures in the osteoporotic patients from the *CER1* sequence variations, adjusted for age, sex, smoking, BMI, years since menopause, and calcium intake, confirming our previous report [32]. Homozygotes or heterozygotes for the above SNP were at a higher risk of hip fracture (1.98-fold) and any fracture (1.38-fold). On the other hand, no significant association between *DKK1* and BMD, age, BMI, smoking, years since menopause, calcium intake, or fracture was found.

Bone markers
Among the studied bone markers previously referred, the serum levels of leptin did not change between osteoporotic patients and controls at any *CER1* variation. Compared to controls' values as well as to normal values' range per bone marker, a statistically significant number of osteoporotic patients with minor alleles of rs3747532 and rs7022304 had higher serum levels of PTH (mean = 78.4, standard deviation (SD) = 41.23) and CT (mean = 10.1, SD = 4.13) and lower serum levels of OC (mean = 4.9, SD = 3.52) and IGF-1 (mean = 80.2, SD = 62.62) (Figure 1). In addition, only serum OC levels and patients with hip fractures were significantly correlated and were found to be lower than total osteoporotic and control groups (p = 0.012), supporting the previous reports of Akesson et al. [34,35]. No significant association was found between the aforementioned bone markers and the age of menopause.

Menopause
Postmenopausal women with osteoporosis seem to suffer menopause 2 years earlier than healthy women ($p \leq$ 0.05) as it is presented in Table 3. In addition, patients with hip fractures suffered menopause significantly earlier compared to the control group. However, our results

did not verify an association between sequence variations of *DKK1* and *CER1* genes and bone marker serum levels or menopause age in the osteoporotic or in the total cohort group (osteoporotic and control).

Discussion
Most genetic studies on osteoporosis, until now, have focused on the regulation of BMD. A number of them suggest an important genetic component in the determination of peak bone mass and, in some instances, in the susceptibility to subsequent fractures.

In our study, we investigated the possible association of two important susceptibility genes for osteoporosis, *DKK1* and *CER1*, that participate in Wnt and TGFbeta signaling pathways, respectively, and are known for their functional role in bone mass regulation. The *DKK1* gene is able to modulate canonical Wnt signaling, and because of the established role of this pathway in the regulation of bone strength, this study aimed at understanding the influence of common genetic variations in *DKK1* and *CER1* genes on BMD, bone markers, and age of menopause. In a large genome-wide linkage scan, Ralston et al. [36] already suggested that the chromosomal region 10q21 containing the *DKK1* gene was specifically associated with the regulation of BMD in men.

Our findings for two *DKK1* variations, rs2241529 and rs1569198, support the previous report of Piters et al. [18] in the male population, while our report is the first in Caucasian women. In addition, the recent meta-analysis of GWAS of Estrada et al. revealed no correlation with any variation inside the *DKK1* gene sequence, although a variation upstream of the *DKK1* gene was significantly associated with FN-BMD (p = 1.3×10^{-5}) and LS-BMD ($p = 3.2 \times 10^{-4}$) as well as with fractures [14].

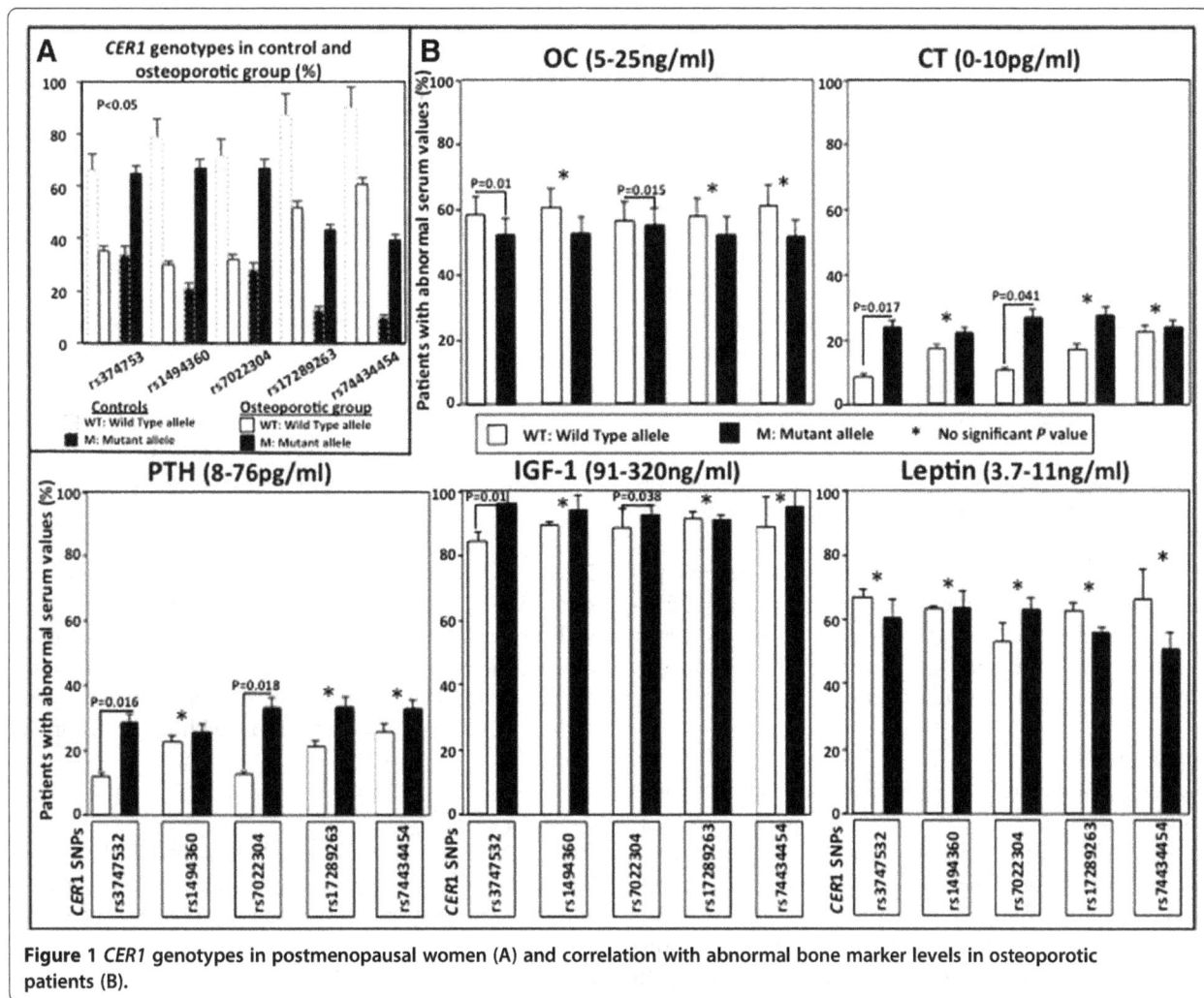

Figure 1 *CER1* genotypes in postmenopausal women (A) and correlation with abnormal bone marker levels in osteoporotic patients (B).

Our previously reported findings in Caucasians [32] as well as the report of Tang et al. in southern Chinese women [33] suggest a significant association between *CER1* variations and BMD and/or fragility risk. Among all *CER1* sequence variations studied, only the rs3747532 SNP, located in exon 1, results in an Ala>Gly amino acid change, but both amino acids are classified as nonpolar. Both rs1494360 and rs7022304 SNPs are located in introns, rs74434454 is located in the 3′UTR, and the synonymous rs17289263 SNP is located in exon 2. Moreover,

mice studies suggested that the *CER1* gene is an inhibitor of BMPs. BMP signaling is very important in bone development; it is not surprising that variations in BMP antagonists may affect skeletogenesis and BMD variations in humans (e.g., the sclerosteosis/van Buchem disease gene, which is caused by mutations in SOST) [37].

This is the first report on the correlation of rs3747532, rs1494360, rs7022304, rs17289263, and rs74434454 *CER1* variations with early menopause and bone markers. When *CER1* variations were correlated with the age of

Table 3 Menopause age and serum OC value correlation with control and osteoporotic (total, hip/vertebral fracture) groups

Study group	Age of menopause			OC (5–25 ng/ml)		
	Mean	SD	P	Mean	SD	P
Control	51.2	0.95	0.041	6.2	3.38	0.752
Osteoporotic total	49.1	12.33		5.4	4.75	
Osteoporotic hip fracture	48.6	5.85	0.024	4.7	4.57	0.012
Osteoporotic vertebral fracture	50.0	4.13	0.072	5.3	2.97	0.197

menopause, they were found to be independent while osteoporotic women with hip fracture were found to suffer menopause approximately 2.5 years earlier than the control group. Osteoporotic patients with rs3747532 or rs7022304 *CER1* variations were found to have significantly higher serum levels of PTH and CT, compared to both controls' and normal values per bone marker. Higher PTH levels in osteoporotic patients are in accordance with the indirect stimulation of bone resorption by PTH through osteoclasts. A further pharmacogenomic analysis of the above variations with different osteoporotic treatments could be of great interest in order to understand their mechanism. Both rs3747532 and rs7022304 variations were associated with low levels of OC and IGF-1 in osteoporotic postmenopausal women. Furthermore, low serum values of OC were associated with osteoporotic hip fractures, concluding that bone formation, as assessed by OC, is apparently lower in elderly women who sustain a hip fracture. Follow-up measurements in osteoporotic patients' serum samples, after 6 months and 1 year of fracture or starting therapy, will possibly show a stronger correlation with the *CER1* gene, leading to a new insight into personalized therapy of osteoporosis.

Conclusions

Our study underlines a significant association of two sequence variations of the *CER1* gene with PTH, CT, OC, and IGF-1 in a Hellenic cohort of postmenopausal women. The studied *DKK1* SNPs seem to have no correlation with either the bone markers or the age of menopause, while the association of the *CER1* gene with bone markers supports its previously reported correlation with osteoporosis and suggests its potential role as a predictive marker of osteoporosis and hip fracture in postmenopausal women. In further GWAS, both the studied *CER1* and *DKK1* variations should be included in order to evaluate their biological role in osteoporosis.

Methods

Subjects

In this case–control study, peripheral blood samples were collected from 700 postmenopausal Greek women, who were treated at the Department of Orthopaedic Surgery of the University Hospital of Thessalia in Larissa, Greece, and gave their informed consent prior to their inclusion in the study. All the subjects of the present study underwent a physical examination and were interviewed using a structured questionnaire to obtain information on age, BMD, age of menopause, fracture, family history of osteoporosis and fracture, medical and reproductive history, smoking, alcohol intake, physical activity, and other secondary causes. Subjects were excluded from this study if they had diseases known to affect bone metabolism, were premature to menopause

(absence of menstruation for at least 12 months, age <45 years), or had a history of drug use that could affect bone turnover and BMD. Moreover, high-trauma fractures including major trauma occurring during a motor vehicle accident or a fall from more than the standing height were excluded. Therefore, only 655 postmenopausal women met the inclusion criteria, of which 457 individuals were osteoporotic and 150 were normal, according to their dual-energy X-ray absorptiometry (DXA) findings (Table 1). In order to avoid misclassification and a potential effect dilution, 48 subjects with 'gray-zone' *T*-scores ranging between –1 and –2.5 were excluded from the study; thus, the study included 607 subjects. The study was approved by the Ethics Committee of the University of Thessalia, Larissa, Greece, and conducted according to the Declaration of Helsinki.

T-score

BMD was measured at the femoral neck and at the lumbar spine (L2 to L4) by DXA. Cases were defined as subjects with a low BMD (*T*-score ≤–2.5) at either the spine or the hip, which was equivalent to osteoporosis according to the WHO definition [38]; control subjects were individuals with normal BMD (*T*-score >–1) without a history of fracture.

Bone markers

Patients and controls were fasted for at least 12 h. Venous blood samples were drawn in the morning between 8:00 and 9:00 a.m., and patients' samples were measured within a mean of 12 h (±5 h) of fracture and before starting treatment. The samples were immediately centrifuged and stored at –80°C for further analysis. Total serum leptin and IGF-1 levels were measured using human radioimmunoassay (RIA) diagnostic kits (KIPMR44 and KIP1588, respectively, DIASource Europe SA, Louvain-La-Neuve, Belgium). The leptin kit is suited for human leptin, and no cross-reactivity has been found with other proteins such as insulin or IGF-1. The sensitivity of the leptin assay is 0.1 ng/ml, with a calibrators' range of 0–64 ng/ml. The IGF-1 kit has a sensitivity of 3.4 ng/ml and a calibrators' range of 0–1,529 ng/ml, with no cross-reactivity to insulin and growth hormone. Serum human intact osteocalcin, parathyroid hormone, and calcitonin values were measured using human immunoradiometric assay (IRMA) diagnostic kits (KIP1381, KIP1491, and KIP0429, respectively, DIASource Europe SA, Louvain-La-Neuve, Belgium). The OC kit has a sensitivity of 0.22 ng/ml and a calibrators' range of 0–69 ng/ml, with no cross-reactivity to N-terminal and C-terminal fragments. The PTH kit has a sensitivity of 4.1 pg/ml and a calibrators' range of 0–973 pg/ml and does not cross-react with PTH fragments and PTH-related proteins. The CT kit has a sensitivity of 0.9 pg/ml and a calibrators' range of 0–674 pg/ml. No

Table 4 Primers for PCR and sequencing of the *DKK1* gene

Localization (nucleotide)	Forward primer 5′ → 3′ sequence	Reverse primer 5′ → 3′ sequence
1–490	GCAGAGCTCTGTGCTCCCT	ACCGCACACATTCAGCACG
396–1,043	AGGTGAGAGGGGTCGGGCAC	CGGAGGAAGATAAGGACCTC
963–1,470	CGCTGAAGTATCTTCATTGCA	GGAGACCTCTTTAGCTGTCT
1,407–2,010	AGCACAGATCCACTAACTT	GGAAGCAGGAAATAGTGATT
1,945–2,481	GCCACTGTCACAGCTGTTA	TGGTGATCTTTCTGTATCCG
2,427–2,965	CCAGCGTTGTTACTGTGGA	CCAAGAGATCCTTGCGTTC
2,950–3,377	CGCAAGGATCTCTTGGAATGA	TAGGTATTATTAATTTATTGG

significant interference has been found (at concentrations up to 100 ng/ml) with calcitonin gene-related peptide (CGRP), salmon calcitonin, katacalcin (PDN-21), and procalcitonin N-terminal. Moreover, all RIA and IRMA kits are calibrated against valid international standards. The radiotracer used in all kits is iodine-125 (^{125}I, half-life $t_{1/2}$ 60 days, 35.5-keV gamma radiation, 27–32-keV X-rays, no beta radiation). All sample assays were performed in duplicate and were included in the same run for each biological parameter. If the difference between duplicate results of a sample was more than 5%, the sample assay was repeated, and the in-run coefficients of variation were 3.9% for leptin, 3.4% for IGF-1, 2.9% for OC, 3.1% for PTH, and 2.8% for CT. An automatic gamma counter (Cobra II/5010, Packard, Conroe, TX, USA) was used to count the radioactivity and calculate the results.

Amplification and resequencing of the human *CER1* and *DKK1* genes

Genomic DNA was isolated using QIAamp DNA Blood Mini Kit (QIAGEN, Venlo, Netherlands). *CER1* and *DKK1* genes were polymerase chain reaction (PCR)-amplified and resequenced to identify the underlying sequence variation. Eleven pairs of primers, four pairs for *CER1* and seven pairs for *DKK1* (Table 4), were designed in order to cover the five variants of the *CER1* gene previously reported by Koromila et al. [32] as well as the entire sequence of the *DKK1* gene (3,377 bp) (Figure 2). Sequencing was performed twice per sample (two independent PCR products) in both forward and reverse orientations. Genomic DNA information was obtained from GenBank wild-type sequences [*CER1*: chromosome 9, NC_000009.11 (14719731..14722715), MIM: 603777,

ID: 9350; *DKK1*: chromosome 10, NC_000010.10 (54074041..54077417), MIM: 605189, ID: 22943]. Sequence variants were verified using the MegaBACE 1000 DNA Sequencing System (Amersham Biosciences, Piscataway, NJ, USA). Six variants, rs2241529, rs11001560, rs11815201, rs112910014, rs1569198, and rs74711339, in the *DKK1* gene were detected (Figure 2), and five common variants were analyzed in *CER1* by multiple sequence alignments using Chromas Lite 2.01 software and BLAST analysis in the cohort. Five common SNPs in the *CER1* gene, namely rs3747532 (c.194C>G, exon 1), rs1494360 (c.507+506G>T, intron), rs7022304 (c.508-182A>G, intron), rs17289263 (c.531A>G, exon 2), and rs74434454 (c.*121T>C, 3′UTR), were resequenced. Among these five *CER1* SNPs, only rs3747532, which is located in exon 1, causes an amino acid change from Ala to Gly. Six SNPs in *DKK1* (Figure 2) were analyzed as well through direct resequencing. Among the *DKK1* SNPs, only rs2241529 is exon-located (c.318A>G, exon 2) and causes an amino acid substitution, while rs11001560, rs11815201, rs112910014, and rs1569198 are located in introns and rs74711339 in 3′UTR. An association between *DKK1* variations and BMD could not be attempted in our dataset (osteoporotic and control).

Statistical analysis

Continuous variables are presented as mean and SD, while categorical variables are presented as absolute and relative frequencies. The Hardy-Weinberg equilibrium (HWE) was assessed in the control samples by applying an exact test. Deviation from HWE was considered nominally statistically significant at the $p < 0.05$ level [39,40]. Genotype frequency differences between cases and controls were tested using unconditional logistic

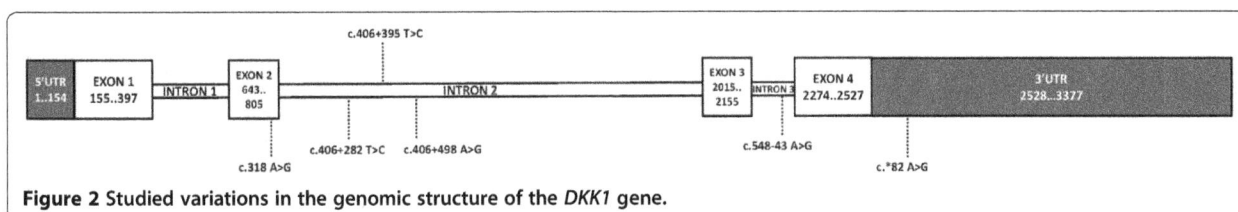

Figure 2 Studied variations in the genomic structure of the *DKK1* gene.

regression without any adjustments. Odds ratios (ORs) and 95% confidence intervals (CIs) were estimated under the log-additive model, using the major allele in the Greek control population as reference. Odds ratios thus represent the risk conferred per copy of the minor allele. Secondary analyses also examined recessive and dominant models of inheritance. Pearson's correlation coefficient (r) was used to estimate the correlations in minor allele frequencies between our study and the HapMap CEU population [26]. The overall correlation between ORs in the Greek population and the GWAS population where each SNP was first discovered was also calculated. The power of the study to detect ORs similar to those previously found in the GWAS, given the allele frequencies observed in the Greek population, was estimated at an α value of 0.05.

Statistical analyses were run in Stata, version 10.1 (College Station, TX, USA). P values for association are two-tailed and not adjusted for multiple comparisons since this is a replication effort for associations that already have robust statistical support. Student's t tests were used for the comparison of mean values between osteoporotic and control groups. Analyses were conducted using SPSS statistical software (version 17.0). Design and reporting follow the STREGA guidance [41].

Abbreviations

BMD: bone mineral density; BMI: body mass index; BMP: bone morphogenetic protein; CER1: cerberus 1; CT: calcitonin; DKK1: dickkopf Wnt signaling pathway inhibitor 1; DXA: dual-energy X-ray absorptiometry; GWAS: genome-wide association studies; IGF-1: insulin-like growth factor-1; IRMA: immunoradiometric assay; NS: not significant; OC: osteocalcin; OR: odds ratio; PTH: parathyroid hormone; RIA: radioimmunoassay; SNP: single nucleotide polymorphism; SOST: sclerostin; TGFbeta: transforming growth factor beta; UTR: untranslated region; WHO: World Health Organization.

Competing interests

The authors declare that they have no competing interests.

Authors' contributions

TK and PK designed the experiments. ZD provided the human blood and tissue samples. PG carried out the analysis of bone markers. TK, SS, and CC collected the samples and clinical data. TK and EEN prepared the statistical analysis. TK, PK, ZD, and VAM prepared the manuscript. All authors read and approved the final manuscript.

Acknowledgements

Part of this work was supported by a research grant ('Kapodistrias 2009,' University of Athens, Greece) to Dr. P. Kollia.

Author details

[1]Laboratory of Human Genetics, Department of Genetics & Biotechnology, Faculty of Biology, National and Kapodistrian University of Athens, Athens 15701, Greece. [2]Department of Nuclear Medicine, University Hospital of Larissa, School of Medicine, University of Thessaly, Larissa 41110, Greece. [3]Department of Orthopaedic Surgery, School of Medicine, University of Thessaly, Larissa 41110, Greece. [4]Department of Hygiene and Epidemiology,

References

1. Laliberte MC, Perreault S, Jouini G, Shea BJ, Lalonde L: Effectiveness of interventions to improve the detection and treatment of osteoporosis in primary care settings: a systematic review and meta-analysis. *Osteoporos Int* 2011, 22:2743–2768. doi:10.1007/s00198-011-1557-6.
2. Ferrari SL, Deutsch S, Antonarakis SE: Pathogenic mutations and polymorphisms in the lipoprotein receptor-related protein 5 reveal a new biological pathway for the control of bone mass. *Curr Opin Lipidol* 2005, 16:207–214.
3. Baldock PA, Eisman JA: Genetic determinants of bone mass. *Curr Opin Rheumatol* 2004, 16:450–456.
4. Albagha OM, Ralston SH: Genetic determinants of susceptibility to osteoporosis. *Endocrinol Metab Clin North Am* 2003, 32:vi–81.
5. Blain H, Vuillemin A, Guillemin F, Durant R, Hanesse B, de Talance N, Doucet B, Jeandel C: Serum leptin level is a predictor of bone mineral density in postmenopausal women. *J Clin Endocrinol Metab* 2002, 87:1030–1035.
6. Lee NK, Sowa H, Hinoi E, Ferron M, Ahn JD, Confavreux C, Dacquin R, Mee PJ, McKee MD, Jung DY, Zhang Z, Kim JK, Mauvais-Jarvis F, Ducy P, Karsenty G: Endocrine regulation of energy metabolism by the skeleton. *Cell* 2007, 130:456–469. doi:10.1016/j.cell.2007.05.047.
7. Gerdhem P, Ivaska KK, Alatalo SL, Halleen JM, Hellman J, Isaksson A, Pettersson K, Vaananen HK, Akesson K, Obrant KJ: Biochemical markers of bone metabolism and prediction of fracture in elderly women. *J Bone Miner Res* 2004, 19:386–393. doi:10.1359/JBMR.0301244.
8. Delmas PD, Eastell R, Garnero P, Seibel MJ, Stepan J: The use of biochemical markers of bone turnover in osteoporosis. Committee of Scientific Advisors of the International Osteoporosis Foundation. *Osteoporos Int* 2000, 11(Suppl 6):S2–S17.
9. Magni P, Dozio E, Galliera E, Ruscica M, Corsi MM: Molecular aspects of adipokine-bone interactions. *Curr Mol Med* 2010, 10:522–532.
10. Reid IR: Relationships between fat and bone. *Osteoporos Int* 2008, 19:595–606. doi:10.1007/s00198-007-0492-z.
11. Cirmanova V, Bayer M, Starka L, Zajickova K: The effect of leptin on bone: an evolving concept of action. *Physiol Res* 2008, 57(Suppl 1):S143–151.
12. Giustina A, Mazziotti G, Canalis E: Growth hormone, insulin-like growth factors, and the skeleton. *Endocr Rev* 2008, 29:535–559. doi:10.1210/er.2007-0036.
13. Duncan EL, Danoy P, Kemp JP, Leo PJ, McCloskey E, Nicholson GC, Eastell R, Prince RL, Eisman JA, Jones G, Sambrook PN, Reid IR, Dennison EM, Wark J, Richards JB, Uitterlinden AG, Spector TD, Esapa C, Cox RD, Brown SD, Thakker RV, Addison KA, Bradbury LA, Center JR, Cooper C, Cremin C, Estrada K, Felsenberg D, Gluer CC, Hadler J, Henry MJ, Hofman A, Kotowicz MA, Makovey J, Nguyen SC, Nguyen TV, Pasco JA, Pryce K, Reid DM, Rivadeneira F, Roux C, Stefansson K, Styrkarsdottir U, Thorleifsson G, Tichawangana R, Evans DM, Brown MA: Genome-wide association study using extreme truncate selection identifies novel genes affecting bone mineral density and fracture risk. *PLoS Genet* 2011, 7:e1001372. doi:10.1371/journal.pgen.1001372.
14. Estrada K, Styrkarsdottir U, Evangelou E, Hsu YH, Duncan EL, Ntzani EE, Oei L, Albagha OM, Amin N, Kemp JP, Koller DL, Li G, Liu CT, Minster RL, Moayyeri A, Vandenput L, Willner D, Xiao SM, Yerges-Armstrong LM, Zheng HF, Alonso N, Eriksson J, Kammerer CM, Kaptoge SK, Leo PJ, Thorleifsson G, Wilson SG, Wilson JF, Aalto V, Alen M, et al: Genome-wide meta-analysis identifies 56 bone mineral density loci and reveals 14 loci associated with risk of fracture. *Nat Genet* 2012, 44:491–501. doi:10.1038/ng.2249.
15. Bafico A, Liu G, Yaniv A, Gazit A, Aaronson SA: Novel mechanism of Wnt signalling inhibition mediated by Dickkopf-1 interaction with LRP6/Arrow. *Nat Cell Biol* 2001, 3:683–686. doi:10.1038/35083081.
16. Mao B, Wu W, Li Y, Hoppe D, Stannek P, Glinka A, Niehrs C: LDL-receptor-related protein 6 is a receptor for Dickkopf proteins. *Nature* 2001, 411:321–325. doi:10.1038/35077108.
17. Zhang Y, Wang Y, Li X, Zhang J, Mao J, Li Z, Zheng J, Li L, Harris S, Wu D: The LRP5 high-bone-mass G171V mutation disrupts LRP5 interaction with Mesd. *Mol Cell Biol* 2004, 24:4677–4684. doi:10.1128/MCB.24.11.4677-4684.2004.
18. Piters E, Balemans W, Nielsen TL, Andersen M, Boudin E, Brixen K, Van Hul W: Common genetic variation in the DKK1 gene is associated with hip axis length but not with bone mineral density and bone turnover markers in young adult men: results from the Odense Androgen Study. *Calcif Tissue Int* 2010, 86:271–281. doi:10.1007/s00223-010-9334-7.
19. Li J, Sarosi I, Cattley RC, Pretorius J, Asuncion F, Grisanti M, Morony S, Adamu S, Geng Z, Qiu W, Kostenuik P, Lacey DL, Simonet WS, Bolon B, Qian

X, Shalhoub V, Ominsky MS, Zhu Ke H, Li X, Richards WG: **Dkk1-mediated inhibition of Wnt signaling in bone results in osteopenia.** *Bone* 2006, **39**:754–766. doi:10.1016/j.bone.2006.03.017.

20. Ohnaka K, Taniguchi H, Kawate H, Nawata H, Takayanagi R: **Glucocorticoid enhances the expression of dickkopf-1 in human osteoblasts: novel mechanism of glucocorticoid-induced osteoporosis.** *Biochem Biophys Res Commun* 2004, **318**:259–264. doi:10.1016/j.bbrc.2004.04.025.

21. Ohnaka K, Tanabe M, Kawate H, Nawata H, Takayanagi R: **Glucocorticoid suppresses the canonical Wnt signal in cultured human osteoblasts.** *Biochem Biophys Res Comm* 2005, **329**:177–181. doi:10.1016/j.bbrc.2005.01.117.

22. Lee N, Smolarz AJ, Olson S, David O, Reiser J, Kutner R, Daw NC, Prockop DJ, Horwitz EM, Gregory CA: **A potential role for Dkk-1 in the pathogenesis of osteosarcoma predicts novel diagnostic and treatment strategies.** *Br J Cancer* 2007, **97**:1552–1559. doi:10.1038/sj.bjc.6604069.

23. Tian E, Zhan F, Walker R, Rasmussen E, Ma Y, Barlogie B, Shaughnessy JD Jr: **The role of the Wnt-signaling antagonist DKK1 in the development of osteolytic lesions in multiple myeloma.** *N Engl J Med* 2003, **349**:2483–2494. doi:10.1056/NEJMoa030847.

24. Yaccoby S, Ling W, Zhan F, Walker R, Barlogie B, Shaughnessy JD Jr: **Antibody-based inhibition of DKK1 suppresses tumor-induced bone resorption and multiple myeloma growth in vivo.** *Blood* 2007, **109**:2106–2111. doi:10.1182/blood-2006-09-047712.

25. Chen D, Zhao M, Mundy GR: **Bone morphogenetic proteins.** *Growth Factors* 2004, **22**:233–241. doi:10.1080/08977190412331279890.

26. Glinka A, Wu W, Onichtchouk D, Blumenstock C, Niehrs C: **Head induction by simultaneous repression of Bmp and Wnt signalling in *Xenopus*.** *Nature* 1997, **389**:517–519. doi:10.1038/39092.

27. Katoh M, Katoh M: **CER1 is a common target of WNT and NODAL signaling pathways in human embryonic stem cells.** *Int J Mol Med* 2006, **17**:795–799.

28. Belo JA, Bouwmeester T, Leyns L, Kertesz N, Gallo M, Follettie M, De Robertis EM: **Cerberus-like is a secreted factor with neutralizing activity expressed in the anterior primitive endoderm of the mouse gastrula.** *Mech Dev* 1997, **68**:45–57.

29. Biben C, Stanley E, Fabri L, Kotecha S, Rhinn M, Drinkwater C, Lah M, Wang CC, Nash A, Hilton D, Ang SL, Mohun T, Harvey RP: **Murine cerberus homologue mCer-1: a candidate anterior patterning molecule.** *Dev Biol* 1998, **194**:135–151. doi:10.1006/dbio.1997.8812.

30. Shawlot W, Deng JM, Behringer RR: **Expression of the mouse cerberus-related gene, Cerr1, suggests a role in anterior neural induction and somitogenesis.** *Proc Natl Acad Sci U S A* 1998, **95**:6198–6203.

31. Piccolo S, Agius E, Leyns L, Bhattacharyya S, Grunz H, Bouwmeester T, De Robertis EM: **The head inducer Cerberus is a multifunctional antagonist of Nodal, BMP and Wnt signals.** *Nature* 1999, **397**:707–710. doi:10.1038/17820.

32. Koromila T, Dailiana Z, Samara S, Chassanidis C, Tzavara C, Patrinos GP, Aleporou-Marinou V, Kollia P: **Novel sequence variations in the CER1 gene are strongly associated with low bone mineral density and risk of osteoporotic fracture in postmenopausal women.** *Calcif Tissue Int* 2012, **91**:15–23. doi:10.1007/s00223-012-9602-9.

33. Tang PL, Cheung CL, Sham PC, McClurg P, Lee B, Chan SY, Smith DK, Tanner JA, Su AI, Cheah KS, Kung AW, Song YQ: **Genome-wide haplotype association mapping in mice identifies a genetic variant in CER1 associated with BMD and fracture in southern Chinese women.** *J Bone Miner Res* 2009, **24**:1013–1021. doi:10.1359/jbmr.081258.

34. Akesson K, Vergnaud P, Delmas PD, Obrant KJ: **Serum osteocalcin increases during fracture healing in elderly women with hip fracture.** *Bone* 1995, **16**:427–430.

35. Akesson K, Vergnaud P, Gineyts E, Delmas PD, Obrant KJ: **Impairment of bone turnover in elderly women with hip fracture.** *Calcif Tissue Int* 1993, **53**:162–169.

36. Ralston SH, Galwey N, MacKay I, Albagha OM, Cardon L, Compston JE, Cooper C, Duncan E, Keen R, Langdahl B, McLellan A, O'Riordan J, Pols HA, Reid DM, Uitterlinden AG, Wass J, Bennett ST: **Loci for regulation of bone mineral density in men and women identified by genome wide linkage scan: the FAMOS study.** *Hum Mol Genet* 2005, **14**:943–951. doi:10.1093/hmg/ddi088.

37. Hsu DR, Economides AN, Wang X, Eimon PM, Harland RM: **The *Xenopus* dorsalizing factor Gremlin identifies a novel family of secreted proteins that antagonize BMP activities.** *Mol Cell* 1998, **1**:673–683.

38. Kanis JA, Johnell O, Oden A, Jonsson B, Dawson A, Dere W: **Risk of hip fracture derived from relative risks: an analysis applied to the population of Sweden.** *Osteoporos Int* 2000, **11**:120–127.

39. Finner H, Strassburger K, Heid IM, Herder C, Rathmann W, Giani G, Dickhaus T, Lichtner P, Meitinger T, Wichmann HE, Illig T, Gieger C: **How to link call rate and p-values for Hardy-Weinberg equilibrium as measures of genome-wide SNP data quality.** *Stat Med* 2010, **29**:2347–2358. doi:10.1002/sim.4004.

40. Panagiotou OA, Evangelou E, Ioannidis JP: **Genome-wide significant associations for variants with minor allele frequency of 5% or less—an overview: a HuGE review.** *Am J Epidemiol* 2010, **172**:869–889. doi:10.1093/aje/kwq234.

41. Ioannidis JP, Gwinn M, Little J, Higgins JP, Bernstein JL, Boffetta P, Bondy M, Bray MS, Brenchley PE, Buffler PA, Casas JP, Chokkalingam A, Danesh J, Smith GD, Dolan S, Duncan R, Gruis NA, Hartge P, Hashibe M, Hunter DJ, Jarvelin MR, Malmer B, Maraganore DM, Newton-Bishop JA, O'Brien TR, Petersen G, Riboli E, Salanti G, Seminara D, Smeeth L, Taioli E, Timpson N, Uitterlinden AG, Vineis P, Wareham N, Winn DM, Zimmern R, Khoury MJ: **A road map for efficient and reliable human genome epidemiology.** *Nat Genet* 2006, **38**:3–5.

Permissions

All chapters in this book were first published in JNI, by BioMed Central; hereby published with permission under the Creative Commons Attribution License or equivalent. Every chapter published in this book has been scrutinized by our experts. Their significance has been extensively debated. The topics covered herein carry significant findings which will fuel the growth of the discipline. They may even be implemented as practical applications or may be referred to as a beginning point for another development.

The contributors of this book come from diverse backgrounds, making this book a truly international effort. This book will bring forth new frontiers with its revolutionizing research information and detailed analysis of the nascent developments around the world.

We would like to thank all the contributing authors for lending their expertise to make the book truly unique. They have played a crucial role in the development of this book. Without their invaluable contributions this book wouldn't have been possible. They have made vital efforts to compile up to date information on the varied aspects of this subject to make this book a valuable addition to the collection of many professionals and students.

This book was conceptualized with the vision of imparting up-to-date information and advanced data in this field. To ensure the same, a matchless editorial board was set up. Every individual on the board went through rigorous rounds of assessment to prove their worth. After which they invested a large part of their time researching and compiling the most relevant data for our readers.

The editorial board has been involved in producing this book since its inception. They have spent rigorous hours researching and exploring the diverse topics which have resulted in the successful publishing of this book. They have passed on their knowledge of decades through this book. To expedite this challenging task, the publisher supported the team at every step. A small team of assistant editors was also appointed to further simplify the editing procedure and attain best results for the readers.

Apart from the editorial board, the designing team has also invested a significant amount of their time in understanding the subject and creating the most relevant covers. They scrutinized every image to scout for the most suitable representation of the subject and create an appropriate cover for the book.

The publishing team has been an ardent support to the editorial, designing and production team. Their endless efforts to recruit the best for this project, has resulted in the accomplishment of this book. They are a veteran in the field of academics and their pool of knowledge is as vast as their experience in printing. Their expertise and guidance has proved useful at every step. Their uncompromising quality standards have made this book an exceptional effort. Their encouragement from time to time has been an inspiration for everyone.

The publisher and the editorial board hope that this book will prove to be a valuable piece of knowledge for researchers, students, practitioners and scholars across the globe.

List of Contributors

Amber L Beitelshees
Division of Endocrinology, Diabetes and Nutrition, University of Maryland School of Medicine, 660 W. Redwood St, HH469, Baltimore, MD 21201, USA

Christina L Aquilante
Department of Pharmaceutical Sciences, University of Colorado Skaggs School of Pharmacy and Pharmaceutical Sciences, Aurora, CO 80045, USA

Hooman Allayee
Department of Preventive Medicine and Institute for Genetic Medicine, Keck School of Medicine, University of Southern California, Los Angeles, CA 90089, USA

Taimour Y Langaee, Gregory J Welder and Issam Zineh
Department of Pharmacotherapy and Translational Research, Center for Pharmacogenomics, University of Florida College of Pharmacy, Gainesville, FL 32610, USA

Richard S Schofield
Division of Cardiovascular Medicine and Department of Veterans Affairs Medical Center, University of Florida College of Medicine, Gainesville, FL 32603, USA

M. Kathryn Liszewski and John P. Atkinson
Division of Rheumatology, Department of Medicine, Washington University School of Medicine, 660 South Euclid, Saint Louis, MO 63110, USA

Argyro Sgourou and Vassilis Fotopoulos
School of Science and Technology, Hellenic Open University, Patras 262 22, Greece

Vassilis Kontos and Adamantia Papachatzopoulou
Laboratory of General Biology, Faculty of Medicine, University of Patras, Patras 265 04, Greece

George P Patrinos
Department of Pharmacy, School of Health Sciences, University of Patras, Patras 265 04, Greece

Luisa Azevedo, Catarina Serrano and Antonio Amorim
Instituto de Investigação e Inovação em Saúde, Universidade do Porto, Porto, Portugal

IPATIMUP-Institute of Molecular Pathology and Immunology, University of Porto, Rua Dr. Roberto Frias s/n, 4200-465, Porto, Portugal
Department of Biology, Faculty of Sciences, University of Porto, Rua do Campo Alegre, s/n, 4169-007, Porto, Portugal

David N. Cooper
Institute of Medical Genetics, School of Medicine, Cardiff University, Heath Park, Cardiff, CF14 4XN, UK

Yichuan Liu, Yun Li, Michael E. March, Kenny Nguyen, Kexiang Xu, Fengxiang Wang, Yiran Guo, Brendan Keating, Joseph Glessner and Hakon Hakonarson
Center for Applied Genomics, The Children's Hospital of Philadelphia, 1014H, 3615 Civic Center Blvd, Abramson Building, Philadelphia, PA 19104, USA

Jiankang Li, Jianguo Zhang and Xun Xu
Beijing Genomics Institute, Shenzhen, China

Theodore J. Ganley
Center for Sports Medicine and Performance, The Children's Hospital of Philadelphia, Philadelphia, PA, USA

Matthew A. Deardorff
Individualized Medical Genetics Center, The Children's Hospital of Philadelphia, Philadelphia, PA, USA

Cyril Cyrus, Chittibabu Vatte, Shahanas Chathoth and Amein Al-Ali
Institute for Research and Medical Consultation, University of Dammam, Dammam 31441, Kingdom of Saudi Arabia

Rudaynah Al-Ali, Abdullah Al-Shehri, Mohammed Shakil Akhtar, Mohammed Almansori, Fahad Al-Muhanna and Awatif Al-Nafie
King Fahd Hospital of the University, University of Dammam, Al-Khobar 31952, Kingdom of Saudi Arabia

Brendan Keating
Department of Pediatrics, Perelman School of Medicine, University of Pennsylvania, Philadelphia, PA, USA

Danielle Carpenter, Laura M. Mitchell and John A. L. Armour
School of Life Sciences, University of Nottingham, Nottingham NG7 2UH, UK

Gorjana Robevska and Jocelyn A. van den Bergen
Murdoch Childrens Research Institute, Melbourne, Victoria, Australia

Katie L. Ayers and Andrew H. Sinclair
Murdoch Childrens Research Institute, Melbourne, Victoria, Australia
Department of Paediatrics, University of Melbourne, Melbourne, Victoria, Australia

Aurore Bouty
Murdoch Childrens Research Institute, Melbourne, Victoria, Australia
The Royal Children's Hospital, Melbourne, Victoria, Australia

Achmad Zulfa Juniarto, Nurin Aisyiyah Listyasari and Sultana M. H. Faradz
Division of Human Genetics, Centre for Biomedical Research, Faculty of Medicine, Diponegoro University (FMDU), JL. Prof. H. Soedarto, SH, Tembalang, Semarang 50275, Central Java, Indonesia

Wen-Bin Zou and Hao Wu
Department of Gastroenterology, Changhai Hospital, Second Military Medical University, Shanghai, China
Institut National de la Santé et de la Recherche Médicale (INSERM), U1078 Brest, France
Etablissement Français du Sang (EFS)–Bretagne, Brest, France
Shanghai Institute of Pancreatic Diseases, Shanghai, China

Arnaud Boulling
Institut National de la Santé et de la Recherche Médicale (INSERM), U1078 Brest, France
Etablissement Français du Sang (EFS)–Bretagne, Brest, France

David N. Cooper
Institute of Medical Genetics, School of Medicine, Cardiff University, Cardiff, UK

Zhao-Shen Li and Zhuan Liao
Department of Gastroenterology, Changhai Hospital, Second Military Medical University, Shanghai, China
Shanghai Institute of Pancreatic Diseases, Shanghai, China

Jian-Min Chen
Institut National de la Santé et de la Recherche Médicale (INSERM), U1078 Brest, France
Etablissement Français du Sang (EFS)–Bretagne, Brest, France
Faculté de Médecine et des Sciences de la Santé, Université de Bretagne Occidentale (UBO), Brest, France

Claude Férec
Institut National de la Santé et de la Recherche Médicale (INSERM), U1078 Brest, France
Etablissement Français du Sang (EFS)–Bretagne, Brest, France
Faculté de Médecine et des Sciences de la Santé, Université de Bretagne Occidentale (UBO), Brest, France
Laboratoire de Génétique Moléculaire et d'Histocompatibilité, Centre Hospitalier Universitaire (CHU) Brest, Hôpital Morvan, Brest, France

Teresa Requena and Alvaro Gallego-Martinez
Otology & Neurotology Group CTS495, Department of Genomic Medicine, GENYO - Centre for Genomics and Oncological Research – Pfizer/University of Granada/Junta de Andalucía, PTS, 18016 Granada, Spain

Jose A. Lopez-Escamez
Otology & Neurotology Group CTS495, Department of Genomic Medicine, GENYO - Centre for Genomics and Oncological Research – Pfizer/University of Granada/Junta de Andalucía, PTS, 18016 Granada, Spain
Department of Otolaryngology, Complejo Hospitalario Universidad de Granada (CHUGRA), ibs.granada, 18014 Granada, Spain

Jubin Osei-Mensah and Yusif Mubarik
Kumasi Centre for Collaborative Research in Tropical Medicine, Kumasi, Ghana

Tim Becker and Christine Herold
Institute for Medical Biometry, Informatics and Epidemiology, University of Bonn, Bonn, Germany

Linda Batsa Debrah
Kumasi Centre for Collaborative Research in Tropical Medicine, Kumasi, Ghana
Department of Clinical Microbiology, Kwame Nkrumah University of Science and Technology, Kumasi, Ghana

Anna Albers, Andrea Hofmann, Achim Hoerauf and Kenneth Pfarr
Institute for Medical Microbiology, Immunology and Parasitology, University Hospital Bonn, Sigmund-Freud-Str. 25, 53127 Bonn, Germany

Alexander Yaw Debrah
Faculty of Allied Health Sciences of Kwame Nkrumah University of Science and Technology, Kumasi, Ghana

Felix F. Brockschmidt
Institute of Human Genetics, University of Bonn, Bonn, Germany
Department of Genomics, Life and Brain Center, University of Bonn, Bonn, Germany

Holger Fröhlich
Bonn-Aachen International Center for Information Technology (B-IT), University of Bonn, Bonn, Germany

Hung-Lun Chiang
Institute of Clinical Medicine, National Yang-Ming University, Taipei, Taiwan
Institute of Biomedical Sciences, Academia Sinica, Taipei, Taiwan

Yuan-Tsong Chen
Institute of Clinical Medicine, National Yang-Ming University, Taipei, Taiwan
Institute of Biomedical Sciences, Academia Sinica, Taipei, Taiwan
Department of Pediatrics, Duke University Medical Center, Durham, USA

Jer-Yuarn Wu
Institute of Biomedical Sciences, Academia Sinica, Taipei, Taiwan
Graduate Institute of Chinese Medical Science, China Medical University, Taichung, Taiwan

Hye Jin Yoo
National Leading Research Laboratory of Clinical Nutrigenetics/Nutrigenomics, Department of Food and Nutrition, College of Human Ecology, Yonsei University, 50 Yonsei-ro, Seodaemun-gu, Seoul 03722, South Korea

Department of Food and Nutrition, Brain Korea 21 PLUS Project, College of Human Ecology, Yonsei University, Seoul 03722, South Korea

Jong Ho Lee
National Leading Research Laboratory of Clinical Nutrigenetics/Nutrigenomics, Department of Food and Nutrition, College of Human Ecology, Yonsei University, 50 Yonsei-ro, Seodaemun-gu, Seoul 03722, South Korea
Department of Food and Nutrition, Brain Korea 21 PLUS Project, College of Human Ecology, Yonsei University, Seoul 03722, South Korea
Research Center for Silver Science, Institute of Symbiotic Life-TECH, Yonsei University, Seoul 03722, South Korea

Minjoo Kim, Minkyung Kim and Jey Sook Chae
Research Center for Silver Science, Institute of Symbiotic Life-TECH, Yonsei University, Seoul 03722, South Korea

Sang-Hyun Lee
Department of Family Practice, National Health Insurance Corporation, Ilsan Hospital, Goyang 10444, South Korea

Ye Wang, Songqing Deng, Yanmin Luo and Qun Fang
Fetal Medicine Centre, Department of Obstetrics and Gynaecology, The First Affiliated Hospital of Sun Yat-Sen University, Guangzhou, China

Xueli Wu
Department of Dermatology, Guangzhou Institute of Dermatology, Guangzhou, China

Liu Du, Ju Zheng and Hongning Xie
Department of Ultrasonic Medicine, The First Affiliated Hospital of Sun Yat-Sen University, Guangzhou, China

Xin Bi
Guangzhou KingMed Center for Clinical Laboratory, Guangzhou, China

Qiuyan Chen
Dongguan Women and Children's Hospital, Dongguan, China

Claude Férec
UMR1078 "Génétique, Génomique Fonctionnelle et Biotechnologies", INSERM, EFS - Bretagne, Université de Brest, CHRU Brest, Brest, France

David N. Cooper
Institute of Medical Genetics, School of Medicine, Cardiff University, Cardiff, UK

Jian-Min Chen
INSERM UMR1078, EFS, UBO, 22 avenue Camille Desmoulins, 29238 Brest, France

Luluah Alhusain and Alaaeldin M. Hafez
College of Computer and Information Sciences, King Saud University, Riyadh, Saudi Arabia

Minjoo Kim and Minkyung Kim
Research Center for Silver Science, Institute of Symbiotic Life-TECH, Yonsei University, Seoul 03722, Korea

Hye Jin Yoo
Department of Food and Nutrition, Brain Korea 21 PLUS Project, College of Human Ecology, Yonsei University, 50 Yonsei-ro, Seodaemun-gu, Seoul 03722, Korea

Jayoung Shon
Department of Food and Nutrition, Brain Korea 21 PLUS Project, College of Human Ecology, Yonsei University, 50 Yonsei-ro, Seodaemun-gu, Seoul 03722, Korea
Department of Food and Nutrition, National Leading Research Laboratory of Clinical Nutrigenetics/Nutrigenomics, College of Human Ecology, Yonsei University, Seoul 03722, Korea

Jong Ho Lee
Research Center for Silver Science, Institute of Symbiotic Life-TECH, Yonsei University, Seoul 03722, Korea
Department of Food and Nutrition, Brain Korea 21 PLUS Project, College of Human Ecology, Yonsei University, 50 Yonsei-ro, Seodaemun-gu, Seoul 03722, Korea
Department of Food and Nutrition, National Leading Research Laboratory of Clinical Nutrigenetics/Nutrigenomics, College of Human Ecology, Yonsei University, Seoul 03722, Korea

Xiaowei Fan, Lifeng Ma, Zhiying Zhang, Zhipeng Zhao, Yiduo Zhao, Fang Liu, Lijun Liu, Peng Cai, Yansong Li and Longli Kang
Key Laboratory for Molecular Genetic Mechanisms and Intervention Research on High Altitude Disease of Tibet Autonomous Region, School of Medicine, Xizang Minzu University, Xianyang 712082, Shaanxi, China

Key Laboratory of High Altitude Environment and Genes Related to Diseases of Tibet Autonomous Region, School of Medicine, Xizang Minzu University, Xianyang 712082, Shaanxi, China

Meng Hao
Ministry of Education Key Laboratory of Fan et al. Human Genomics, Contemporary Anthropology, Collaborative Innovation Center for Genetics and Development, School of Life Sciences, Fudan University, Shanghai 200433, China

Xingguang Luo
Division of Human Genetics, Department of Psychiatry, Yale University School of Medicine, New Haven, CT 06510, USA

Yi Li
Ministry of Education Key Laboratory of Fan et al. Human Genomics, Contemporary Anthropology, Collaborative Innovation Center for Genetics and Development, School of Life Sciences, Fudan University, Shanghai 200433, China
Six Industrial Research Institute, Fudan University, Shanghai 200433, China

Yu Jin
NUS Graduate School for Integrative Sciences and Engineering, National Division of Medical Sciences, National Cancer Centre, Singapore 169610, Singapore

Jingbo Wang
Department of Biochemistry, National University of Singapore, Singapore 119077, Singapore

Maulana Bachtiar
Department of Biochemistry, National University of Singapore, Singapore 119077, Singapore
Division of Medical Sciences, National Cancer Centre, Singapore 169610, Singapore

Samuel S. Chong
Department of Paediatrics, Yong Loo Lin School of Medicine, National University of Singapore, Singapore 119228, Singapore

Julian Gough
Department of Computer Science, University of Bristol, The Merchant Venturers Building, Bristol BS8 1UB, UK

Caroline G. L. Lee
NUS Graduate School for Integrative Sciences and Engineering, National University of Singapore, Singapore 117456, Singapore
Department of Biochemistry, National University of Singapore, Singapore 119077, Singapore
Division of Medical Sciences, National Cancer Centre, Singapore 169610, Singapore
Duke-NUS Graduate Medical School, Singapore 169547, Singapore

Mingming Liu and Liqing Zhang
Department of Computer Science, Virginia Polytechnic Institute and State University, Blacksburg, VA, USA

Layne T. Watson
Department of Computer Science, Virginia Polytechnic Institute and State University, Blacksburg, VA, USA
Department of Mathematics, Virginia Polytechnic Institute and State University, Blacksburg, VA, USA
Department of Aerospace and Ocean Engineering, Virginia Polytechnic Institute and State University, Blacksburg, VA, USA

Hashem A Shihab, Ian NM Day and Tom R Gaunt
Bristol Centre for Systems Biomedicine and MRC Integrative Epidemiology Unit, School of Social and Community Medicine, University of Bristol, Oakfield House, Oakfield Grove, Bristol BS8 2BN, UK

Matthew Mort and David N Cooper
Institute of Medical Genetics, School of Medicine, Cardiff University, Cardiff CF14 4XN, UK

Nzioka P Muiya, Salma M Wakil, Asma I Tahir, Samya Hagos, Mohammed Najai, Daisy Gueco, Nada Al-Tassan, Editha Andres, Nejat Mazher, Brian F Meyer and Nduna Dzimiri
Genetics Department, King Faisal Specialist Hospital and Research Centre, Riyadh 11211, Saudi Arabia

Theodora Koromila, Vassiliki Aleporou-Marinou and Panagoula Kollia
Laboratory of Human Genetics, Department of Genetics & Biotechnology, Faculty of Biology, National and Kapodistrian University of Athens, Athens 15701, Greece

Panagiotis Georgoulias
Department of Nuclear Medicine, University Hospital of Larissa, School of Medicine, University of Thessaly, Larissa 41110, Greece

Zoe Dailiana, Stavroula Samara and Chris Chassanidis
Department of Orthopaedic Surgery, School of Medicine, University of Thessaly, Larissa 41110, Greece

Evangelia E Ntzani
Department of Hygiene and Epidemiology, University of Ioannina School of Medicine, Ioannina 45110, Greece

Index

A

Age-related Macular Degeneration, 17-18, 20, 158

Allele-sharing Distance, 118, 123, 127-128

Amylase Gene Cluster, 49-51

Ancestry Informative Markers, 127-128

Atrial Natriuretic Peptide, 102, 108-109

Atypical Hemolytic Uremic Syndrome, 9, 18-21

C

C-reactive Protein, 7, 132, 136

Calcitonin, 184-185, 188-190

Carcinoembryonic Antigen-related Cell Adhesion Molecule-1, 83, 85, 92

Cardiac Hypertrophy, 183

Centro Nuclear Myopathy, 73, 80

Centrosomal Protein 57 Kda-like Protein 1, 37

Cholesteryl Ester Transfer Protein, 44, 48

Chromosomal Microarray Analysis, 111-112, 114, 116

Combined Annotation-dependent Depletion, 80

Congenital Hypogonadotrophic Hypogonadism, 55, 63

Copy Number Variation, 34, 42, 49, 52-54

Cxcl5 Polymorphisms, 1, 4-5

D

Dickkopf Wnt Signaling Pathway Inhibitor 1, 185

E

Epithelial Neutrophil Activator-78, 1

Erythropoiesis, 141-142, 144, 146

Exome Hidden Markov Model, 36

Exon Splicing Enhancer, 157

F

Fibroblast Growth Factor Receptor, 63-64

G

Gaussian Mixture Model, 122

Gene Ontology, 80, 170, 172

Gonadotrophin-releasing Hormone, 55, 63-64

H

Hardy-weinberg Equilibrium, 3, 44, 85, 92, 104, 108, 119-121, 128, 132, 136, 189, 191

High-altitude Polycythemia, 138-139, 145

Homeostatic Model Assessment, 104-105, 107-108, 131, 136

Human Chorionic Gonadotrophin, 56, 63

Human Gene Mutation Database, 35, 42, 171-172

Hypospadias, 55-59, 61-64

I

Immunoradiometric Assay, 103, 131, 188, 190

Insulin-like Growth Factor-1, 184, 190

Intellectual Disability, 42, 57, 111

Intronic Splicing Regulatory Element, 150, 157

K

Kallmann Syndrome, 55, 63-64

L

Linkage Equilibrium, 121

Lymphatic Filariasis, 83-84, 86, 92-94

Lymphedema, 83-87, 89-94

M

Matrix Metalloprotease-2, 83, 85, 89, 92

Matrix Metalloproteinases, 13, 19, 94

Matthew's Correlation Coefficient, 169

Meniere's Disease, 73, 81-82

Metalloprotease-2, 83, 85, 89, 92

Mfold Analysis, 114, 116

Microcephalic Osteodysplastic Primordial Dwarfism Type-1, 110, 112

Microfilariae, 84, 91, 93

Minor Allele Frequency, 35, 56, 82, 104, 111, 120, 127, 132, 145, 174, 191

Mirna Binding Site, 150

N

Neighbor-joining Tree, 123, 125, 128-129

Next-generation Sequencing, 81-82, 111, 116

Nonsense-mediated Mrna Decay, 66, 70-71

O

Osteocalcin, 184-185, 188, 190-191

P

Paralogue Ratio Test, 52

Parathyroid Hormone, 184-185, 190 Monooxygenase, 102, 106-108

Peroxisome Proliferator-activated, 130, 136-137

Phosphatase and Tensin Homolog, 160-161

Phosphoinositide 3-kinase, 141, 146

Plasma Malondialdehyde, 130

Polymerase Chain Reaction, 26, 35, 48, 189

Prospero Homeobox Protein-1, 91-92

Q

Quantitative Pcr, 34, 36-37, 39

R

Rs5882 Polymorphism, 43-44, 46-47

S

Sanger Sequencing, 97, 114

Sclerostin, 190

Single Nucleotide Polymorphisms, 33, 48, 83-84, 96, 102, 118-119, 127, 158

Single-nucleotide Polymorphisms, 94, 130-131, 166, 173-174

Singular Value Decomposition, 121, 123, 128-129

Standard G-banding Karyotyping, 111, 116

Super Paramagnetic Clustering, 124-125, 129

T

Trans-species Polymorphisms, 28-29, 31

Transcription Factor Binding Site, 157

Tumor Suppressor Genes, 160-162, 165

U

Ultra-conserved Elements, 147, 150, 156

V

Vascular Endothelial Growth Factor Receptor-3, 83, 85, 92, 94

W

Whole-exome Sequencing, 42, 72, 81, 110-111, 116

www.ingramcontent.com/pod-product-compliance
Lightning Source LLC
Chambersburg PA
CBHW080244230326
41458CB00097B/3207